Advanced Sciences and Technologies for Security Applications

The series Advanced Sciences and Technologies for Security Applications comprises interdisciplinary research covering the theory, foundations and domain-specific topics pertaining to security. Publications within the series are peer-reviewed monographs and edited works in the areas of:

- biological and chemical threat recognition and detection (e.g., biosensors, aerosols, forensics)
- crisis and disaster management
- terrorism
- cyber security and secure information systems (e.g., encryption, optical and photonic systems)
- traditional and non-traditional security
- energy, food and resource security
- economic security and securitization (including associated infrastructures)
- transnational crime
- human security and health security
- social, political and psychological aspects of security
- recognition and identification (e.g., optical imaging, biometrics, authentication and verification)
- smart surveillance systems
- applications of theoretical frameworks and methodologies (e.g., grounded theory, complexity, network sciences, modelling and simulation)

Together, the high-quality contributions to this series provide a cross-disciplinary overview of forefront research endeavours aiming to make the world a safer place.

The editors encourage prospective authors to correspond with them in advance of submitting a manuscript. Submission of manuscripts should be made to the Editor-in-Chief or one of the Editors.

More information about this series at http://www.springer.com/series/5540

Anthony J. Masys
Editor

Disaster Forensics

Understanding Root Cause and Complex Causality

Springer

Editor
Anthony J. Masys
Centre for Security Science
Ottawa, ON
Canada

ISSN 1613-5113 ISSN 2363-9466 (electronic)
Advanced Sciences and Technologies for Security Applications
ISBN 978-3-319-41847-6 ISBN 978-3-319-41849-0 (eBook)
DOI 10.1007/978-3-319-41849-0

Library of Congress Control Number: 2016944903

Printed on acid-free paper

This Springer imprint is published by Springer Nature
The registered company is Springer International Publishing AG Switzerland

Contents

Introduction

Recently the global threat landscape has seen the emergence of high-impact, low-probability events. Events like Hurricane Katrina, the Great Japan Earthquake and Tsunami, Hurricane Sandy, Super Typhoon Haiyan, global terrorist activities, aviation and critical infrastructure disasters have become the new normal. Extreme events challenge our understanding regarding the interdependencies and complexity of the disaster etiology and are often referred to as Black Swans. As described in UNISDR [31] '...between 2002 and 2011, there were 4130 disasters recorded, resulting from natural hazards around the world where 1,117,527 people perished and a minimum of US$1195 billion was recorded in losses. In the year 2011 alone, 302 disasters claimed 29,782 lives; affected 206 million people and inflicted damages worth an estimated US$366 billion.' This book opens the black box of disasters by presenting 'forensic analysis approaches' to disaster investigations and analysis, thereby revealing the complex causality that characterizes them. In so doing it identifies 'new and innovative' strategies in analyzing accidents and disasters.

The anatomy of disasters and accidents depicts an etiology that reflects an inherent complexity that involves elements beyond the temporally and spatially proximate, thereby supporting a holistic or systemic view of disasters and accidents. A systems perspective of accident etiology recognizes, as Hollnagel [13] remarks '... how functions depend on each other and can therefore show how unexpected connections may suddenly appear.' Urry [32] describes how complexity recognizes the emergent properties that result from the dynamic interaction within a system, thereby developing collective properties that are not reflected in the individual components. As such, complexity argues against reductionism. As noted in Styhre [30], the complexity perspective recognizes that changes result from a multiplicity of interconnected causes and effects. Within the context of understanding accident etiology, the systems approach as a guiding methodology informed by complexity theory facilitates a break from '...mechanistic, linear, and causal methods of analysis towards viewing interdependence and interrelation rather than linearity and exclusion' [8: 140]. To capture and address the complexity inherent within socio-technical accidents, Dekker [6: 78] argues '...it is critical to capture the relational dynamics and longer term socio-organizational trends behind system failure.'

As argued in Dekker [4: 103], 'Were we to really trace the cause of failure, the causal network would fan out immediately, like cracks in a window, with only our own judgment to help us determine when and where to stop looking, because the evidence would not do it for us.' Accident analysis utilizing fault/event-based approaches tends toward an explanation of the etiology with a concentration on the proximate events and actors immediately preceding the loss [16].

Leveson [16: 25] characterizes event-based models as best suited for component failures rather than explaining systemic factors such as '…structural deficiencies in the organization, management deficiencies, and flaws in the safety culture of the company or industry.' Leveson [16: 25] argues that 'new models that are more effective for accidents in complex systems will need to account for social and organizational factors, system accidents and dysfunctional interactions, human error and flawed decision making, software errors, and adaptation.'

Perrow [24] coined the phrase 'systems accident' to describe an etiology that resides within complex relationships between elements comprising a system. The complexity that resides in current systems creates what Perrow [24] refers to as 'normal accidents.' Perrow [25: 12] remarks that:

> We have produced designs so complicated that we cannot anticipate all the possible interactions of the inevitable failures; we add safety devices that are deceived or avoided or defeated by hidden paths in the systems. The systems have become more complicated because either they are dealing with more deadly substances, or we demand they function in ever more hostile environments or with ever greater speed and volume.

In what Perrow [24] classifies as high-risk systems, accidents are inevitable or normal stemming from the way failures interact and tie a system together. His introduction of the term 'normal accident' refers to the inherent characteristics of the system.

Understanding such complex disaster and accident etiology thereby requires novel and innovative approaches to analysis.

Forensics

The Integrated Research on Disaster Risk (IRDR) strategic plan (2013–2017) identifies 'reducing risk and curbing losses through knowledge-based actions' as a key goal. Building upon that, the Forensic Investigations of Disasters (FORIN) project proposed an approach that aims to uncover the root causes of disasters through in-depth investigations that go beyond the typical reports and case studies conducted post-disaster events [14]. The evidence-based approach is rooted in the traditional conceptualization of Forensics where in the early twentieth century, Dr. Edmond Locard, a forensic science pioneer in France, formulated the theory which states, 'Every contact leaves a trace.' This became known as Locard's exchange principle and is the basis for all forensic science as we know it today.

FORIN [10: 6] asks the question: why, when so much more is known about the science of natural events, including extremes, and when technological capacity is so much stronger, are large-scale and even small- and medium-scale disasters apparently becoming more frequent and the losses continuing to increase at a rapid rate (IRDR 2009; White et al. 2001)?

1. Understanding Disaster Etiology

As part of the Springer book series: Advanced Sciences and Technologies for Security Applications, this edited volume, **Disaster Forensics: Understanding Root Cause and Complex Causality**, introduces novel perspectives and innovative approaches that reveal the complexity associated with disaster etiology.

The 16 chapters in this book reflect contributions from various experts and through case studies and research reveal many perspectives, tools, and approaches to support disaster forensic analysis. The value added through disaster forensics can enable resilience and help support disaster mitigation, prevention, response, and recovery efforts.

2. Content

Polinpapilinho F. Katina in his Chapter 'Individual and Societal Risk (RiskIS): Beyond Probability and Consequence During Hurricane Katrina' argues that, individuals can have varying understanding of risk which in turn affects their decisions and actions. The varying understanding and actions stem from deep-seated fundamental assumptions (i.e., beliefs and predispositions). However, deep-seated fundamental assumptions are often not included in traditional risk measures of probability and consequences. This chapter attempts to close this gap by developing a risk framework, RiskIS, that includes individual and society measures influencing decisions and actions. These measures are developed by examining literature and contrasting the resulting measures with a well-known event: Hurricane Katrina in New Orleans. A synthesis of this research provides a wider array of measures that influence decisions and actions (i.e., norms and personal attitudes, organizational structures, knowledge base, and social context, degree of connectivity, race and ethnicity, mass media, and national ideology). The proposed framework provides a basis for inclusion of a contextual frame of reference that influences actions beyond probability and consequence. Implications for those involved in disaster management are provided.

Ivan Taylor in his Chapter 'Application of Problem Inversion to Cascading Critical Infrastructure Failure' argues for the need to prepare for frequent future natural disasters and find ways to mitigate the potential death and destruction they cause. This chapter will discuss a novel method for making preparations to avoid the problem of cascading disasters created by a single natural event. This approach will adapt a knowledge-based technique from manufacturing, called Ideation Failure Analysis™, to correct deficiencies in critical infrastructure. The technique involves a number of approaches that are combined into a comprehensive process.

First, a simple direct approach is attempted with the assistance of a knowledge base. If the problem cannot be resolved in the direct manner, then an indirect approach is suggested. A detailed Failure Analysis Questionnaire is used to assist in model building. A model of the failure network is developed. However, instead of working directly toward failure correction, an inversion process is conducted. That is, in order to facilitate greater creativity, the analysis team is asked to imagine ways to produce the failure. The creative work is then assisted by a knowledge base. The analysis team is able to prioritize the likelihood that a cause might have resulted in the failure. The next step is to find ways to prevent, eliminate, or reduce the impact of the failure. Again to assist the creative process, a knowledge base provides suggestions for correction techniques that can be prioritized in a hierarchical fashion. Finally, the results are evaluated to avoid negative side effects or drawbacks in the suggested ways to correct the failure. This chapter will conclude by providing some recommendations and an evaluation of the potential value of this technique.

Jonathan Gao and Sidney Dekker in their Chapter 'Heroes and Villains in Complex Socio-technical Systems' highlight how the history of efforts to reduce 'human errors' across workplaces and industries suggests that people (or their weaknesses) are seen as a problem to control [2, 11, 17, 26] (Woods and Cook 2002). However, some have proposed that humans can be heroes as they can adapt and compensate for weaknesses within a system and direct it away from potential catastrophes [26]. But the existence of heroes would suggest that villains (i.e., humans who cause a disaster) exist as well [2] and that it might well be the outcome that determines which human becomes which (Baron and Hershey 1988). The purpose of this chapter is to examine whether complex socio-technical systems would allow for the existence of heroes and villains, as outcomes in such systems are usually thought to be the product of interactions rather than a single factor [16]. The chapter will first examine if the properties of complex systems as suggested by Dekker et al. [7] would allow for heroes and villains to exist. These include the following: (a) synthesis and holism; (b) emergence; (c) foreseeability of probabilities, not certainties; (d) time-irreversibility; and (e) perpetual incompleteness and uncertainty of knowledge, before concluding with a discussion of the implications of the (non) existence of heroes and villains in complex systems for the way we conduct investigations when something goes wrong inside of those systems.

Anthony J. Masys in his Chapter 'Patient Safety and Disaster Forensics: Understanding Complex Causality Through Actor Network Ethnography' argues that human error is often cited as a major contributing factor or cause of incidents and accidents. For example, in the aviation domain, accident surveys have attributed 70 % of incidents to crew error citing pilot error as the root cause of an aviation accident [12: 781; 29: 60, 34: 2]. Similarly in the healthcare domain, medical errors are reported to be a major cause of morbidity and mortality. Aggarwal et al. (2010: i3) argues that '…an increasing awareness of medical injuries has been paralleled by the rise in technology, and the increasing complexity it causes.'

According to Woods et al. [34], human error can be characterized either as a cause of failure or as a symptom of a failure. The label 'human error' as reported by Woods et al. [34] is considered prejudicial and unspecific. They argue that the label 'human error' retards rather than advances our understanding of how complex systems fail and the role of the human in both successful and unsuccessful system operations. This is contextualized and further supported in the healthcare domain (clinical encounters) by Artino et al. [1]. A systems view of the problem space regards human error as a symptom of '…contradictions, pressures and resource limitations deeper inside the system' [3: 2]. Actor Network Theory provides a conceptual foundation and lens to facilitate a systems-thinking-based analysis [19, 33] to examine the key dynamics that reside in the black box of human error pertaining to patient safety.

This chapter frames patient safety and human error through the lens of Actor Network Theory by leveraging insights from accidents and disasters such as the 2003 US/Canada Blackout [20] and the Uberlingen mid-air crash [18]. An example of the medical errors associated with the use of an Emergency Department Information Systems (EDIS) in a clinical situation is given. This ANT facilitated 'disaster forensic analysis' reveals a complex causality that is rhizomal rather than linear thereby challenging our notion of human error and highlights where intervention strategies can be focused to support patient safety.

Allan Bonner in his Chapter 'The Fog of Battle in Risk and Crisis Communication: Towards the Goal of Interoperability in the Digital Age' conducts a 'forensic' examination of disaster communications in the effort to unearth key gaps and contradictions. He argues that clear communication which produces an attitudinal or behavioral change is crucial in business, government, and politics. Selling products, providing entitlements, and obtaining consent from voters rely on communication. This is more crucial when communicating about a risk or during a crisis—recalling product, providing health information, or preparing an urban area for an emergency, for example. The risk communication literature provides guidance in these crucial cases.

Dorte Jessen in her Chapter 'Disaster Forensics: Governance, Adaptivity and Social Innovation' seeks to contribute to the discourse on disaster forensics, by arguing that the root cause and complex causality is ultimately governance, ideally cultivating the collective ability to navigate disasters rather than to command control. The focus will be on the social dimension and its impact on disasters. Governance theory, combined with complex adaptive systems theory [9], will provide the analytical foundation for the examination of Hurricane Katrina and the Fukushima Daiichi nuclear disaster. The theoretical deconstruction will reveal that the traditional virtues embedded in the social amplification of risk [15] remain at the heart of complex causality. With this insight, it is observed that social innovation, with its inherent positive connotation [23], is expanding the horizon for how social divisions, vulnerabilities, and resilience are measured. Optimistically, it is suggested that social innovation, driven by civil society, may prove a vital component in the creation of a new social narrative.

Simon Bennett in his Chapter 'Disasters and Mishaps: The Merits of Taking a Global View' examines the case of the Germanwings Flight 4U9525. On March 24, 2015, Andreas Lubitz, the First Officer of Germanwings Flight 4U9525, committed suicide by aircraft. Following the disaster, there was a 'rush to blame,' with Lubitz painted as the sole villain. Few reviewed the wider circumstances. While accepting the primacy of Lubitz's actions in the destruction of Germanwings Flight 4U9525, this chapter scans the horizon for contributory factors. In doing so it demonstrates the contribution systems-thinking can make to understanding failure in complex, transnational socio-technical systems (like commercial aviation). The chapter offers a counterweight to the fundamental attribution error. It offers an antidote to blamism. It references the work of Ross, Reason, Turner and Fiske and Taylor. While blame and punishment satisfy our baser instincts (the urge to hurt those who have hurt us is hard to resist), they generally undermine safety. The chapter argues that, from a safety standpoint, blamism is an inappropriate response to mishap and disaster.

Dmitry Chernov and Didier Sornette in their Chapter 'Dynamics of Information Flow Before Major Crises: Lessons from the Collapse of Enron, the Subprime Mortgage Crisis and Other High Impact Disasters in the Industrial Sector' present an analysis of the two largest financial disasters in the USA so far in the first decade of this century—the collapse of Enron in 2001 and the subprime mortgage crisis of 2007–2008—suggests that the huge scale of these disasters stemmed from a lack of timely information. Chernov and Sornette present extensive evidence that regulators, investors, and associates were not informed of the conditions and risks associated with the activities of Enron management in the first case, or with the assessment and underwriting of collateralized debt obligations (CDOs) in the second; and with little understanding of the 'whole picture' of risks, they could not intervene decisively to prevent or minimize disaster. Moreover, Chernov and Sornette identify similar obstacles to the transmission of reliable risk information in past cases such as the Barings Bank Crash, the Deepwater Horizon Oil Spill, the nuclear accidents at Chernobyl, and Fukushima Daiichi as well as in the current development of the US shale energy industry. Based on the careful observation of events before the moment of collapse in three financial events (Barings, Enron and subprime crisis), one mixed financial-industrial case and three industrial catastrophes, Chernov and Sornette document and discuss how the inadequate transmission or outright concealment of risk information constitutes a powerful engine of disasters.

Ross Prizzia in his Chapter 'Climate Change and Disaster Forensics' relates and applies forensic theory, insight, and analysis to disaster-related research and practice. It explores, describes, and explains human causality of climate change-related disasters and their impact on human and environmental losses. The chapter also identifies and describes new and innovative methodologies and strategies to analyze climate-related disasters, reduce disaster risk, and improve disaster mitigation, adaption, and management. Emphasis is given to vigilant monitoring and assessment of Intended Nationally Determined Contributions (INDC) to reduce greenhouse gas (GHG) emissions to limit global warming to 2 °C by 2030, a critical target set to prevent some of the worst impacts of climate change.

D. Elaine Pressman in her Chapter 'The Complex Dynamic Causality of Violent Extremism: Applications of the VERA-2 Risk Assessment Method to CVE Initiatives' describes the complexity inherent within radicalization. Attacks by violent extremists have been occurring with increasing frequency over the past years from a spectrum of ideological objectives. Incidents have occurred in the USA, Canada, the UK, Spain, France, Denmark, Norway, Australia, Pakistan, Iraq, Afghanistan, and other countries around the globe. In 2013, more than 9700 terrorist incidents were recorded in 93 countries. These incidents claimed more than 18,000 lives and 33,000 injured. In 2014, a rise in lone offender attacks was observed. Many of these attacks were inspired by ISIL, Al-Qaeda, and other extremist groups. Others do not appear to have been specifically directed by a terrorist organization, extremist group, or their affiliates. In the future, centralized leadership of terrorist organizations may be less important than the radicalization process itself, the individual identity of the perpetrator, and the narratives believed. This chapter examines the dynamic causality of violent extremism through the application of the VERA-2 method.

Gisela Bichler and Stacy Bush in their Chapter 'Staying Alive in the Business of Terror' examine the domain of terrorism. As it pertains to terrorists, they argue that staying alive (and at large) is a career advantage when you manage an insurgent group. If instead, your objective was to detonate a suicide bomb, success would be measured differently. These divergent goals must be considered when examining the social network within which individual actors are embedded, as each outcome may require a different supporting structure, warranting the application of different theory and associated metrics. Breaking from the extant literature that is principally concerned with assessing the cellular structure of attack groups and the centrality of actors, this chapter applies a business model of competitive advantage to examine how varied egonet structures correlate with the operational success of command staff—here the objective is to stay alive. Investigating the utility of Burt's (1992; 1997) theory of structural holes, we find that the communication patterns of central leaders of Al Qa'ida and the Islamic State of Iraq (ISI), who were active since 2006 and survived at-large until November 2015, involved smaller egonets that had fewer non-redundant ties, lower density, and were significantly less likely to involve reliance on a central actor for information. In short, less social capital and lower constraint improved the likelihood of survival.

Anthony J. Masys in his Chapter 'Counter-Terrorism and Design Thinking: Supporting Strategic Insights and Influencing Operations' describes how the recent terrorist attacks in Paris and Jakarta, Ankara, Ivory Coast, and Brussels in 2015 and 2016, respectively, highlight the complexity and challenges associated with counter-terrorist operations. The words of Rosenhead and Minger [28: 4–5] resonate with the complex space of counter-terrorism and these recent incidents. They argue that we '…are not confronted with problems that are independent of each other, but with dynamic situations that consist of complex systems of changing problems that interact with each other.' Such a complex problem space can be viewed as 'wicked problems' or 'messes' [27]. Rosenhead and Mingers [28: 4–5] describe 'messy problems' as that which have inherent complex interdependencies and dynamic complexity. They argue that 'Individual problems may be solved. But if they are components of a mess, the solutions to individual problems cannot be added, since those solutions will interact.'

Within the context of counter-terrorism, deep uncertainty is the source of surprises and shocks in a system and the main cause of discontinuity in the strategic space of a system. Regions in the EU have been identified as key nodes of radicalization and violent extremism. It highlights the complex social factors that require an empathic approach to uncover the connectivity and processes [22] supporting this convergence of violent extremism. The problem space transcends domain-specific analysis to require a more inclusive approach that draws upon insights from sociology, economics, political science, humanities in the problem framing [21]. New methods and methodologies have evolved to address such inherent complexity in problem spaces. This chapter examines the counter-terrorism problem space leveraging the epidemiological model to illustrate how design thinking can be applied to develop analysis methodologies and intervention strategies to support counter-terrorism and resilience.

Jason Levy, Peiyong Yu and Ross Prizzia in their Chapter 'Economic Disruptions, Business Continuity Planning and Disaster Forensic Analysis: The Hawaii Business Recovery Center (HIBRC) Project' propose that modern disaster forensics can reduce supply chain disruptions, enhance disaster resilience, and promote a more robust economy. This chapter examines the root causes of economic disruptions by presenting 'forensic analysis approaches' to disasters that impact the economy of the US island state of Hawaii. Supply chain disruptions and investigations pertaining to business disruptions are undertaken with an emphasis on modeling, understanding, and characterizing the complex causality that defines them. In so doing this chapter uncovers creative, timely and important strategies for analyzing accidents and disasters that impact the economy of Hawaii. In order to promote business continuity planning and disaster forensics in Hawaii, the twenty-eighth Hawaii State Legislature enacted, and the Governor of Hawaii has signed, House Bill (HB) 1343 which provides funds for new state-of-the-art Hawaii Business Recovery Center (HIBRC), a joint partnership between the State of Hawaii Emergency Management Agency (HIEMA), the State of Hawaii Department of Business, Economic Development and Tourism (DBEDT), the State of Hawaii State Procurement Office (SPO), and the University of Hawaii West Oahu (UHWO). This designated business recovery hub will provide both outreach and dissemination of business recovery resources in addition to serving as a center for presenting disaster forensics approaches to disaster investigations in Hawaii, thereby uncovering the complex causality that underlies them. The center will help inform businesses of the importance of disaster preparedness, assist with post-disaster business recovery efforts, and create a robust business recovery network that shares the highest-level of management and governance with business leaders and strives for best disaster management practices and continuous improvement.

Jason Levy in his Chapter 'An Event-Driven, Scalable and Real-Time Geospatial Disaster Forensics Architecture: Decision Support for Integrated Disaster Risk Reduction' examines water resources disasters and their impact on humans, the built environment and natural systems. The chapter also identifies and describes timely and innovative decision support architectures to analyze climate-related disasters, enhance emergency preparedness, reduce disaster risk, promote disaster resilience, and improve disaster mitigation, adaption, and management. The root causes of water resources disasters are explored, and a distributed, scalable and real-time disaster forensics architecture with event-driven messaging and advanced geomatics engineering capabilities is put forth. Emphasis is given to vigilant monitoring, assessment, response, and recovery of floods and oil and molasses spills in the US state of Hawaii. The decision support and situational awareness advances found in this chapter complement the recent success of water resources disaster risk management and disaster forensics in Europe and elsewhere. The herein proposed disaster forensics architecture helps managers uncover creative, timely and important strategies for analyzing water resources accidents and disasters. In this manner, professionals have additional tools to model the complex causality of disasters and are better equipped to apply disaster forensics theory to the promotion of a more holistic, sustainable relationship between society and the environment. Specifically, this contribution provides theoretical insights and practical examples to manage water resources disasters under uncertainty.

Jason Levy and Peiyong Yu in their Chapter 'Advances in Economics and Disaster Forensics: A Multi-criteria Disaster Forensics Analysis (MCDFA) of the 2012 Kahuku Wind Farm Battery Fire on Oahu, Hawaii' illustrate how the discipline of economics and its many sub- and closely related disciplines offer valuable modeling techniques to relate and apply forensic theory, insight and analysis to disaster-related research. We herein propose advances in economics and disaster forensics to also reduce disaster risk and assess the direct and indirect impacts of disasters. This chapter constitutes a landmark attempt to address, comprehensively and in-depth, many timely and important issues associated with using the field of economics to build a culture of disaster prevention and to understand the root cause and complex causality of disasters. In particular, advances in microeconomic, macroeconomic, and forensic analyses are used to assess the causes and consequences of energy-related disasters. A timely, original and valuable Multi-Criteria Disaster Forensics Analysis (MCDFA) approach for the forensic analyses of disasters is put forth, and the 2012 battery room fire at the Kahuku wind energy storage farm on Oahu, Hawaii, is used as a case study to illustrate the proposed approach. Modeling identifies dynamic volt-amp reactive (D-VAR) technology as a preferred alternative over lead acid batteries for the Kahuku Wind Farm.

Jason Levy in his Chapter 'Complexity and Disaster Forensics: Paradigms, Models and Approaches for Natural Hazards Management in the Pacific Island Region' assesses and applies complex systems theory, modeling and analysis to disaster forensics policy and research. To better understand the root cause and complex causality of disasters, complex systems theory, with roots in the fields of statistical physics, information theory, and nonlinear dynamics, and systems analysis, is applied to help communities and nations achieve important social development goals, reduce institutional brittleness, and increase disaster resilience by promoting positive transformations in the coevolving and mutually dependent human-environmental condition, and by capitalizing on opportunities provided by human creativity, diplomatic openings, technologic capacities and environmental change. The case studies, investigations and models outlined in this chapter collectively demonstrate the quality, breadth and depth of complex systems and disaster forensics methodologies. Game-theoretic ('Small World') decision analyses and complex systems ('Large World') models of mutually interactive game design are put forth to capture the complexity of climate-related disasters and to reduce the threat of climate refugees in the Pacific Island region. The resulting risk management lessons learned were applied to communities in the Pacific Island of Vanuatu, the most natural disaster-prone country in the world.

References

1. Artino AR, Durning SJ, Waechter DM, Leary KL, Gilliland WR (2012) Broadening our understanding of clinical quality: from attribution error to situated cognition. Clin Pharmacol Ther 91(2)
2. Clarke DM (2010) The human contribution: unsafe acts, accidents and heroic recoveries. J. Risk Res 14:144–145
3. Dekker S (2002) The re-invention of human error, Technical Report 2002-01, Lund University School of Aviation
4. Dekker SWA (2003) Illusions of explanation: a critical essay on error classification. Int J Aviat Psychol 13(2):95–106
5. Dekker S (2004) The hindsight bias is not a bias and not about history, Technical Report 2004-01, Lund University School of Aviation
6. Dekker S (2006) Resilience engineering: chronicling the emergence of confused consensus. In: Hollnagel E, Woods DD, Leveson N (eds) Resilience engineering: concepts and precepts. Ashgate Publishing Ltd., Aldershot, pp 77–92
7. Dekker S, Cilliers P, Hofmeyr J-H (2011) The complexity of failure: implications of complexity theory for safety investigations. Saf Sci 49:939–945
8. Dennis K (2007) Time in the age of complexity. Time Soc 2007(16):139–155
9. Duit and Galaz (2008) Governance and complexity: Emerging issues for governance theory. Gov Int J Policy Adm Inst 21(3):311–335
10. FORIN (2011) http://www.irdrinternational.org/wp-content/uploads/2012/06/FORIN-REPORT_web.pdf
11. Haq H (2013) How human error can cause a plane crash. BBC Travel [Internet]. UK; 2013. Available from: http://www.bbc.com/travel/story/20130521-how-human-error-can-cause-a-plane-crash
12. Helmreich RL (2000) 'On error management: lessons from aviation' Br Med J 320:781–785
13. Hollnagel E (2004) Barriers and accident prevention. Ashgate Publishing, Hampshire, England
14. IRDR (2013) http://www.irdrinternational.org/projects/forin/
15. Kasperson J, Kasperson R (2005) The social contours of risk, Volume I: publics, risk communication & the social amplification of risk. Earthscan, London

16. Leveson NG (2002) System safety engineering: back to the future. Aeronautics and Astronautics Department, Massachusetts Institute of Technology, Cambridge, MA: MIT. http://sunnyday.mit.edu/book2.pdf
17. Levitt P (2014) When medical errors kill: American hospitals have embraced a systems solution that doesn't solve the problem. Los Angeles Times [Internet]. Los Angeles; 2014. Available from: http://articles.latimes.com/2014/mar/15/opinion/la-oe-levitt-doctors-hospital-errors-20140316
18. Masys AJ (2005) A systemic perspective of situation awareness: an analysis of the 2002 mid-air collision over Uberlingen, Germany. Int J Disaster Prev Manage 14(4):548–557
19. Masys AJ (2010) Fratricide in air operations: opening the black box- revealing the social. Ph. D. Dissertation, June 2010, University of Leicester, UK
20. Masys AJ (2014) Critical infrastructure and vulnerability: a relational analysis through actor network theory. In: Masys AJ (ed) Networks and network analysis for defence and security. Springer, Berlin
21. Masys AJ (ed) (2015) Applications of systems thinking and soft operations research in managing complexity. Springer, Berlin
22. Masys AJ (ed) (2016) Exploring the security landscape- non-traditional security challenges. Springer, Berlin
23. Matei A, Antonie C (2015) Complexity theory and the development of the social innovation. Soc Behav Sci 185: 61–66
24. Perrow C (1984) Normal accidents: living with high-risk technologies. Basic Books, Inc., New York, NY
25. Perrow C (1999) Normal accidents: living with high-risk technologies, 2nd edn., Princeton University Press, Princeton, NJ
26. Reason J (2008) The human contribution: unsafe acts, accidents and heroic recoveries. Ashgate Publishing, Surrey
27. Rittell HW, Weber MM (1973) Dilemmas in general theory of planning. Policy Sci 4: 155–169
28. Rosenhead J, Mingers J (2001) A new paradigm of analysis. In: Rosenhead J, Mingers J (eds) Rational analysis for a problematic world revisited. Wiley, West Sussex, England
29. Shappell SA, Wiegmann DA (2001) Applying reason: the human factors analysis and classification system (HFACS). Hum Factors Aerosp Saf 1(1):59–86
30. Styhre A (2002) Non-linear change in organizations: organization change management informed by complexity theory. Leadersh Organ Dev J 23(6):343–351
31. UNISDR (2012) Towards a post-2015 framework for disaster risk reduction. http://www.unisdr.org/files/25129_towardsapost2015frameworkfordisaste.pdf
32. Urry J (2003) Global complexity. Polity Press, Cambridge, UK
33. Wickramasinghe N, Tumu S, Bali RK, Tatnall A (2007) Using actor network theory (ANT) as an analytic tool in order to effect superior PACS implementation. Int J Networking Virtual Organ 4(3):257–279
34. Woods D, Johannesen L, Cook R, Sarter N (1994) Behind human error: cognitive systems, computers, and hindsight. Crew Systems Ergonomics Information Analysis Center (CSERIAC), Wright-Patterson Air Force Base, Ohio

Individual and Societal Risk (RiskIS): Beyond Probability and Consequence During Hurricane Katrina

Polinpapilinho F. Katina

Abstract The classical definition of risk revolves around probability and consequence. However, individuals can have varying understanding of risk which in turn affects their decisions and actions. The varying understanding and actions stem from to deep-seated fundamental assumptions (i.e., beliefs and predispositions). However, deep-seated fundamental assumptions are often not included in traditional risk measures of probability and consequences. This chapter attempts to close this gap by developing a risk framework, RiskIS, that includes individual and society measures influencing decisions and actions. These measures are developed by examining literature and contrasting the resulting measures with a well-known event: Hurricane Katrina in New Orleans. A synthesis of this research provides a wider array of measures that influence decisions and actions (i.e., norms and personal attitudes, organizational structures, knowledge base, and social context, degree of connectivity, race and ethnicity, mass media, and national ideology). The proposed framework provides a basis for inclusion of a contextual frame of reference that influence actions beyond probability and consequence. Implications for those involved in disaster management are provided.

Keywords Hurricane Katrina · Individual · New Orleans · Risk individual—society (RiskIS) · Society · Traditional risk

1 Introduction

Hurricane Katrina has the distinction of being the "costliest and one of the five deadliest hurricanes to ever strike the United States" [24, p. 1]. The damages associated with Hurricane Katrina are estimated to have been 108 billion dollars (US) with

P.F. Katina (✉)
Old Dominon University, Norfolk, USA
e-mail: pkatina@odu.edu

A.J. Masys (ed.), *Disaster Forensics*, Advanced Sciences and Technologies for Security Applications, DOI 10.1007/978-3-319-41849-0_1

over 1800 deaths [24, 49]. It is natural to stop and reflect on why there was so much damage and loss of life from a hurricane which had been publicized with prior warnings by federal, state, and local levels/authorities especially related to aspects of making landfall and potential consequences for the city as well as the residents of the city of New Orleans [4]. Arguably, a response might be found in the notion that past experiences should not be the basis for accurate prediction of future events. More succinctly, this notion suggests that "experiences of the past, encourages anticipation of the wrong kind of risk" [6, p. 330]. The irony is in the fact that there is a pervasive use of past information in many aspects of decision and actions in human-life. Certainly this is the case when someone decides to 'ride-out' a storm since they weren't harmed by a previous storm. This form of thinking is often tacit and therefore difficult to be made explicit and almost impossible to quantify. When one revisits this situation from a perspective of risk, the conclusion is easily reached by assuming that the probability of occurrence associated with an impending disaster and the related consequences will have the same outcomes as in previous similar situations.

Discernibly, this type of thinking has implications on actions and decisions that one makes in association with an impending disaster. For those involved in disaster type-situations (i.e., potential victims and rescuers), one area of concern might be how to incorporate the range of tacit beliefs in risk formulations to enable better decision-making. This issue is especially essential since the traditional formulation of risk [4] considers two measures: (1) probability that an event will occur and (2) articulation of consequences of occurrence of such an event. However, this does not account for tacit and implicit beliefs held by people in a disaster situation. The supposition at hand is that it is possible to make a decision and take actions contrary to what is expected given probability and consequence of a disaster. An examination of *why* that might be the case provides insights and could be instrumental in the development of techniques and tools instrumental in minimizing damage and loss of lives in disaster scenarios.

There is wide recognition of the importance and the role of risk and its measures of probability and consequence. However, these measures alone are not sufficient for decision-making and action-taking for an increasingly emerging low probability high impact events of the 21st century. In these events individuals might make decisions and act based on unarticulated core and tacitly held values and beliefs. It is from this perspective that this chapter is developed to propose a risk-based framework that incorporates individual and society factors influencing decisions and actions. These measures are supported by literature and the events leading up to, during, and after Hurricane Katrina in New Orleans in 2005.

To fulfill the purpose of this chapter, the remainder of this chapter is organized as follows: Sect. 2 explores the concepts of culture, society, and engineering risk. The aim of this section is to indicate how the concept of risk changes based on different contexts. Section 3 uses Hurricane Katrina as a case application. This section contrasts the relationship between classical risk and the core values/tacit knowledge that might have influenced how residents of the city of New Orleans viewed Hurricane Katrina. Section 4 provides an alternative approach for risk incorporating individual and societal measures: RiskIS. The proposed view of risk incorporates several measures that attempt to refine risk and could be used to better understand actions and decisions of those in disaster stricken areas. Future research directions are provided.

2 Foundations: Culture, Society, and Risk

Modern times have been characterized as tumultuous [26, 43]. This characterization is manifested in the ambiguity, complexity, emergency, interdependencies, and the uncertainties of the last three centuries [23]. There is *ambiguity* associated with an increasing lack of clarity and situational understanding, *complexity* associated with large numbers of richly and dynamically interacting complex systems and subsystems with behavior difficult to predict. Moreover, there is *emergence* with respect to the inability to deduce behavior, structure, or performance of current critical systems as a function of their constituent subsystems and elements and *interdependency* associated with mutual influence among different complex systems through which the state of a system influences, and is influenced by, the state of other interconnected systems. Uncertainty appears to be the norm in which there is always incomplete knowledge about situations and therefore casting doubt for decision/action consequences [21–23].

This landscape aligns with the notions of 'messes' [2] as well as 'wicked problems' [37]. The increasing prevalence of these conditions suggests a need to question classical elements of our understanding of how the world operates. With respect to risk formulation, Beck [6] notes that the increasing ubiquitous computing and pervasive computing technologies should force policymakers as well as researchers to abandon how risk has been approached. The call for a revision of the predominant risk paradigm is also evident across areas of disaster management [27], critical infrastructures [8, 25, 48], terrorism and homeland security [10, 19], system of systems [35] and cultural studies [45, 47, 50]. These studies acknowledge that the classical view of risk that past experiences can be quantified and then used as the basis for accurate prediction measures of probability and consequences. However, they also acknowledge that recent events, natural (e.g., hurricanes) or man-made (e.g., the 9/11 attacks), continue to defy and call into question the logic of this classical view of risk.

The need to understand and 'control' risk events fits within the purview of engineering managers and systems engineers. Therefore, there is benefit from examination of risk from this an extended perspective that extends beyond more classical views.

2.1 Engineering Risk

Often associated with risk analysis and decision-making processes, engineering risk is a science of risk and its probabilities based on statistical analysis of available data. From the engineering perspective, the aim of risk analysis is to establish future risk estimates based on independent sampling and past experiences [34]. The term 'engineering' serves as an indicator that one is concerned with an engineered system. Engineered systems are characterized as having been 'designed' by someone [17]. Such systems have "a high degree of technical complexity, social intricacy, and elaborate processes, aimed at fulfilling important functions in society" [11, p. 31]. Examples of engineered systems include healthcare systems (i.e., hospitals), banking and finance, and energy systems [42].

Moreover, there is an increasing realization that the modern world is more integrated. This integration, evident by interdependencies and connectivity among engineered and 'non-engineered' systems creates a situation in which the term 'engineering' is not a necessary requirement in risk analysis. This is especially the case when one considers that there is a need for integration of different disciplines to better understand and deal with risk events in modern times. Coincidentally, the very characterization of engineered systems—existing in the real world, artificiality, dynamic, hybrid state, and involving human control [11], supports this thesis of the limitations imposed by a singular formulation of 'engineering' risk. It is thus obvious that the term *engineering risk* carries multiple connotations in risk analysis that may in fact be limiting given the current state of risks in modern systems.

In this chapter, there is a deliberate effort to move away from risk as defined by any specific field and more towards emerging perspectives on risk [19, 22]. These emerging perspectives suggest risk can arise from designed and natural systems. This might be attributed to system deficiencies, and regardless as to whether such deficiencies are known, unknown, unpredictable, or undetectable, they still have the possibility of generating loss/injury and consequences to system users. A cautionary tale is provided by Beck [6]. His research provides several examples in science, government, business, and defense where this emerging paradigm is evident. He suggests, for example that, the discovery of chlorofluorocarbon (CFC) and its potential benefits should have been examined for known and unknown implications—especially those associated with ozone layer impacts and contributions to global warning [30]. Beck's [6] thesis is that the time is right for modern society to move away from the paradigm of 'science'. This is especially the case based on its assumption that the future is predictable and controllable based on past experience. In the case of CFC, one could argue that the promises of beneficial science have in fact generated threats to the society it had well-meaning intentions to improve.

The preceding discussion has implications for the concepts of risk engineering and risk itself. First, it suggests that there is no succinct difference between engineering risk and risk. While most engineering projects deal with engineered systems; such systems interact with natural systems—making them more of *hybrids* in nature. Risk analysis, for such systems, requires a consideration of a system of interest as well as the interacting systems. Second, while the measures of probability of occurrence of an event and the magnitude of the resulting consequences are considered fundamental [4, 20], there are scenarios (e.g., unknown unknowns [33] and black swans [44]) in which these measures are not sufficient to enable 'anticipation' [6] of risk and support for subsequent actions and decisions based on historically based predictions. The consideration of measures of probability and consequences, while necessary, is not sufficient to account for the multitude of measures that influence actions and decisions. The hybrid nature of complex systems, coupled with the increasing unpredictability of such systems supports the need to revisit risk measures and their influence on decision-making.

2.2 *Culture and Risk Analysis*

One measure affecting decision and action is *culture*. Webster's New Explore Encyclopedic Dictionary defines culture as "a set of shared attitudes, values, goals, and practices that characterizes a company or corporation" [29, p. 442]. When shared, a culture can influence collective action regarding specific issues—including responsive behavior. This issue is well noted in Beck's [6] discussions related to how certain issues are prioritized and addressed. This is the case for terrorism as well as climate change, where global warming is considered a critical issue for the European Union and terrorism as the most pressing issue in the US [6]. Cultural differences are also attributed to establishment of two different risk philosophies for addressing issues. Table 1 draws on Beck [6] to illustrate two different risk philosophies that have been attributed to culture.

In the context of risk analysis and management of risk, culture could be defined as emanating from a *group of individuals who share a set of specific beliefs and those beliefs are manifested in the attitudes, values, goals, visions, and objectives of the group*. This definition comes with a caveat from Schein [38]: culture is one of the most difficult 'organizational' attribute to change. The uniqueness of an

Table 1 Risk philosophies attributed to culture

Applicable philosophy	Description	Risk implications—example
Laissez-faire	It is safe, as long as it has not been proven to be dangerous	Policymakers are inclined to accept genetically modified foods (GMF) without fully understanding the consequences of such foods—a GMF is safe since it has no known bad consequence
Precautionary principle	Nothing is safe, as long as it has not been proven harmless	Policymakers are induced to reject GMFs in favor of more research to establish the fact that they are safe—a GMF not proven harmless renders it unsafe not given definitive evidence to the contrary

organization lies with the concept of culture [31]. Arguably, there is utility in consideration of culture in the articulation of risk events especially as people of the same 'culture' could be expected to have similar approaches and mindsets related to the analysis, modeling, and management of problematic issues.

2.3 Society and Risk Analysis

A related measure that can influence risk is *society*. If one takes the perspective that society is "an enduring and cooperating social group whose members have developed organized patterns of relationships through interaction with one another," then society can be thought of as "a community, nation, or broad grouping of people having common traditions, institutions, and collective activities and interests" [29, p. 1747]. This perspective is portrayed by modern society being characterized by an increasing demand for quality services, long-term sustainability, increasing trans-boundary dependencies, rapid technological changes, shifting to privatization of public sector [46]. Arguably, these characterizations involve having a set of *common interests* which in turn can create and shape society. A good example can be found in the field of critical infrastructures where an increased need for more secure and predictable electrical power infrastructures led to the emerging paradigm of 'electricity plus information' (E + I) in which information is viewed to be as relevant as the electricity itself [18]. This is a realization common in heavily industrialized societies.

In addition to having a set of common interests, societal behavior can also be influenced by a *common tradition*, *institutions*, and *collective activities and interests*. Again, and certainly within the context of the European Union, it has been argued that deregulation, internalization, liberalization, and unbundling of the European Electric Power System can be attributed to interests in use of digital electronics, better measurements, quicker operations, powerful control schemas and broadband access [18]. These perspectives alone seem to indicate that the modern world is interconnected as a result of increasing cooperation which could be attributed to common activities, needs, or demands.

Consequently, the term *society* encompasses traditions and institutional needs, demands, activities, and interests beyond those of a culture, which we have defined within the context of a given organization or a profession. In our current discussion *society* could be defined in terms of *a larger group of people (across time and space) with common traditions, institutions, activities and interests*. In context of risk analysis and management of risk, there is utility in consideration of culture in the articulation of risk events especially as people of the same 'society' are expected to have similar mindsets concerning appropriate approaches related to addressing problematic situations.

The concepts of culture and society are purposefully selected and contrasted to engineering risk to illustrate how the classical view of risk (i.e., measures of probability and consequence) evolve in today's more complex operating landscape.

In this landscape, there is an increasing realization that promises of modern systems free of risk from science, state, business, and defense are failing [6, 16]. Society and institutions themselves continue to change with rapid technological innovations—as a result of increasing demand for cheaper, higher quality, and safer systems as well as higher levels of outputs (i.e., products, goods, and services) from such systems. Moreover, these changes continue to transform society's current penchant for creating systems not bounded by traditional geo-political boundaries. It is from this worldview that engineering managers and system engineers must address the current hybrids of systems, spanning the manmade and natural formulations. In risk and especially disaster-type scenarios, risk based efforts can benefit by consideration of several measures and factors that might influence how people make decisions and take actions in risky-type situations. These measures are the basis of the discussion in the following section.

3 A Review of Literature: The Influence of Culture and Society on Risk Analysis

This section provides a review of literature supporting the supposition that culture and society influence activities and approaches that might be used in dealing with complex situations. Moreover, it has been suggested that how one deals with a situation can be related to deep-seated dispositions, inclinations, and tendencies [22]. This is an issue of extensive research in the social sciences.

In cultural theory it is widely contended that the humanity based environments contain social 'norms' that drive individual and societal perceptions on health, risk, and safety [12, 45]. Cultural theory is essential in risk analysis and management since it can be used in attempts to explain how having shared values, goals, and practices (cultures) and common traditions and interests (society) influences how individuals and large groups approach risk events. Although studies in cultural theory are not known for their attempts to distinguish culture and society, they reinforce the idea that *culture* can be viewed in terms of groups and individuals. The group level is concerned with inside and out of the group while the individual level is concerned with being influenced by or attempting to influence a group from an individual perspective.

Interestingly, discussions about risk in cultural theory are not restricted solely to issues of safety in reducing magnitude and probability. Tansey and O'Riordan [45] suggests that risk is best understood when it is viewed in terms societal issues of *power*, *justice*, and *legitimacy*. This is based in the notion that traditional risk quantification measures of probability and consequences is too simplistic—technically driven, based on utility theory, and heavily intertwined with the assumption that all humans make rational decisions. This issue is supported by McKinnon's [28] research which highlighted weaknesses in one of World Bank's social protection frameworks. McKinnon observed that social risk management for the World Bank did not account for contextual issues "preferencing the primacy of individual responsibility before collective action…[and] challenges the aspirational and redistributive policy agenda of social security in its pursuit of 'social justice' for all" [28, p. 22]. In other words, there is need to consider *irrational* actions of preferencing to act based on self-sacrifice—deferring to the interests of others beyond self.

Personal attitude also influences different approaches to risk. This might be more evident in *modern society* where there exists a belief that science, state, business, and the military have failed to deliver on their promises of a modern world free of risks. Arguably, free will is a major driver in how individuals view risk-related aspects including the identification, mitigation, and communication related to risk. In culture theory, this categorization of risk is defined in terms of *individual risk approach*—how individuals perceive and measure risk events. To illustrate: consider personal interpretations of risks associated with GMF and global warming. These interpretations induce people to view the associated risks in a particular manner. An alternative approach is a *cultural risk approach* in which social solidarities, judgments (about fairness and reliability), and risk communication are more important than the individual view of risk [45]. A cultural risk approach has been instrumental in the emerging worldviews of how to address global risks as suggested by Aaron and Dunlap [1] and Beck [6].

Another issue to consider is *social context.* Tansey and O'riordan [45] note that social context "impose[s] order on reality in particular ways" (p. 73). In other words, the need to maintain social identity might invoke particular preferred approaches to dealing with phenomena. In order to maintain societal identity and reduce risk, Tansey and O'riordan [45] note that some *cultures* may promote certain 'norms' as explained in an example where women are encouraged to avoid contact with cattle for the sake of not causing death. While this approach might appear 'primitive' to some, it turns out that even 'non-primitive' societies are equally influenced by societal 'norms.' Examples in which *culture* influences risk includes issues of "immorality" and "promiscuity" [45], the fight against terrorism, global warning, and GMFs.

Tansey and O'Riordan [45] have also postulated that culture and society affect risk analysis via the *degree of connectivity* which is explained in terms of 'center' and 'border' of an issue at hand. In the 'center' of the culture, Tansey and O'riordan [45] notes that people feel more attached to the beliefs held by the group as opposed to people on the borders. This would then suggest that people in the center of the group will have a higher degree of legitimacy and influence on those on the border. The ones on the border are more susceptible to following the center's approach an issue such as risk and approach [45]. Hence, one can reasonably conclude that a boundary associated with risk, which will be reflected in people and their tendencies, can be a manifestation of influence of a given culture. This idea seems to be supported by Douglas and Wildavsky [13] who suggested that being close to the beliefs of a group plays a major role in evolution, transformation, and quantification of risk. Certainly, this can have a significant implication for how to conduct analysis, modeling, and management of engineered systems. There are two major implications stemming from this discussion: (1) an issue that may have been considered relevant could become irrelevant and largely ignored and (2) the group or individual is able to identify new relevant issues—risk events. To illustrate, consider 'new' risks that are being associated with the field of *critical infrastructures*. This field involves identification and protection of essential systems (physical and/or information) whose destruction can have debilitating impact on public well-being. This emerging view of these systems is bringing to light new risks and vulnerabilities [25]. Moreover, it should be noted that discovering these 'new' risks could also be a basis that creates isolation among groups and individuals especially in scenarios where some groups fail to 'see' the new risks.

McKinnon [28] observed yet another way in which culture and society influences risk: *through powerful organizational structures*. This idea becomes more evident when one examines the role and influence of political powers, organizational agendas, and economic interests of certain entities such as those involved in risk assessment of nuclear energy programs and numerous accidents such as Three Mile Island, Exxon Valdez oil spill, and the Challenger accident. Tierney [47] notes that organizations such as the Atomic Energy Commission, the Joint Committee on Atomic Energy, and the Nuclear Regulatory Commission can influence risk analysis and management by promoting their ideas. For example, despite mounting evidence from the Brookhaven National Laboratory, regarding safety issues in power plants, the Atomic Energy Commission, the Joint Committee on Atomic Energy, and the Nuclear Regulatory Commission "consistently pushed the idea that nuclear power plants were necessary, safe, and economical" [47, p. 223]. Alternatively, political powers and organizational structure use mass media to influence risk and how it should be perceived. Tierney [47] espouses that "people's perceptions on risk are shaped by the ways in which risk-related information is communicated to them by these sources" and that "perceptions are also influenced significantly by the trust people have in organizations, including the producers of hazards, the organizations providing risk information, and the organizations responsible for protecting the public" (p. 234). Recall that the issue at hand is not whether the way of influencing people on risk is 'correct' or 'incorrect,' but rather, a recognition that this is one of the ways factors that influence risk should be noted when addressing how people might view risk. This issue of powerful organizational structure is closely related to Tierney's [47] knowledge base. *Knowledge base* is related to specific means that might be used in the calculations of risk which might involve a knowledge base as established by engineers, scientists, and certain government agencies.

Literature also establishes that *race and ethnicity* can be major influences concerning risk. Using an example of industrial pollution of the 1970s, [47, p. 232] notes that "states, poor, minority, and less politically organized communities end up as the 'hosts' for such [toxic and polluting] facilities." While noting that correlation does not imply causality, Tierney [47] argues that most low-income and minority communities could not simply resist becoming exposed to toxic wastes since they could rationalize the risks in terms of employment provisions. This can be compared to accepting risk, given in terms of freedom, which could be associated with giving up personal freedoms—freedom of expression (e.g., posting a message on WWW) despite possible ruminations.

In the perspective of *national origin*, it is stipulated that people's background can also have an influence on different risk perception. For example, Weber and

Hsee's [51] research illustrates that there are differences in how American students and their counterparts from China perceive risk. After using a survey, Weber and Hsee [51] concluded that since Chinese culture emphasized collectivism and inter-dependence on others in one's family and the community as a whole, there is a tendency to engage in 'risky' behavior. This is the opposite for highly individualized nations like the U.S. In effect, people in the Chinese culture are able to see 'big risks' as small because of their collectivism ideology [51].

A final perspective is that of allocating risk by the default of *prestige*. Beck [6] espouses that this is an issue more prevalent when one does a comparison between the west and east. According to Beck [6], Western nations are always seen to portray a symbol of success and safety because they appear to have succeeded in "bringing under control contingencies and uncertainties" (p. 332) involving issues such as accidents, violence, and sickness. Coincidentally, this view allows developing nations to engage in "lax [rules] or nonexistent land-use regulation, the proliferation of squatter camps in hazardous areas, environmental degradation, insufficient infrastructure to support the population and provide protection should a disaster occur, and governments that are only too willing to allow risks to be imposed on the poor for the benefit of elites" [47, p. 235]. Nations in this category, may see that allocation of risk by the 'elite' nations as the only way forward. After all, when 'amateur' nations are unable to recover toward 'elitism,' they are often considered to be living in pathological conditions, undeserving, and beyond help [28].

This discussion attempts to draw out indicators of *primary influences* of culture and society in the analysis and management of risk in engineered systems. These include individual 'norms' informing personal attitudes, organizational structures (i.e., power, justice, and legitimacy), knowledge base, and social context (i.e., environment). Additionally, degree of connectivity (i.e., being at the 'center' verses being at the 'border' of issues), race and ethnicity, and media and communications affect risk perception. The reviewed literature also suggests that national ideology (i.e., individualism versus collectivism) affects how risk events and incidents are perceived. This includes the amount of risk people are willing to be exposed to—which is related to understanding risk, management, and vulnerabilities.

There are two implications of interest related to our current development efforts concerning risk. First, an engineer and/or policymaker has to realize that in addressing current issues, one cannot simply rely on traditional risk measures of *probability* of occurrence and *potential* of harm. Such an approach assumes that man makes rational decisions. Second, the proceeding section indicates that risk analysis and management approaches should consider contextual issues that might influence actions and decisions. A reflexive modernization philosophy (transformation) in which the society as a whole uses reflex insights to reform and shape its future using principles of adaptability, sustainability, and precautionary principles is suggested by Beck [6]. This is especially the case since "disaster [emerges] from what we do not know and cannot calculate" [6, p. 330]. It is from these insights that the analogy (Fig. 1) of an onion peel level of issues was developed. The accompanying table (Table 2) elaborates on each level attempting to create a relationship between risk, individual, culture, and society.

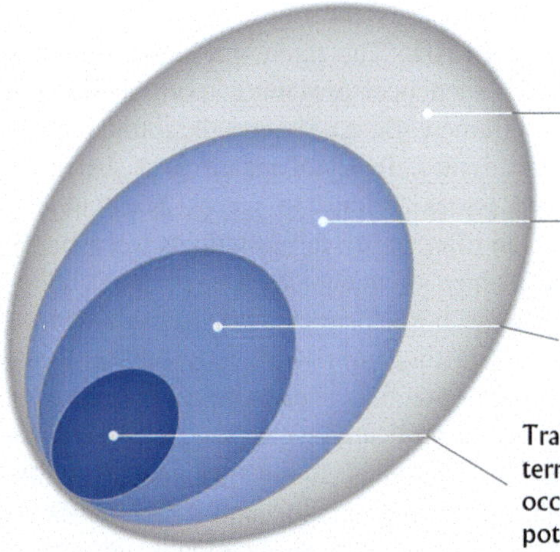

Fig. 1 An onion peel analogy of levels of issues to consider

Table 2 An expanded mapping of different levels to support individual and group perceptions

Level of quantification	Level descriptions	Relevant observations
1: Traditional risk	Founded on the basis of probability of occurrence (O_p) and potential loss ($L_{p.}$)	Besides O_p and L_p, risk calculations may involve measures of exposure, communication, perception, and vulnerability [50]
2: Individual level	A consideration of risk (i.e., $O_{p.}$ and L_p) and influence of individual beliefs	The individual perception measures might involve understanding the role of choice, free will, predispositions, and personal beliefs
3: Culture level	Understanding risk, individual beliefs, and a consideration of culture at the level of the system (organization) of interest	The culture at the location of system/issue at hand may be discerned through ***shared attitudes, values, goals, and practices*** involving organizational structures, knowledge base, degree of connectivity, norms, mass media, and race/ethnicity and dominant ***philosophical paradigms*** (e.g. laissez-faire and precautionary)
4: Societal level	Understanding risk, individual beliefs, culture, and influences at the societal level	Societal influences might be discerned through examination of ***shared attitudes, values, goals, and practices*** at a society level. The scale of operations for societal is at a global level involving elements of time, space and magnitude (e.g., organizational culture versus national ideology)

The proposed risk framework has three phases (i.e., peels) beyond the core which is defined based on classical view of risk. The three 'peels' are meant to ensure a more holistic view of risk with each peel providing more structure to understanding through risk analysis. As a framework, it only lays the foundation for understanding and dealing with complex issues. It is not meant as a systematic process to solving complex risk related issues. Rather, it serves as a guide to support the present topic. Moreover, it can offer utility to individuals interested in deeper understanding and appreciation of how their choices and personal beliefs influence how risk is perceived. It also offers a basis for seeing effects of culture and society on individual and risk events. While this framework is not a full-blown methodology, it provides a glimpse into issues that may need to be considered when addressing risk events involving individuals and large groups (i.e., culture and society) with implicit predispositions. The presented framework is only a first iteration and as such must be prone to inconsistencies and incompleteness. Refinement will be achieved through applications. A case application is presented is the following section.

4 Case Application: Hurricane Katrina

4.1 Context of Hurricane Katrina

New Orleans is located in southeastern part of the state of Louisiana, straddling the Mississippi River. The city and Orleans Parish boundaries are extended to parishes of St. Tammany to the north, St. Bernard to the east, Plaquemines to the south and Jefferson to the south and west. City limits also include Lake Pontchartrain (north) and Lake Borgne lies to the east. The geography of New Orleans is dominated by water.

During the 2005 Atlantic hurricane season, New Orleans was engulfed by floods from Hurricane Katrina causing loss of life and huge socio-economic damages. It is also fair to describe the New Orleans area as below sea-level and therefore "prone to flooding from the river, the lakes, and the Gulf of Mexico" [49, p. 35]. Plans to protect the city from flooding have always been in place and they date back to 1927. The Great Mississippi Flood of 1927 led the U.S. Army Corps of Engineers (USACE) to construct the world's longest levee system under the Flood Control Act of 1928 [5]. Additionally, the New Orleans Flood and Hurricane Protection System was created to protect the city from flooding. It consisted of "350 miles of levees which are embankments, usually earthen, that serve as flood barriers... floodwalls, hundreds of bridges, closable gates, culverts and canals that facilitate transportation in and out of the system. It is comprised of a series of four main compartmented basins designed to limit the flooding impacts on the entire system resulting from individual failures of levees and floodwalls. In addition, large pump stations are used to pump out and redirect water from the city. These pumps are designed to mitigate flooding that results from significant rainfall and can, over time, remove water from moderate overtoppings" [49, p. 35].

Prior to Hurricane Katrina, the levee system in place was constructed to protect the city from at up to 18 ft. above sea level. The levee system was also designed to withstand Category 2 hurricanes. However, as Hurricane Katrina made landfall in Louisiana on August 29, 2005, the hurricane was classified as a Category 3 storm. A Category 3 storm has 111–130 mph sustained winds on the Saffir-Simpson Hurricane Scale as compared to 96–110 mph for a Category 2 storm. In essence, the city was not prepared for the corresponding storm surge, extreme amount of rain, and high winds, which in retrospect were responsible for most of the damage in the city, including inoperability of the information and communication systems [49]. Figure 2 depicts the path of Hurricane Katrina. With over 80 % of New Orleans flooded, Hurricane Katrina asserted itself as the most destructive hurricane US history [4]. The levee system failed to prevent storm surge and this failure has been described as the "most costly failure of an engineered system in U.S. history" with an estimated damage cost of over $100 billion [40, p. 702].

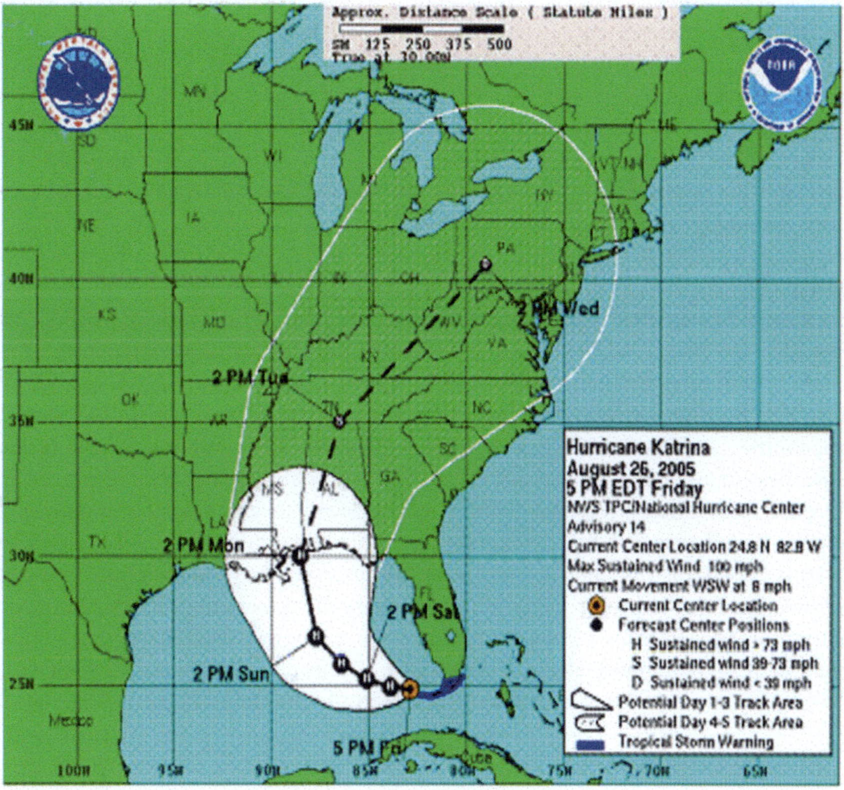

Fig. 2 Timeline and path of Hurricane Katrina [49, p. 23]

After Katrina, the federal, state, and local governments issued several reports on the levee system as well as lessons learned. First, it was widely-accepted that the system built to protect the city failed miserably. This failure was partly attributed to choices and actions of USACE especially the design and construction materials that were supposed to prevent breaching and overtopping. Sills et al.'s [41] research is more specific: the "Levees that had been constructed using hydraulic fill and higher silt and sand content were severely damaged…on the other hand, rolled fill levees that were constructed of cohesive materials, for the most part, were able to survive without breaching" (p. 560). These findings are echoed by Seed et al. [40] and *The Federal Response to Hurricane Katrina: Lessons Learned* [49].

Secondly, review of literature supports the idea that different groupings of society, culture, and individual played important roles in decision-making of those affected by the disaster event (i.e., Hurricane Katrina). This is especially the case for decision-making involving the issue of evacuation. Specifically, it has been suggested that the *social characteristics* of New Orleans played a major role in how the city evaluated, coped, and responded to the threat of Hurricane Katrina [9]. Moreover, Townsend [49] notes that the levee system in New Orleans (pre-storm) had been designed to meet the specification of past storms and not the storm and the surge that took down the levees. In essence, there was an anticipation of the wrong kind of threat. Interestingly, when the levees failed during the Great Flood of 1927, the USACE opposed building of spillways and floodways which could have enabled the levees to withstand the increased volume of water [5].

A synthesis of literature also reveals that a large number of people did not evacuate New Orleans because of how they perceived risk associated with the hurricane, the potential consequence, and the warning mechanisms [14, 15, 40, 49]. More succinctly, Elder et al. [15, p. S113] notes that a "collective memory of past hurricanes combined with distrust of authorities led to minimization of their perceived risk associated with Hurricane Katrina." Similarly, Burnside et al.'s [7] research indicates that the decision to evacuate, during disaster-like events, is dependent on perceived risk which could be heightened by dissemination of consistent information and understanding of subcultures that might exist in an affected area. This information would seem to suggest a need to understand risk in terms of probabilities and consequence as well as how individuals might be viewing the same disaster and the influence of local culture and society as a whole. However, in the case of Hurricane Katrina there is also a need to examine the engineering culture. Recall that it is widely accepted in the engineering community to design systems based on previous events using engineering base knowledge. Clearly, failures associated with the landfall and the subsequent consequences of Hurricane Katrina in New Orleans cannot be attributed to a single issue. There was a combination of several interrelated factors at play. To this end, the American Society of Civil Engineers noted that four major factors in the flooding [4, pp. 9–10]:

- First, experts knew that a hurricane like Katrina was inevitable, and that when it occurred the city would be flooded. No one heeded their warnings or effectively communicated the risks to decision makers, government officials, or the people of who lived in New Orleans.
- Second, although the southeast Louisiana hurricane-protection system was a complex assemblage of earth levees, concrete and steel walls…it was a system in name only. It was not designed as a system, nor operated as one. It was planned, designed, constructed, and operated without a system-wide approach or integration with land use, emergency evacuation, or recovery plans.
- Third, everyone was in charge, and yet no one was in charge…No single agency or organization was empowered to provide system-wide oversight or a focus on critical life-safety issues. The result was management by committee, and no one could say, 'The buck stops here.'
- …the hurricane-protection system was designed and constructed over a 40-year period with little adjustment to changing regional conditions. Despite new meteorological information, the standard project hurricane (the design hurricane) was not updated…And despite field test data that showed unacceptable deflections in I-walls resisting loads from the floods, I-wall designs were not revised.

This section provides necessary and sufficient background information to begin to understand issues regarding Hurricane Katrina landfall and effects on residents of New Orleans. This is an appropriate place to examine how tacit knowledge of individuals and groups (i.e., in terms of culture and society) might have influence actions in considering Hurricane Katrina as a risk event.

4.2 Action Taking: The Case for New Orleans During Katrina

The working assumption of this chapter is that organizational (power and legitimacy) structures, knowledge base, degree of connectivity and belief in themes, national origins, societal norms, mass media, race and ethnicity, and personal attitudes influence decision-making and actions at the individual, culture, and societal levels. According to Reason [36], it is essential to recall too that making a decision and taking actions are related to mental cognitive processes. In the scenario of Hurricane Katrina, taking action might involve a range of issues including selection of levee design, when and how to disseminate information, and certainly deciding to or not to evacuate.

In the proposed risk framework, the core is the traditional risk formulation. In the context of the current discussion, this is provided by examination of the risk event—Hurricane Katrina. Meteorologists often use satellite imagery to map out positions of the large air masses circling the Earth and are able to predict weather patterns with a relatively high degree of accuracy. Such information contains probability of occurrence of certain events and the type of damage that is typically associated with

Table 3 Category types and associated damages

Category	Sustained winds	Types of damage due to hurricane winds
1	74–95 mph	It suggests that very dangerous winds will produce some damage. Well-constructed frame homes could have damage to roof, shingles, vinyl siding and gutters. Large branches of trees will snap and shallowly rooted trees may be toppled. Extensive damage to power lines and poles likely will result in power outages that could last a few to several days
2	96–110 mph	It suggests that extremely dangerous winds will cause extensive damage. Well-constructed frame homes could sustain major roof and siding damage. Many shallowly rooted trees will be snapped or uprooted and block numerous roads. Near-total power loss is expected with outages that could last from several days to weeks
3	111–129 mph	This is considered a major category. It suggests that devastating damage will occur. Well-built framed homes may incur major damage or removal of roof decking and gable ends. Many trees will be snapped or uprooted, blocking numerous roads. Electricity and water will be unavailable for several days to weeks after the storm passes
4	130–156 mph	A major category hurricane suggesting that catastrophic damage will occur. Well-built framed homes can sustain severe damage with loss of most of the roof structure and/or some exterior walls. Most trees will be snapped or uprooted and power poles downed. Fallen trees and power poles will isolate residential areas. Power outages will last weeks to possibly months. Most of the area will be uninhabitable for weeks or months
5	157 mph or higher	A major category suggesting that catastrophic damage will occur. A high percentage of framed homes will be destroyed, with total roof failure and wall collapse. Fallen trees and power poles will isolate residential areas. Power outages will last for weeks to possibly months. Most of the area will be uninhabitable for weeks or months

such events. This was the case for Hurricane Katrina even though it landed as a Category 3 after being predicted as a Category 2 [49]. Table 3 depicts effects associated with different types of hurricane categories [32]. The key issue here is that probability of occurrence was known along with its associated potential impact.

This view of risk formulation only offers a partial view of why there might have been so much damage and loss of lives in the New Orleans area. According to the proposed framework, there is a need to consider both individuals and groups. The individual level involves the consideration of factors that might have influenced how individuals perceived threat associated with Hurricane Katrina. There is also a need to examine culture and society levels (i.e., shared attitudes, values, and practices) that might have influenced actions and decisions.

For New Orleans and Hurricane Katrina, culture could be described in terms of different players including organizations (e.g., USACE) while society could be

Table 4 Going beyond probability and consequences

Level of consideration	Factors	Example of supporting text
Individual	Perception of a risk event as they relate to choice, predispositions, and personal beliefs	…[p]articipants integrated media messages with the communications of friends, family, neighbors, and church members in deciding whether to evacuate… [14, p. S112]
Culture	Shared attitudes, values, goals, and practices	ASACE standard practice of inspecting the levee system revealed that the levee system had an 'acceptable rating.' This ensured that nothing much could be done to improve the state of the levee system [39]
Society	Shared attitudes, values, goals, and practices	Institutional values of the Federal government, state, and local authorities (i.e., power, justice, and legitimacy) influence actions that were undertaken by President Bush in the declaration of 'State of Emergency, Governor Blanco and Major Nagin's issue for 'Mandatory Evacuations' due to threat to public wellbeing

described in relation to *physical boundaries* of New Orleans, the State of Louisiana, or the at the Federal level. Interestingly, a simple examination of events leading up to, during, and the aftermath of Hurricane Katrina suggests that going beyond probabilities and consequences provides insights into actions and decisions of different players (i.e., residents and rescuers). Table 4 is an attempt to examine the suggested three levels, beyond the traditional risk formulation, that could influence decision and action.

Even though Hurricane Katrina had been classified as a powerful Category 2 storm, choice, free will, predispositions, and personal beliefs regarding a threat event were still influencing actions of individuals. In fact, some individuals opted to stay in New Orleans despite 'mandatory evacuations orders.' This suggests a need to know the attitude of people towards a risk event. By some estimates, only 73.8 % of residents of New Orleans had evacuated at least once in the past 10 years before Katrina [7]. One of the major factors was the 'crying wolf' hypothesis. In this case, value associated with the threat of landfall of Katrina diminished.

It has also been suggested that the culture of New Orleans was a major influence on events surrounding Hurricane Katrina. If one takes *culture* to be a set of shared attitudes, values, and goals, then Campbell's [9] research provides a basis for defining the culture in New Orleans as that classified by repeated exposure to past hurricanes, lack of trust in government, environmental injustice, low income and poor education system, and differing philosophies in regards to dealing with New Orleans protection. While there is no succinct information regarding how cultural philosophies of *laissez-faire* and *precautionary* principle might have played a role, applying these philosophies could provide further insights into the culture of the city as well as discovery of deep systemic issues that might have influenced decisions and actions.

Society was defined in terms of common tradition, institutions, and collective activities and interests. If this definition is extended to different institutions and organizations, then it is possible to acknowledge the role of local, state, and federal institutions such as Federal Emergency Management Agency (FEMA) and USACE. This is especially the case since these organizations are involved in the monitoring, mitigation, rescue, and recovery efforts during disaster events [49]. Such entities have legitimate powers instrumental in threat assessment, risk communication, risk perception, vulnerability assessment, and assignment of probabilities and consequences. These entities have a shared vision based on belief in protecting citizens from natural events which was manifested through actions of building the levee system, providing shelter, and issuing evacuation directions.

Specific influences related to Hurricane Katrina also included *knowledge base* and *mass media*. Knowledge base is related to data that is previously collected, shared, and organized into information to enable decision-making and action. In the case of Hurricane Katrina, it has been suggested that knowledge base played a major role, especially during levee design and evacuation. For example, the knowledge base of USACE and its local sponsors was influential in characterizing the levee system as having an 'acceptable' risk to the city of New Orleans prior to storm surge [39]. In years and months leading to the levee collapse, annual inspections revealed that the levee system had an acceptable rating [39]. Furthermore, the Federal government, USACE, and other authorities had accumulated data over the past six times hurricane events affected New Orleans [39, 40]. This view is also supported by Elder et al.'s [15] research which suggests that some individuals never left the city because of thc knowledge they had regarding past hurricanes.

Media technologies including the internet, radio, cinema, television, blogs, RSS feds, podcasts, mobile, newspapers, magazines, books, and outdoor media were used to reach large audiences *en* mass. Campbell's [9] work suggests that mass media can have a positive and influential impact on a given phenomenon. In the case of Katrina, mass media exposed the threat of a deadly event and their consequence to residents [39]. It is also credited with saving many lives during the storm, especially since the National Hurricane Center "disseminated warnings and forecasts via NOAA Radio and the internet, operating in conjunction with the EAS [Emergency Alert System]" [49, p. 28]. The importance of mass media is also expressed by Elder et al.'s [15] research which notes that residents of New Orleans "expressed trust in media reporting of the size and strength of the hurricane, and almost all information on the hurricane was received from television reporting" (p. S127).

Thus far, two conclusions can be drawn for this research. First, the example of Hurricane Katrina offers compelling support for the notion that knowing probabilities and consequences of an event is essential and fundamental to risk formulation. However, there is a need to also consider those deep seated fundamental assumptions that might influence actions of individuals in disastrous situations. The classical view of risk does not adequately appreciate complexity inherent in the

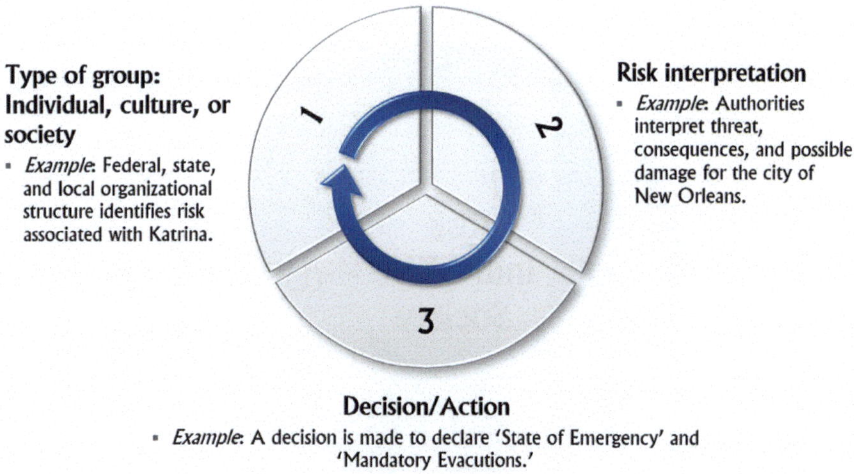

Fig. 3 An example of how actions could be linked to risk interpretation and group

nature of Katrina-like events, the tacit nature of people, or the context of their environment. Second, the proposed framework's attempt to more holistically consider the influences of culture and society to risk might provide a means to think more critically about risk formulation. Specifically, how culture and society influence decisions and actions and as such are necessary in modeling, analysis, and management of risk engineering. It is possible to draw a direct link between a risk event, actions, and society. This is illustrated in Fig. 3 where actions (e.g., declarations) are developed based on risk perception (e.g., flooding) and influenced by society (i.e., a knowledge based on risk to public). This supports the notion that one might act based on where they are within a group and thus being influenced by the group [45].

If this is the case, then there becomes a need to acknowledge individual and group knowledge (implicit and/or explicit) in risk, especially in cases where people might need saving. This approach might involve incorporating issues such as consideration of knowledge bases, effective risk communication, risk perception, and evaluation of system vulnerabilities, susceptibility, and fragility in addition to core measures of probability and consequences to develop more robust frameworks for risk assessment and management.

4.3 Limitations of RiskIS Framework

The presented framework (i.e., RiskIS) seeks to include individual and society in risk analysis. At the individual level, RiskIS enables a consideration of a vast number of individual issues (i.e., choice, free will, predispositions, and personal beliefs) that influence decisions and actions. The society aspect attempts to ensure risk formulation

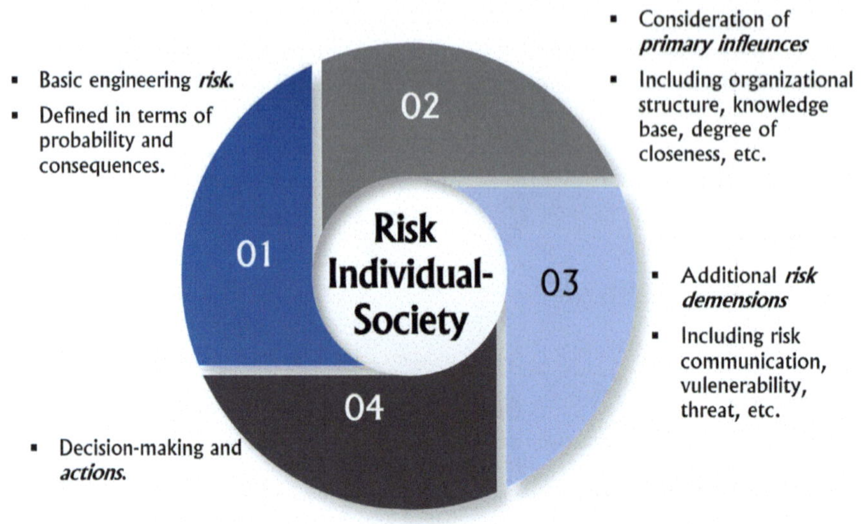

Fig. 4 Elements of revised RiskIS framework

inclusive of shared attitudes, values, goals, and standard practices that might shape decisions and actions. However, the application of this framework to Hurricane Katrina as a risk event provides means to refine the suggested framework. It was suggested that classical risk formation lies at the core of the framework and that there are three additional peels (i.e., individual, culture, and society). In light of the application to Hurricane Katrina, there appears to be no significant difference between culture and society when it comes to primary influences. In fact, *organizations* which is part of 'culture' and institutions which is part of 'society' actions are heavily influenced by similar issues i.e., organizational structures and knowledge base). This would seem to support the notion that society is simply a *larger* group than a culture, as previously defined. Thus, *culture* and *society* are taken to be the same in the context of RiskIS. A revised framework is presented in Fig. 4.

This revised framework also omits national origin, race, and ethnicity since in the Katrina scenario there is no evidence these were major factors influencing decisions and actions. However, there is evidence that race and ethnicity might have played a role in actions of the policymakers and potential evacuees [3, 9, 15]. However, it appears that this was not a systemic and wide-spread issue.

5 Conclusions and Further Research

In a global society, dealing with risk requires holistic approaches capable of addressing different aspects of emerging issues. In this research, classical risk formulation based on the measures of probability and consequences are presented as foundational. However, there is a need to consider fundamental issues that might influence individuals and groups to make decisions and take actions. Literature suggests that decision-making and action-taking is influenced by choice, free will, predispositions, and personal beliefs regarding a risk event at a personal level. At the society level, decision-making and actions are influenced by the core beliefs of the group. People at the 'center' of the group will have a higher degree of legitimacy and influence on those on the 'border' of the group. This information could be essential in disaster stricken areas especially when attempting to develop effective communications systems. More specifically, simply providing information about an impending risk event might not be necessary and sufficient to enable better decision-making and actions such as those related to evacuation. There is still a need to understand the tacit issues and role of society in such scenarios.

This research also provides a set of *primary influences* affecting decisions and actions of those in disaster-type scenarios. These include individual and societal 'norms', personal attitudes, organizational (power and legitimacy) structures, knowledge base, and social context (i.e., environment), degree of connectivity, race and ethnicity, mass media, and national ideology. A consideration of these issues in analysis, modeling, and management of engineered systems might provide insights into issues that would otherwise be left untapped. For example, the consideration of *degree of connectivity* could reveal who is at the 'center' of a group and who is at the 'border.' Strategies to influence these groups of people could then be developed for the targeted audience.

Cleary, and as illustrated by the case of Hurricane Katrina, ***different people view eminent risk differently***. Accepting as a given that people view a risk even in a similar manner is a challenge that remains open-ended. Models that attempt to represent risk events from a classical risk view and then incorporating individual and society layers might need to be developed on a case by case basis. This might present challenges associated with capturing tacit information of individuals and that of the society at large. Moreover, one might fail to see the utility of such models since they are not intended to be used on similar disaster. Thus, there is also a challenge of building a distinct model for each given scenario.

Finally, the developed RiskIS framework is only a framework. It is developed out of need for reformulation of how risk should be addressed in context of disasters with a simple case application to Hurricane Katrina. Its underpinnings, especially the measures that influence decisions and actions at the individual and society level need to be 'tested' in other scenarios. Nevertheless, the proposed research provides a basis for including personal beliefs and groups into risk formulation.

Acknowledgment This research was partly supported by a grant from the National Centers for System of Systems Engineering (NCSOSE) and Emergent Risk Initiative (ERI) @ Old Dominion University. NCSOSE is a university Enterprise Research Center of the Batten College of Engineering and Technology focused on decision making for multidisciplinary problems. The vision of *ERI@ODU* is to create the next generation body of knowledge in risk management for current and future systems and organizations characterized by uncertainty, emergence, complexity, and interdependence.

References

1. Aaron MM, Dunlap RE (2000) Challenging global warming as a social problem: an analysis of the conservative movement's counter-claims. Soc Probl 47(4):499–522
2. Ackoff RL (1974) Systems, messes, and interactive planning. In: Redesigning the future: systems approach to societal problems. Wiley, New York, NY, pp 20–33
3. Adeola FO (2007) Nativity and environmental risk perception: an empirical study of native-born and foreign-born residents of the USA. Hum Ecol Rev 14(1):13–25
4. ASCE (2009) Guiding principles for the nation's critical infrastructure. American Society of Civil Engineers, Reston
5. Barry JM (1997) Rising tide: the great Mississippi flood of 1927 and how it changed America. Simon and Schuster, New York
6. Beck U (2006) Living in the world risk society. Econ Soc 35(3):329–345
7. Burnside R, Miller DMS, Rivera JD (2007) The impact of information and risk perception on the hurricane evacuation decision-making of greater New Orleans residents. Sociol Spectr 27(6):727–740
8. Calida BY, Katina PF (2012) Regional industries as critical infrastructures: a tale of two modern cities. Int J Crit Infrastruct 8(1):74–90
9. Campbell NM (2009) Hurricane Katrina and the perception of risk: incorporating the local context. University of Central Florida, Orlando
10. Chatterjee S, Abkowitz MD (2011) A methodology for modeling regional terrorism risk. Risk Anal 31(7):1133–1140. doi:10.1111/j.1539-6924.2010.01565.x
11. de Weck OL, Roos D, Magee CL (2011) Engineering systems: meeting human needs in a complex technological world. MIT Press, Cambridge
12. Douglas M (1992) Risk and blame: essays in cultural theory. Routledge, New York
13. Douglas M, Wildavsky A (1982) Risk and culture: an essay on the selection of technical and environmental dangers. University of California Press, Berkeley
14. Eisenman DP, Cordasco KM, Asch S, Golden JF, Glik D (2007) Disaster planning and risk communication with vulnerable communities: lessons from Hurricane Katrina. Am J Public Health 97(Supplement_1):109–115
15. Elder K, Xirasagar S, Miller N, Bowen SA, Glover S, Piper C (2007) African Americans' decisions not to evacuate New Orleans before Hurricane Katrina: A qualitative study. Am J Public Health 97(Supplement 1):S124–S129
16. Escobar, A. (2004). Beyond the third world: imperial globality, global coloniality and anti-globalisation social movements. Third World Q 25(1):207–230. doi:10.1080/0143659042000185417
17. Gheorghe AV, Katina PF (2014) Editorial: Resiliency and engineering systems—Research trends and challenges. Int J Crit Infrastruct 10(3/4):193–199
18. Gheorghe AV, Masera M, Weijnen MPC, De Vries JL (eds) (2006) Critical infrastructures at risk: securing the European electric power system, vol 9. Springer, Dordrecht
19. Hill KN (2012) Risk quadruplet: integrating assessments of threat, vulnerability, consequence, and perception for homeland security and homeland defense (Ph.D.). Old Dominion University, United States—Virginia
20. INCOSE (2011) In: Cecilia H (ed) Systems engineering handbook: a guide for system life cycle processes and activities, 3.2 ed. INCOSE, San Diego
21. Katina PF (2015) Systems theory-based construct for identifying metasystem pathologies for complex system governance (Ph.D.). Old Dominion University, United States—Virginia
22. Katina PF, Keating CB, Jaradat RM (2014) System requirements engineering in complex situations. Requirements Eng 19(1):45–62
23. Keating CB, Katina PF, Bradley JM (2014) Complex system governance: concept, challenges, and emerging research. Int J Syst Syst Eng 5(3):263–288
24. Knabb RD, Rhome JR, Brown DP (2011) Tropical Cyclone Report Hurricane Katrina 23–30 August 2005. National Oceanic and Atmospheric Administration, Washington, pp 1–43

25. Kröger W, Zio E (2011) Vulnerable systems. Springer, London
26. Martin J (2006) The meaning of the 21st century: a vital blueprint for ensuring our future. Riverhead Books, New York
27. Masys A (ed) (2015) Disaster management: enabling resilience. Springer International Publishing, New York
28. McKinnon R (2002) Social risk management: a conceptual fallacy of composition. Risk Manag 4(2):21–31
29. Merriam-Webster (2006) Webster's new explorer encyclopedic dictionary. Federal Street Press, Springfield
30. Metz B, Kuijpers L, Solomon S, Anderson SO, Davidson O, Pons J, Meyer L et al (eds) (2005) Safeguarding the ozone layer and the global climate system: issues related to Hydrofluorocarbons and Perfluorocarbons. Cambridge University Press, New York
31. Needle D (2004) Business in context: an introduction to business and its environment, 4th edn. Cengage Learning Business Press, London
32. NHC (2015) Saffir-Simpson Hurricane wind scale. Retrieved 16 Dec 2015, from http://www.nhc.noaa.gov/aboutsshws.php
33. Parsons VS (2007) Searching for "Unknown Unknowns." Eng Manag J 19(1):43–46. doi:10.1080/10429247.2007.11431721
34. Paté-Cornell ME, Dillon RL (2006) The respective roles of risk and decision analyses in decision support. Decis Anal 3(4):220–232
35. Pinto CA, McShane MK, Bozkurt I (2012) System of systems perspective on risk: towards a unified concept. Int J Syst Syst Eng 3(1):33–46. doi:10.1504/IJSSE.2012.046558
36. Reason J (1990) Human error. Cambridge University Press, Cambridge, UK
37. Rittel HWJ, Webber MM (1973) Dilemmas in a general theory of planning. Policy Sci 4 (2):155–169. doi:10.1007/BF01405730
38. Schein E (1992) Organizational culture and leadership: a dynamic view. Jossey-Bass, San Francisco
39. Seed RB, Bea RG, Abdelmalak RI, Athanasopoulos A., Boutwell GP Jr, Bray JD, Cohen-Waeber J et al (2006) Investigation of the performance of the New Orleans Flood Protection System in Hurricane Katrina on 29 Aug 2005. University of California at Berkeley, Berkeley, CA
40. Seed RB, Bea RG, Abdelmalak RI, Athanasopoulos-Zekkos A, Boutwell GP, Briaud JL, Govindasamy AV et al (2008) New Orleans and hurricane Katrina. I: introduction, overview, and the east flank. J Geotech Geoenviron Eng 134.701
41. Sills GL, Vroman PE, Wahl RE, Schwanz PE (2008) Overview of New Orleans levee failures: lessons learned and their impact on national levee design and assessment. J Geotech Geoenviron Eng 134:556
42. Suh NP (2005) Complexity: theory and applications. Oxford University Press, Cambridge
43. Tainter JA (1988) The collapse of complex societies. Cambridge University Press, New York
44. Taleb NN (2010) The black swan: the impact of the highly improbable. Random House Trade Paperbacks Edition, New York
45. Tansey J, O'riordan T (1999) Cultural theory and risk: a review. Health Risk Soc 1(1):71–90. doi:10.1080/13698579908407008
46. Thissen WA, Herder PM (2003) Critical infrastructures: state of the art in research and application. Kluwer Academic Publishers, Boston
47. Tierney KJ (1999) Toward a critical sociology of risk. Sociol Forum 14(2):215–242
48. Tokgoz BE (2012) Probabilistic resilience quantification and visualization building performance to hurricane wind speeds (Ph.D.). Old Dominion University, United States—Virginia
49. Townsend FF (2006) The federal response to Hurricane Katrina: lessons learned. US Government Printing Office, Washington
50. Vamanu BI, Gheorghe AV, Katina PF (2016). Critical Infrastructures: Risk and vulnerability assessment in transportation of dangerous goods, Vol 31. Cham: Springer International Publishing
51. Weber EU, Hsee C (1998) Cross-cultural differences in risk perception, but cross-cultural similarities in attitudes towards perceived risk. Manage Sci 44(9):1205–1217

Application of Problem Inversion to Cascading Critical Infrastructure Failure

Ivan Taylor

Abstract Between 1994 and 2013, over 6800 natural disasters have occurred worldwide, claiming 1.35 million lives. Flooding accounted for over 40 % of these disasters displacing nearly 2.5 billion people. Storms were the second most frequent type of disaster: killing more than 240,000 people and costing over US$900 billion in damage to infrastructure. It is commonly felt that the number of severe natural disasters has been increasing in recent years because of climate change. However, the problem of climate change may not be solved in the near future. So governments need to prepare for frequent future natural disasters and find ways to mitigate the potential death and destruction they cause. This chapter will discuss a novel method for making preparations to avoid the problem of cascading disasters created by a single natural event. This approach will adapt a knowledge based technique from manufacturing, called Ideation Failure Analysis™, to correct deficiencies in critical infrastructure. The technique involves a number of approaches that are combined into a comprehensive process. First, a simple direct approach is attempted with the assistance of a knowledge base. If the problem can not be resolved in the direct manner, then an indirect approach is suggested. A detailed Failure Analysis Questionnaire is used to assist in model building. A model of the failure network is developed. However, instead to working directly towards failure correction, an inversion process is conducted. That is, in order to facilitate greater creativity, the analysis team is asked to imagine ways to produce the failure. The creative work is then assisted by a knowledge base. The analysis team is able to prioritize the likelihood that a cause might have resulted in the failure. The next step is to find ways to prevent, eliminate, or reduce the impact of the failure. Again to assist the creative process, a knowledge base provides suggestions for correction techniques that can be prioritized in a hierarchical fashion. Finally, the results are evaluated to avoid negative side-effects or drawbacks in the suggested ways to correct the failure. This chapter will conclude by providing some recommendations and an evaluation of the potential of this technique.

I. Taylor (✉)
Policy Dynamics, New Hamburg, Ontario, Canada
e-mail: Ivanwtaylor@gmail.com

A.J. Masys (ed.), *Disaster Forensics*, Advanced Sciences and Technologies for Security Applications, DOI 10.1007/978-3-319-41849-0_2

Keywords Critical infrastructure · Cascading failure · TRIZ · Ideation failure analysis

1 Introduction

Between 1994 and 2013, 6873 natural disasters have occurred worldwide, claiming 1.35 million lives. Flooding accounted for 43 % of these disasters displacing nearly 2.5 billion people. Storms were the second most frequent type of disaster: killing more than 244,000 people and costing US$936 billion in damage to infrastructure [1]. It is commonly felt that the number of natural disasters has been increasing in recent years because of climate change. However, the problem of climate change may not be solved in the near future. So governments need to prepare for frequent future natural disasters and find ways to mitigate the potential death and destruction they cause.

"Red Teaming" has been used by military planners for many years [2]. In this type of analysis, a war game is simulated in which one group of players called the "red team" try to sabotage or otherwise disrupt the plans of the so-called "blue team". By this challenge method of counter-analysis, plans can be tested under severe conditions to ensure they will hold up when used in real-life.

The author has adapted a version of the Nominal Group Technique [3] which is referred to as structured brainstorming. In a number of cases, this technique has been used with event planners [4]. The planners are asked to generate creative ideas to answer the question: "In what ways could the event fail?" This allows the planners to imagine the worst-case scenario and thus take appropriate actions beforehand to avoid the imagined problems. Klein [5] has used this technique successfully in project management. He appropriately named his approach a "pre-mortem".

The software package, called Ideation Failure Analysis™ [6], employs a similar problem inversion process to enhance the creativity of an analysis team trying to find the root cause of a manufacturing failure. This chapter will discuss the potential of problem inversion to find the reasons for cascading critical infrastructure failure caused by natural disasters. In this paper, we will use the knowledge based technique from manufacturing in the Ideation Failure Analysis™ software. However, we recognize that the knowledge base could be adapted to correct deficiencies in the design of critical infrastructure that lead to cascading failures.

It will be shown that a comprehensive application of the Ideation Failure Analysis™ to correct the failure could prove overwhelming to an individual analysis team or even a team of analysis teams. So, a prioritization technique based on the Analytical Hierarchy Process [7] is suggested to allocate effort optimally in the processes of failure diagnosis and correction.

The Ideation Failure Analysis™ process is quite time consuming when conducted in its full form. Therefore, a "short-cut" direct approach to the failure analysis is suggested that might resolve simple problems. Section 2 will describe this direct approach to the problem of finding the root cause of a cascading infrastructure failure caused by a natural disaster. Although this direct approach is not new, the Ideation Failure Analysis™ software supports the process by providing an extensive knowledge base of typical failure modes to facilitate the creative process conducted by the analysis team.

In Sect. 3, it is assumed that the direct approach did not lead to satisfactory results. Then the analysis team, using the Ideation Failure Analysis™ software, begins an indirect approach by completing a detailed Failure Analysis Questionnaire to support model building. We will provide two models that were developed using the software. These models are similar to network models that attempt to describe the interconnected process in a cascading critical infrastructure failure. These models are used to invert the problem of failure analysis. Using the software, the creative process is facilitated by asking the analysis team to suggest ways the failure could be produced.

The software provides "Directions" for investigation by the analysis team. Each of these Directions also comes with so called "Operators" which are specific examples and case studies of how the Direction might occur. Using this knowledge base of Directions, Operators and Case Studies, the analysis team is encouraged to develop ideas on the root cause or causes of the failure. In Sect. 4, this process is examined.

The process of determining how the failure can be corrected is described in Sect. 5. In this section, the Directions and Operators that are provided by the software to prevent, eliminate or reduce the impact of the failure are outlined.

In Sect. 6, the final steps in the Ideation Failure Analysis™ process are described. These steps involve the methods that can be employed to avoid negative side-effects or drawbacks of the solutions that were developed by the analysis team.

Section 7 provides a brief summary, some concluding remarks, some recommendations and an evaluation of the potential of the Ideation Failure Analysis™ for use by government agencies.

2 The Direct Approach

In the Ideation Failure Analysis™ process, the analysis team should begin by taking a direct approach to finding the root cause of the failure. The first step in this direct approach is to describe the situation in "everyday language" avoiding the use professional terminology. It is believed that this will result in the problem being "generalized," thereby the analysis team will be able to suggest more methods for solving it.

We can describe the problem situation as follows: "How do we avoid a cascading failure in the critical infrastructure of a city when a natural disaster occurs?"

The analysis team can now apply the knowledge base provided in the software directly. Below is a list of typical failures:

(a) Explosion,
(b) Combustion,
(c) Corrosion,
(d) Malfunction of electric or electronic device,
(e) Deformation or destruction,
(f) Disappearance of a useful object or material,
(g) Appearance of a harmful object or material,
(h) Disruption of useful system functioning, and
(i) Appearance of a harmful effect in the system.

Of course, more than one of these typical failure modes might be applicable. Then some process of prioritization could be conducted. One way this could be done is using the Analytical Hierarchy Process [7]. In this case, pair-wise comparisons of the importance of each of these failure modes could be turned into percentage values. These percentage values can then be used to allocate effort to resolving these failure modes starting with the failure mode with the highest percentage value. Table 1 shows the results of an example run of the Analytical Hierarchy Process for these typical failure modes.

The investigation of these failure modes by the analysis team is supported by a knowledge base in the Ideation Failure Analysis™ software. One of the features of the knowledge base is that it is highly hierarchical. In the next paragraph, we will outline the types of information available in the knowledge base by looking at the particular failure mode of an explosion.

The knowledge base contains detailed information of the potential causes of chemical, thermal and mechanical explosions. If none of these types of explosions seem applicable, the analysis team need not stop there. The knowledge base outlines how some object or material might enter the system to create an explosion. There are two aspects that need to be considered to make this happen: a driving force and a transport path. The knowledge base contains specific details on how large or small objects could enter the system to cause an explosion as well as how liquids or gases could enter the system to create an explosion. One way small objects, liquids or gases could enter the system is on "carriers" like system inputs or

Table 1 Prioritization of typical failure modes

Typical failure mode	Priority value (%)
Explosion	21
Combustion	16
Corrosion	0.1
Malfunction of electric or electronic device	10
Deformation or destruction	12
Disappearance of useful object or material	4
Appearance of harmful object or material	13
Disruption of useful system functioning	5
Appearance of a harmful effect in the system	19

auxiliary materials such as lubricants or coolants, through small openings like cracks, broken or damaged seals or through porous materials. As an example of the extensiveness of the knowledge base for the direct failure modes, below is an excerpt concerning damage to connections:

> Damage to connections between parts (movable or stable, permanent or dismountable junctions) can result from: Inappropriate connections such as wrong connection method, wrong dimensions, tolerance, and fit of connecting elements, excessive or insufficient fastening pressure, excessive or insufficient fastening elements (clamps, clasps, clips, bayonet plugs, sockets), excessive distance between fastened locations, wrong material for connecting elements, wrong method of welding, soldering, gluing, riveting, wrong connecting material (glue, solder, electrode material, wire); Impacts to the junction during operation (load, shocks, vibration, temperature deviation) that can cause deformation or shift in the elements that in turn damages the junction; Impacts from the medium on the material of the connected parts – corrosion, impurity saturation, abrasive wear, pollution; Harmful impact from other system parts – in particular, electro-chemical corrosion and diffusion due to the contact of different materials, thermal stress during deviations in temperature; Impact from repeated assembly and disassembly – deformation of parts, part damage; and, Adhesive effects (elements adhering to each other). Additional causes of junction damage are due to the fact that junctions are often inadequately protected from environmental impacts. Any barrier is subject to cracks or micro-cracks, either initially or due to various impacts.

This discussion outlines the high level portions of the knowledge base for the explosion failure mode. For each of these elements, the analysis team can look deeper into the knowledge base to obtain more details and creative ideas.

If this direct approach helps the analysis team identify root cause of the failure, they can look into the knowledge base to find ways to correct the failure.

There are details in the knowledge base on how to eliminate an energy source or a material from the system to correct the failure. Here is an excerpt from the knowledge base:

> The analyst should consider the possibility of preventing the energy supply that causes the harmful effect by: preventing energy production; reducing the energy flow parameters to the point where the harmful effect is eliminated; dividing the energy flow into harmless flows; creating insulation against an undesirable energy; redirecting the energy flow; creating an energy flow that will oppose the undesirable flow of energy; absorbing the undesirable energy. If the energy that causes the harmful effect also serves some useful function, he or she should attempt to find a simple way to resolve the contradiction.

Also there are suggestions on how to separate the cause of the failure from the system in space or time. The analysis team is encouraged to look at specific conditions that might be present to create the failure mode. By removing these specific conditions, the failure mode might be eliminated. There are suggestions in the knowledge base on how to eliminate or reduce the impact of the failure mode or the likelihood of the failure or reduce the negative consequences of the failure.

If after examining these failure modes directly, the analysis team finds that the problem is not typical, the analysis team should perform the following steps.

(a) Complete the Failure Analysis Questionnaire;
(b) Conduct Failure Analysis Modeling;

(c) Develop Failure Mode Prioritization;
(d) Failure Mode Correction Prioritization; and
(e) Evaluate Results.

3 The Failure Analysis Questionnaire

The Failure Analysis Questionnaire allows the analysis team to systematically examine the failure situation, and then build a model to describe it. There are two parts to the questionnaire: information about the system, and information about the failure.

It is important to identify the system in which the failure occurs. The first step is to name the system in a descriptive but concise manner. For our problem, the system can be thought of as "a city". The next step is to describe the main parts of the system. The main parts of a city are its people, its institutions such as the government, financial systems and the emergency services, and finally its infrastructure such as its transportation systems, its water system, its gas and oil system, its electrical system and its communication system. Of course, with a little imagination the analysis team could identify other important systems or sub-systems that would need to be considered. However, for demonstration purposes this list will be enough with which to start.

Then the system environment should be outlined. This environment can be thought of as the super-system that surrounds the system and of which the system is a part. The super-system could be the country or region in which the city is located.

The next aspect of the system that needs to be identified is its primary useful function. The primary purpose of a city is as a place in which people can live and work efficiently. This primary useful function may be supported by other useful sub-functions that should also be listed. The sub-functions of a city are shelter, wealth generation and storage, medical care and social events to name just a few.

The final part of the system description is the potential harmful functions that are created by the system. The potential harmful functions of a city are waste and pollution, and possibly excessive energy and resource use. High levels of population density might also be considered to some extent a harmful function of a city.

It is useful at this point to describe why the failure could not be solved using the direct method. The reason that the problem has not been solved already has many parts. The main reason the problem might not have been solved already is related to the expense in terms of private and public investment in prevention of critical infrastructure failure that is not considered affordable.

The analysis team should answer the question "What about the failure is unclear?" The nature of interconnectedness of the critical infrastructure might be one of the things that is most unclear about the problem.

Another important element of the description is determining what events are associated with the failure. What are the historic events prior to the failure? And what are the reasons they occurred? There are many examples of natural disasters causing failure of critical infrastructure. Most of the recent ones have been analyzed extensively by many national and international agencies.

The analysis team should consider whether other attempts have been made to solve this problem. It is believed that although many attempts have been made to solve this problem, none have been completely successful. However, there have been many useful partial successes created by the general awareness of the need. Too often these successes come after a critical infrastructure failure caused by a natural disaster rather than before. Another result is that as time passes without another natural disaster occurring, the successful strategies are forgotten and the problems return.

Localizing the failure is also an important step. The analysis team should try to determine the specific circumstances in which the failure occurred. The analysis team should identify the last event before the failure occurred. Also the analysis team should identify any concurrent events that happened with the failure or specific conditions associated with the failure. Different cities are susceptible to different types of natural disasters but the most common ones are floods and storms. The most common occurrences to storms and floods are in coastal regions or near rivers. Inland cities might also be commonly susceptible to tornadoes and occasionally susceptible to earthquakes. One of the specific conditions of critical infrastructure failure is lack of preparation and planning and possibly poor architectural design of buildings, ports and roads.

With the answers to the Failure Analysis Questionnaire, the analysis team should then be able to develop a system model. Figure 1 shows a simplified model of the

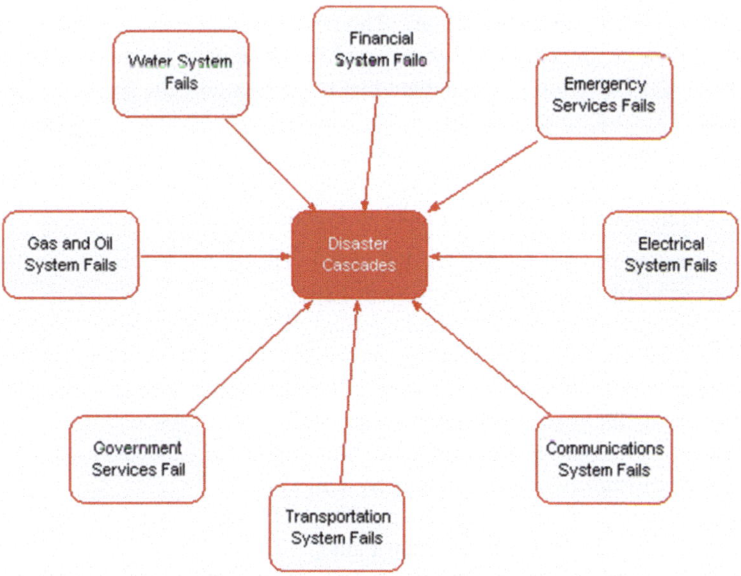

Fig. 1 A simple model of cascading critical infrastructure failure

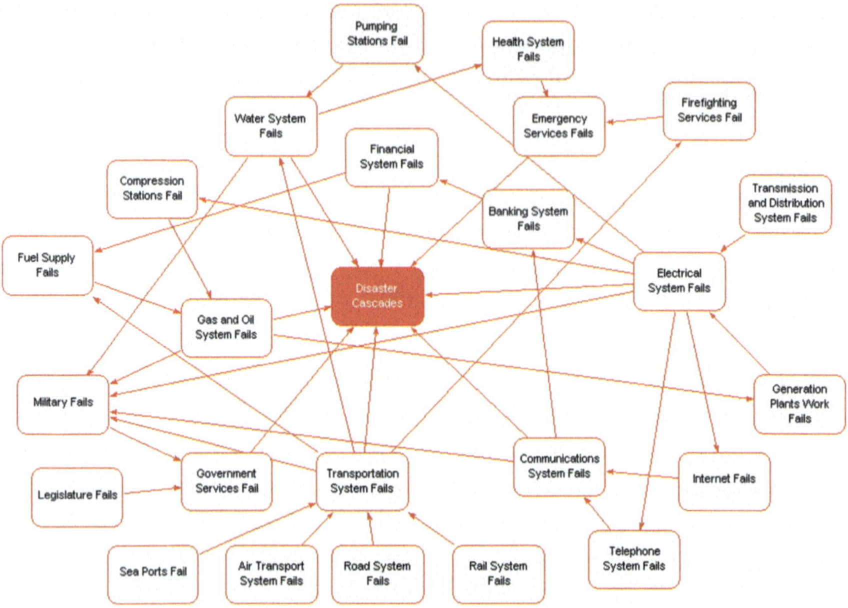

Fig. 2 A more complex model of cascading critical infrastructure failure

problem of cascading failure of critical infrastructure while Fig. 2 shows a more complex model of the problem.

Using this model, the problem can be inverted. The analysis team is encouraged to "find a way to produce the failure". The analysis team then uses an embedded knowledge base of Directions, Operators and Illustrations to create possible causes of the failure. A Direction is a statement that represents how the model can be used to suggest ways to create the failure. Each model can be used to suggest many Directions towards the failure.

Below is a complete list of the Directions created by the software for the inverted problem suggested by the simple model in Fig. 1:

1. Consider opportunities for providing [the] (Disaster Cascades) with help of resources of [the] (Gas and Oil System Fails).
2. Consider opportunities for providing [the] (Disaster Cascades) with help of resources of [the] (Government Services Fail).
3. Consider opportunities for providing [the] (Disaster Cascades) with help of resources of [the] (Transportation System Fails).
4. Consider opportunities for providing [the] (Disaster Cascades) with help of resources of [the] (Communications System Fails).
5. Consider opportunities for providing [the] (Disaster Cascades) with help of resources of [the] (Electrical System Fails).
6. Consider opportunities for providing [the] (Disaster Cascades) with help of resources of [the] (Financial System Fails).

7. Consider opportunities for providing [the] (Disaster Cascades) with help of resources of [the] (Emergency Services Fails).
8. Consider opportunities for providing [the] (Disaster Cascades) with help of resources of [the] (Water System Fails).

For the more complex model shown in Fig. 2, the software generated 25 similar Directions for consideration by the analysis team.

4 Hypothesizing Failure Modes

At this point, it might be worthwhile to divide the problem into its particular sub-problems and employ a number of analysis teams, one for each of the sub-problems. For example, the analysis team could be broken into analysis sub-teams for the gas and oil system, the government services, the transportation system, the communication systems, the electrical system, the financial system, the emergency services system and the water system. It is recommended that these sub-teams for each of these sub-problems be augmented by experts in each of these fields.

There are two possible ways to conduct the analysis from here. One approach involves a top-down comprehensive method of identifying all possible ways to create the failure. This strategy allows for an exhaustive set of possible causes. A second method, involves considering the most promising way to create the failure. Then the process is repeated with the next most promising way, and so on. For each possible cause, the question is whether the necessary resources are available for the cause to be feasible.

As one could imagine, the exhaustive method would take a great deal of time. However, it is the recommended way to conduct the analysis. The second method would seem more efficient although could still be quite time consuming. We will again prioritize the inverted failure modes using the Analytical Hierarchy Process (see Table 2). Then we will look in some detail at one failure mode.

Let's look at the knowledge base concerning the search of resources related to the Gas and Oil System Fails to determine if this cause could create the cascading

Table 2 Prioritization of inverted failure modes

Inverted failure modes	Priority value (%)
Gas and oil system fails	4
Government services fail	11
Transportation system fails	8
Communications system fails	9
Electrical system fails	7
Financial system fails	2
Emergency services fails	43
Water system fails	15

disaster. The question that needs to be answered is: "Are there sufficient resources in or near the Gas and Oil System so that if it fails, a cascading critical infrastructure failure will occur?" Using the Ideation Failure Analysis™ knowledge base, the analysis sub-team who are particularly knowledgeable about the gas and oil system would begin with a direct approach to finding resources.

4.1 Finding Resources Directly

For this purpose, the knowledge base contains information on available: materials resources, energy resources, time resources, space resources, structural resources, functional resources, and information resources. Table 3 shows a subjective prioritization using the Analytical Hierarchy Process that could be used to allocate effort towards the search for these resources. With this allocation in mind, the analysis team would use the knowledge base to look in more detail at each of these resources.

There is extensive information in the knowledge base concerning each of these resources. Here is an excerpt from the knowledge base concerning energy resources:

> The analyst should consider as a resource any kind of energy available in the object or process that might help to provide existing functions or operations, or to perform new ones. The analyst should consider mechanical impacts of solid-state objects, mechanical impacts of liquids and gases, thermal impacts, chemical impacts, electrical fields, magnetic fields, radiation impacts. Mechanical impacts of solid-state objects can be used for: creating a desired shape, creating a dynamic shape, transportation, control of transport, protection, slackening an undesired action, obtaining information, energy accumulation. Mechanical impacts of liquid or gases can be used for: retaining objects, transporting liquids and gases, creating a support, reducing resistance to movement, destroying solid-state objects, sealing, heating, or temperature control. Thermal impact can be used for: changing material properties, creating phase transitions, destroying an object, generating force, joining parts, transporting an object, or controlling chemical processes. Thermal impact can be used to change material properties. In particular, the following can be changed: structure, composition, magnetic properties, mechanical strength, color, shape and dimension. Thermal impacts can be used to create phase transitions, in particular, for: melting – solidifying, evaporation – condensation, sublimation, changing crystal structure. Chemical impacts can be used for: creating a material, changing properties, joining objects, creating force, process

Table 3 Prioritization of failure resources

Typical failure resources	Priority value (%)
Material resources	15
Energy resources	38
Time resources	19
Space resources	19
Structural resources	5
Functional resources	2
Informational resources	2

activation, obtaining information, transferring information. Electrical fields can be used for: transporting liquids or light objects, separating objects, changing an object's structure, changing an object's properties, material decomposition, process control, generating permanent forces, creating shocks, obtaining information. Magnetic fields can be used for: joining parts, integrating powder or granules, creating force, developing a particular structure, material separation, obtaining information about an object. Radiation impacts can be used for: destroying, heating, generating an electrical field, comparing and distinguishing objects, making objects visible, developing a non-visible image, marking parts, image transformation, or documenting information. If the analyst cannot find energy resources within the object or process, he or she needs to consider the resources within the nearby environment. In particular, he or she should consider the energy of the super-system of which the object or process forms a part.

We will briefly go through the steps of determining the various resources required for the gas and oil system to be the cause of the failure.

The analysis team should consider any kind of material available in the system. Obviously, the oil and gas system contains flammable material that could result in explosions. In this case, the analysis team would need to look for material that could act as an igniter. This igniter might be found in the energy resources which will be discussed below.

The analysis team should consider any kind of energy available in the system. The analysis team should consider mechanical impacts of solid-state objects, mechanical impacts of liquids and gases, thermal impacts, chemical impacts, electrical fields, magnetic fields, radiation impacts. Table 4 provides a subjective prioritization of these energy resources using the Analytical Hierarchy Process that can be used to allocate effort to the further analysis of these energy resources.

Mechanical impacts of solid-state objects might include the transportation system of trucks that are used to move the gas and oil. Mechanical impacts of liquid or gases might include the pumping stations and piping that moves the gas and oil to the city. Thermal impact might be the destruction of some parts of the system caused by fires created by explosions. Electrical fields can be used to create sparks from short circuits that cause the ignition of gas and oil in the system. Radiation impacts can be used for heating which might be related to the ignition of a gas or oil system explosion.

Table 4 Prioritization of energy resources

Energy resource	Priority value (%)
Mechanical impact of solid-state objects	20
Mechanical impacts of liquids and gases	18
Thermal impacts	27
Chemical impacts	10
Electrical fields	21
Magnetic fields	2
Radiation impacts	2

The analysis team should consider the time intervals in the system in particular, time before delivery of the gas and oil such as the time in refiners or compressors, time during the delivery such as the delivery of gas and oil in ships, trucks and pipelines, time after the delivery has ended such as at the facilities of the gas and oil users.

Free space might be available in the object. This space can involve the co-location of parts of the gas and oil system with other facilities or the distance that must be traveled between the sources of the gas and oil and the final destination of the gas and oil.

The analysis team should consider the structure of the system. In particular, they should consider the location of subsystems such as the refineries being near coast lines of oceans, lakes or rivers or gas and oil generators being co-located with electrical energy plants.

The analysis team should consider how part of the process could perform additional functions in particular, useful functions, such as heat or energy generation; and harmful functions such as pollution or contamination of water supplies.

The analysis team should consider any information about the system such as rate of flow on pipelines, shiploads, truckloads and barrels of oil and gas delivered.

After this thorough examination of resources, the gas and oil analysis team might conclude that all the resources required for this failure mode exist near the gas and oil system. Then this failure mode needs to be addressed. Or the analysis team may conclude that none of the resources are available near the gas and oil system. Then they can rule out this failure mode. Finally, the analysis team might conclude that some but not all resources are in or around the system. Then the analysis team can find new resources. For the sake of this example, we will examine the process of finding new resources using the Ideation Failure Analysis™ knowledge base.

4.2 *Obtaining New Resources*

The analysis team should consider the methods for changing or modifying the system under consideration in order to obtain: new material resources, new energy resources, new space resources, modifying time, modifying structure. Table 5 shows a prioritization of the various types of new resources that need to be examined by the analysis team.

Table 5 Prioritization of new failure resources

New failure resources	Priority value (%)
Material resources	24
Energy resources	57
Space resources	8
Time resources	8
Structure resources	3

Here is an excerpt from the knowledge base concerning new material resources:

> To obtain new material resources, the analyst should try to modify material available within the object. He or she should consider the following ways of modifying available material properties: material transfer, obtaining new material resources through physical effects, or obtaining new material resources through chemical effects. New materials can be obtained in a system via transfer from another system. The analyst should consider applying the following transfer mechanisms: using a flow of liquid or gas, "capturing" elements, electro-transfer. The analyst should consider the following physical effects capable of changing or modifying material properties and obtaining material resources: by thermal treatment, by fractionating, by decomposition, by mixing, by introducing additives, by ionization/recombination, by using specific effects. Treatment with heat or cold can change the initial properties of a material. In particular, the following can be changed: hardness, shape, aggregate state. Fractionating a material can provide it with new properties, in particular, mobility or controllability. The decomposition or transformation of a material to a mobile state can help in removing the material after it has fulfilled its desired function. Mixing materials can provide several new properties such as the ability to: connect, affix. Additives introduced into a material can provide new properties such as the ability to provide: insulation, expansion. Ionization (recombination) can change material properties to provide: electrical resistance, stages of aggregation.

We will briefly go through the process of looking for new resources in the gas and oil system.

For new material resources in the gas and oil system, there are many ways that the material resources could be transformed such as through additives, decomposition and mixing. Broken piping could allow harmful outside materials to enter the system and interact with the gas and oil. Refining chemicals could initiate harmful material interactions in the gas and oil system.

For new energy resources, the analysis team should consider transferring or delivering energy through the transportation system of ships, trucks and compression systems. New electrical energy could come from downed power lines. The knowledge base provides many more energy sources that could be used to develop analogies about new energy resources.

For new space resources, the team should consider the movement through space such as the space in and around the transportation network. Leaks from the gas and oil system might occur in these new spaces.

For new time resources, the analysis team should consider modifying available time in the processes of the system. The gas and oil system might try to adapt to the natural disaster using time resources. Another way the time resources might be affected is through an interruption of service that creates a cascading failure of the critical infrastructure.

The analysis team should consider ways of changing the structure of the oil and gas system such as integrating two processes. For example, the gas delivery system could interact with the oil transportation system to enhance the failure. Or the failure could be enhanced by the transportation of oil by ship, ocean and port being connected to the transportation of oil by truck and road.

In addition, the analysis team should consider the specific failure resources available in the system as shown below.

4.3 *Specific Failure Resources*

Table 6 shows an example Analytical Hierarchy Process prioritization that could be used to allocate effort in the search for specific failure resources.

As an excerpt from the knowledge base, the following is a description of possible human errors or malicious acts that should be considered:

> Human errors and malicious acts are often resources for system failure. Unpleasant as it is, we must recognize the existence of people who will intentionally damage a system to serve their own interests. Unintentional failures can be caused by humans as well. A person can be dangerous due to his/her specific characteristics or habits, including: a physical or mental disorder such as poor motor coordination, low intelligence, delayed reactions; mental instability that manifests as lack of self-confidence (easily influenced by others), or aggressiveness; a habit of under-estimating the level of risk such as belief that a failure will never take place, a habit of relying on the competence or wisdom of others, a habit of taking risks; the acceptance of harm as an "unavoidable evil" such as remaining passive in a dangerous situation, not learning from accidents; tendency to act "by rote" an in emergency such as following customary (familiar) instructions regardless of the situation, mimicking the actions of others when individual decisions are necessary, relying on personal, everyday experience; tendency to apply a creative approach to any routine situation such as disobeying instructions "for good reasons", perfectionism; having a "guilty conscience" that manifests as illegal actions, anti-social behaviour. An "ordinary" person can be transformed into a source of danger for the system he/she is dealing with due to the following unfavourable conditions: poor professional background such as lack of professional knowledge, lack of attention to auxiliary operations and elements, violating safety regulations, disregard for complicated devices or processes; stressful environment such as life/health threatening environment, stress due to frequent changes, high level of responsibility, repeated failure; fatigue caused by repeated irritation/annoyance, monotonous working conditions, specific physical, biological or chemical impacts; hindered control of a situation such as diverting

Table 6 Prioritization of specific failure resources

Specific failure resources	Priority value (%)
Intentional actions or spontaneous events	8
Differences in some parameter or characteristic	4
Specific characteristics and properties of a system	3
Harmful structures	11
Small failures and disturbances	2
Dangerous adjacent systems or elements	29
Faulty control devices	10
Faulty counter-failure systems	10
Human errors and malicious acts	22
Exhaustion of the useful resources	1

attention, erroneous or incomplete information; misunderstanding due to non-specific words or gestures, cultural differences, falsified information; poor written instruction that are impractical, lack of recommendations for emergency situations, including unreasonable restrictions, conflicting with other recommendations, unclear to the user, modified or appended many times; automatic reaction; irresponsible environment impacting a person's behaviour; group egoism causing a person to intentionally misrepresent information, providing special privileges to certain people, imposing unreasonable restrictions on others, neglecting community interests when making decisions; lack of training for extraordinary situations, not taking protective measures beforehand, lack of training regarding the use of protective means.

The analysis team should consider the consequences of changed location and speed such as in delivery systems for gas and oil. Pressure differentials caused by leaking of gas or oil pipelines might be considered here.

The analysis team should consider harmful structures. In particular, they should consider either in space or in the supply schedules if the gas and oil supply system is disrupted. They should consider specific characteristics of the subsystems such as when sub-systems of the gas and oil system can have large harmful effects. They should consider small failures and disturbances that tend to lead to more dangerous consequences such corrosion in pipelines. They should consider elements that become unfit or harmful under irregular circumstances. In particular, they should consider inflammable materials and toxic materials.

Faulty control devices or protection systems can be sources of danger. In particular, the analysis team should consider electrical or electronic measuring devices located at system inputs/outputs such as pressure monitoring systems. Faulty counter-failure systems, protective means, safety measures can be the sources of high-risk. In particular, the analysis team should consider faulty controlled or automatic valves or emergency switching mechanisms.

Harmful resources can appear by the exhaustion of the useful resources. For example, the gas and oil system could run out of useful transportation services.

5 Correcting Failure Modes

The failure correction stage involves using the model and knowledge base to prevent, eliminate or reduce the harmful effects of the failure.

5.1 Preventing the Failure by Averting Its Causes

For this purpose, the analysis team should identify the events prior to the failure. They should consider going back and modifying the events such as in our case, diversifying the fuel supply system, or hardening or providing backups for the compression stations. These measures might be costly to implement but it might be worthwhile if this would avoid cascading failure of the critical infrastructure.

5.2 Eliminating the Failure's Harmful Effects

The harmful effect might be eliminated by: removing or changing the source of harm; modifying the harmful effect; counteracting the harmful effect; isolating the system from the harmful effect; increasing the system's resistance to the harmful effect; modifying or substituting the effected object. Table 7 provides a prioritization of these ways to eliminate the harmful effects that was developed using the Analytical Hierarchy Process.

To get a feeling for the nature of the failure correction knowledge base, below is an excerpt on isolating the system from the harmful effect:

> The analyst should consider isolating the system from the harmful effect. In particular, he or she should try to isolate the system from: wear, fire, explosion, ambient oxygen, evaporation, thermal impact. To make the isolation more effective, he or she should consider inventive ways to introduce an isolating material. The analyst could consider the following ways of introducing an isolating material: use available materials; use selectively permeable isolation; use an easily-destroyed intermediate layer; transform a mediating element. The analyst could consider using materials available in the system or process as isolating materials. In particular, he or she could make use of materials: involved in causing the harmful effect; available in the system or produced by it; that are inexpensive. The analyst should pay particular attention to the possibility of easily replacing the isolating material if and when it is destroyed. If you need to isolate a system from the environment but also need to maintain access to the system, the analyst should consider using an isolating material or arrangement such as foam, a layer of liquid, gas or air, or a grating (to separate large particles from smaller ones). If two adjacent components that will have to be separated adhere too tightly to each other, the analyst should consider using an intermediate layer. The intermediate layer should hold the two components together but should be easily removed or destroyed.

The analysis team should try to remove the source of danger. For example, hardening of the gas and oil refineries and delivery systems would make the process more secure. They should consider modifying the harmful effect. If the harmful effect takes place at a point, the analysis team should consider changing the point contact to a multiple points of contact. This might involve having many smaller gas and oil storage sites rather than one large site. They should consider using another effect. For this purpose, they should consider neutralizing the harmful effect with a countering effect such as hardening of the gas and oil system. They should consider isolating the system. The analysis team could use available materials such as open spaces.

Table 7 Prioritization of ways of eliminating the failure mode

Ways of eliminating the failure mode	Priority value (%)
Remove or change the source of harm	25
Modify the harmful effect	3
Counteract the harmful effect	7
Isolate the system from the harmful effect	30
Increase the system's resistance to the harmful effect	30
Modify or substitute the effected object	4

5.3 Stopping the Harmful Effects of the Failure

The analysis team should try to make the system more resistant. In particular, they should decrease the sensitivity to a harmful effect by for example providing hardening, backup systems, multiple supply lines, and a diversity of sources of supply. An alternative would also provide in advance for immediate restoration of the system by replacing or repairing portions that are destroyed or damaged. This could involve dedicating resources to come into effect in the case of a natural disaster.

If it is impossible to protect the system, the analysis team should consider substituting the system. In particular, the analysis team could look into substituting the gas and oil system with electrical energy systems or replacing the trucking of gas and oil with pipelines.

If the analysis team has not found a way to resolve the problem, they should try to reduce its harmful effect by: localizing its harmful effect; reducing the effect; facilitating detection; or "sugar coating the pill".

To show the extent of the knowledge base to support the stopping of the harmful effects, below is an excerpt from the knowledge base:

> If it is impossible to eliminate a harmful effect, the analyst should consider the possibility of localizing it. This will help to protect other parts of the system and to assess damage as well. For this purpose, the analyst should consider: confining the effect to a definite location or time interval, or sheltering a material inside another material. The analyst should consider reducing the harmful effect at a specific location and for a specific period of time. In particular, he or she could distribute or dilute the harmful result. If a local defect can not be eliminated, the analyst should consider multiplying this defect so a pattern develops which hides the defect. He or she should see if there is a way to utilize, even temporarily, a harmful material, energy or undesirable system parameter. If it is impossible to eliminate a harmful effect that leads to a search (for lost or damaged systems or for individuals responsible for the harm), the analyst should consider making provisions in advance that will facilitate the search. If it is impossible to eliminate or reduce the harmful effect, the analyst should consider achieving some partially compensating positive effect. This might make things easier to accept, at least.

The analysis team should consider confining the effect to a definite location, or in this case creating smaller gas and oil storage facilities. The solution of many smaller facilities would also have the possibility of reducing the harmful effect. They should consider implementing detection systems such as pressure monitors on pipelines. They should consider providing shelters for citizens to move to in case their gas and oil heating systems fail.

6 Evaluating the Results

The analysis team would now select the most promising of the possible solutions. The most promising possible solutions would be considered "Good, that is, the failure would be completely prevented or eliminated". However, the analysis team should not stop there. Even solutions that can prevent or eliminate the failure might have adverse side-effects. The solution should be examined for adverse side-effects using a simplified form of the Ideation Failure Analysis™ and then these adverse side-effects should be prevented, eliminated or reduced.

The second most promising solutions might be considered "Good, but…, that is, there are one or more minor problems with correcting the failure". Again these minor problems should be avoided, eliminated or reduced. The knowledge base provides specific suggestions that can be used to separate these minor problems from the potential solutions. Thus the "Good, but" solution has a known defect while the so-called "Good" solution has no known defects until the simplified Ideation Failure Analysis™ is conducted on it.

To provide an example of how this might work in practice, we will consider as an example the "Good" solution of replacing the gas and oil system with an electrical energy system. We will consider the "Good but" solution as diversifying the gas and oil supply, storage and delivery system. This solution has the known drawback that it might be more costly.

6.1 Simplified Ideation Failure Analysis™

The "Good" solution completely solves the problem. However, unpleasant surprises can arise. Therefore, the analysis team should conduct a simplified Ideation Failure Analysis™: step 1, invert the problem; step 2, invent possible failures; step 3, prioritize failures to be eliminated.

In step 1, the analysis team should "invert" the problem as follows: "It is necessary to produce all possible failures to [critical infrastructure] and/or its environment with the help of [an electric energy system]".

In step 2, the analysis team can then imagine possible failures that use the following resources: systemic resources; change resources; differential resources; inherent resources; organizational resources; small failures and disturbances; dangerous elements; control devices; protection systems; human errors and malicious acts; exhaustion of useful resources.

It is likely that a natural disaster that is harmful to the gas and oil system would also be harmful to electric energy system. For example, cables might be broken, generators could fail or be over loaded, fires or explosions could be created by short-circuits.

In step 3, the analysis team should list possible problems with the solution. Then they should prioritize the problems based on their consequences and likelihood and determine which problems should be eliminated. It is likely that electric energy generators would fail during a natural disaster such as a flood or a severe storm. Overhead transmission wires and cables would probably break and fall possibly causing fires or explosions.

The analysis team should consider various ways to resolve the problems as shown in Table 7. These drawbacks in the electric energy system might be eliminated by having generator backups and running cables underground instead of overhead.

6.2 *Resolve the Contradictions*

If the solution contains some known drawbacks, the analysis team should formulate a contradiction such as "The concept [of diversifying the supply, storage and delivery system for gas and oil] should provide [avoidance of a cascading failure of the critical infrastructure in the city] … but should avoid [the problem of being costly]". Then using the separation principles they should try to resolve this contradiction.

To demonstrate how the knowledge base can support the resolution of the drawbacks of the potential solutions, the following excerpt from the knowledge base is provided.

> The analyst should try to separate the contradictory requirements: in space, in time, between a whole object and its parts depending on different conditions. The analyst should try to separate opposite requirements in space. For this purpose, he or she should partition the object or assign each contradictory function or condition to a different part. The analyst should try to separate opposite requirements in time, that is, schedule operation so that conflict requirements or functions take effect at different times. The analyst should try to separate opposite qualities between the whole object and its part. For this purpose, he or she should: assign one of the contrary functions or conditions to one or more parts, while the whole object retains the remaining functions and conditions; separate the part(s) with the undesirable qualities from the rest; isolate the part(s) of the system or process that has the undesirable qualities; consider making use of the special properties or features of the object. The analyst should determine what action will manifest these properties or features and apply that action and try to identify a parameter or condition that can change to allow the system to meet one requirement under one condition and the opposite one under another condition.

The analysis team should try to avoid the contradiction: in space, in time, between a whole object and its parts depending on different conditions. For example, one way to separate the contradiction in time is to shift the solution into the future. As the centralized gas and oil system ages and replacement is being considered, steps could be taken to diversify the system, making storage sites smaller and distant from each other in space.

7 Concluding Remarks

In this chapter, we have shown how problem inversion can be used to enhance creativity in failure analysis. We have looked at an example of the problem of cascading infrastructure failure in a city caused by a natural disaster such as a flood or a storm. We have also demonstrated how the processes and the knowledge base in the Ideation Failure Analysis™ software could augment this problem inversion process.

We have demonstrated that solving this problem using this method would take a great deal of effort. The allocation of effort could be distributed to teams of experts because the process breaks the problem into its individual parts and the knowledge base has been developed in a highly hierarchical manner. We suggested that a prioritization method be applied to the allocation of effort to make the process more efficient. In particular, in our example, we used the Analytical Hierarchy Process in this allocation of effort. The Analytical Hierarchy Process seems to be an excellent way to allocate effort because it provides the prioritization in percentage terms where the total amount of effort can be broken down into its separate components. Also although not all of the components will get the same allocation of resources, none of the components will be completely overlooked. Each component will receive some amount of effort allocated to it even if it is small. This is important because failure analysis is often affected by small problems that have big impacts.

This approach does not come without some serious reservations that must be considered. First, as we have already mentioned is the time and effort required to complete the process in an exhaustive manner. The second is the fact that analysts using the knowledge base as a checklist might stop their creative process after finding just one potential failure mode or correction technique for the particular Direction or Operator. It is imperative to push the creative process to attempt to exhaust the ideas that can be generated from each Direction and Operator. The third concern relates to the applicability of the knowledge base in the Ideation Failure Analysis™ software. It could take a great deal of effort to adapt the knowledge base to make it more applicable to the problem of cascading failure of critical infrastructure. Associated with this problem with the method is the availability of expertise. Getting teams of experts together to work through the issues for a great length of time might be difficult or costly. Also it is likely that the solutions found by this method have already been ruled out because they are too costly to implement. If they were not too costly, they would probably have already been implemented. Finally, there may be security concerns. If the failure modes became widely known to the public, there would be a great deal of negative publicity that there are a large number of failure modes being considered by government agencies. There are also the obvious security concerns of keeping this knowledge of the numerous failure modes out of the hands of enemy forces because this method and its results could facilitate sabotage or terrorism.

None of these reservations would seem to be sufficient to discount the value of further analysis of the inversion technique for solving the problem of cascading failure of critical infrastructure caused by a natural disaster.

References

1. Center for Research on the Epidemiology of Disasters (2015) Human cost of natural disasters: a global perspective. World Heath Organization. http://emdat.be/human_cost_natdis. Accessed 24 Nov 2015

2. Mateski M (2009, June) Red teaming; a short introduction. Red Team J. http://redteamjournal.com/papers/AShortIntroductiontoRedTeaming(1dot0).pdf. Accessed 24 Nov 2015
3. Evaluation Research Team (2006, Nov) Gaining consensus among stakeholders through the nominal group technique. Evaluation Brief No. 7, Department of Health and Human Services, Center for Disease Control and Prevention. http://www.cdc.gov/healthyyouth/evaluation/pdf/brief7.pdf. Accessed 24 November 2015
4. Taylor I (1996, Apr) An extension of structured brainstorming using the six thinking hats. Directorate of Logistics Analysis, Research Note 9603. http://cradpdf.drdc-rddc.gc.ca/PDFS/unc42/p498627.pdf. Accessed 24 Nov 2015
5. Klein G (2007, Sept) Performing a project PreMortem. Harvard Business Review. https://hbr.org/2007/09/performing-a-project-premortem. Accessed 24 Nov 2015
6. Kaplan S, Visnepolschi S, Zlotin B, Zusman A (1999) New tools for failure and risk analysis anticipatory failure determination (AFD) and the theory of scenario structuring. Ideation International Corporation. http://www.ideationtriz.com/new/materials/AFDNewToolsbook.pdf. Accessed 24 Nov 2015
7. Saaty TL (2008) Decision making with the analytic hierarchy process. Int J Serv Sci 1(1):83–98. http://www.colorado.edu/geography/leyk/geog_5113/readings/saaty_2008.pdf. Accessed 24 Nov 2015

Heroes and Villains in Complex Socio-technical Systems

Jonathan Gao and Sidney Dekker

Abstract The history of efforts to reduce 'human errors' across workplaces and industries suggests that people (or their weaknesses) are seen as a problem to control [1, 3, 15, 16]. However, some have proposed that humans can be heroes as they can adapt and compensate for weaknesses within a system and direct it away from potential catastrophes [15]. But the existence of heroes would suggest that villains (i.e. humans who cause a disaster) exist as well [16], and that it might well be the outcome that determines which human becomes which. The purpose of this chapter is to examine if complex socio-technical systems would allow for the existence of heroes and villains, as outcomes in such systems are usually thought to be the product of interactions rather than a single factor [17]. The chapter will first examine if the properties of complex systems as suggested by Dekker et al. [18] would allow for heroes and villains to exist. These include: (a) synthesis and holism, (b) emergence, (c) foreseeability of probabilities, not certainties, (d) time-irreversibility and, (e) perpetual incompleteness and uncertainty of knowledge, before concluding with a discussion of the implications of the (non) existence of heroes and villains in complex systems for the way we conduct investigations when something goes wrong inside of those systems.

Keywords Complexity · Heroic recovery · Human error

1 Introduction

The history of efforts to reduce "human errors" across workplaces and industries strongly suggest that people (or their weaknesses) are seen as a problem to control, typically via replacing the "faulty" individual or the addition/enforcement of procedures [1–7]. For instance, the National Transportation Safety Board (NTSB) [6]

J. Gao (✉) · S. Dekker
Griffith University, Brisbane, Australia
e-mail: jonathan.gao@griffithuni.edu.au

A.J. Masys (ed.), *Disaster Forensics*, Advanced Sciences and Technologies for Security Applications, DOI 10.1007/978-3-319-41849-0_3

stated in their Most Wanted List of Transportation Safety Improvements that accidents can be reduced if pilots were to comply with procedures.

Such notions are single-factor explanations that condense accounts of failure to individual human action or inaction. An analysis by Holden [2] showed that between 1999 and 2006, 96 % of investigated US aviation accidents had their cause mostly attributed to the flight crew. In 81 % of these accidents, people were the only reported cause. "Crew failure" or similar terms were used to describe 74 % of probable causes while the remaining cases contained language such as "inadequate planning, judgment and airmanship," "inexperience" and "unnecessary and excessive … inputs." "Violation" of written guidance was stated as cause or contributing factor in a third of all cases.

Contemporary research suggests otherwise [8–12]. In their review of 13 studies, Alper and Karsh [8] discovered that procedural violations were due to a myriad of reasons, such as a lack of information, poor system design, and competing goals. This finding is consistent with work done by de Brito [9], whose study revealed that factors such as interruptions or difficulties in locating the correct protocols during an emergency contributed to pilots violating procedures. Such a phenomenon has been observed in other domains as well. For instance, de Saint Maurice et al. [10] found that anesthetists tend to deviate from procedures when they had unscheduled surgery, or when work demands were high. In other words, present research strongly indicates that the reasons for procedural violations are more complex than intentional non-compliance by humans.

In contrast to the view of humans as a liability, some have advocated for the view of humans as a valuable resource [13–15]. Reason [15] even argues that humans can be heroes as they are able to adapt and compensate for weaknesses within a complex system and hence direct it away from potential disasters. However, the existence of heroes would suggest that villains (i.e. humans who cause a disaster) exist as well [16].

Therefore, the purpose of this chapter is to examine if complex socio-technical systems would allow for the existence of heroes and villains within it, as outcomes in such systems are usually thought to be the product of interactions rather than a single factor [17]. Complexity Theory suggests that we see performance as an emergent property, the result of complex interactions and relationships. This stands in sharp contrast with the perspective from Newtonian Science.

2 Newtonian Science, Cause and "Human Error"

Newtonian Science is one of simplicity [18]. The complexity of the world is thought to be superficial and can be broken down into individual components. Each component can then be analysed to determine its role. Such simplistic thinking is often used in investigative and legal reasoning in the aftermath of accidents to find a cause. Newtonian Science believes that all accidents have a definitive, identifiable cause and a definitive effect, and that cause and effect are symmetrical. In other words, the bigger the effect, the bigger the cause. Identification of a "cause" or "causes" is generally seen as the most important function of an accident investigation. However, this assumes that the physical effects of an accident can be traced back to physical causes. If efforts to determine cause-effect relationships cannot be achieved, then the typical response is to deem the accident analysis method or agency as unworthy [18]. Then Chairman of the NTSB, Jim Hall, raised the spectre of his agency at not being able to find the "faulty" part in TWA 200, which would challenge its entire reputation (p. 119): "What you're dealing with here is much more than an aviation accident… What you have at stake here is the credibility of this agency and the credibility of the government to run an investigation" [19].

3 Complexity Theory

In contrast with Newtonian Science, Complexity Theory proposes that complex behaviour arises because of the interactions between the components of a system [18]. The properties of a system emerge from these interactions, instead of being contained within individual components. External designers are not necessary for a complex system, as new structures can be generated internally. These internally generated structures can adjust in response to changing conditions in the environment. Complexity is a feature of the system, not of components inside it. The knowledge of each component is limited and local, and no component can possess enough capacity to represent the complexity of the entire system. The behaviour of the system cannot be reduced to the behaviour of the components. Complex systems are held together by local relationships only. Each component is ignorant of the behaviour of the system as a whole, and cannot know the full influences of its actions. Components respond locally to information presented to them, and complexity arises from the huge, multiplied webs of relationships and interactions that result from these local actions. Asymmetry or non-linearity in these relationships means that a tiny change in initial conditions can lead to huge differences later. There are no linear, symmetrical cause-effect relationships like that of Newtonian Science.

This chapter will first examine if key properties of complex systems, namely: (a) synthesis, holism, and emergence, (b) foreseeability of probabilities, not certainties, (c) time-irreversibility and, (d) perpetual incompleteness and uncertainty of knowledge, would each allow for heroes and villains to exist. It will then conclude with a discussion of the implications of the (non) existence of heroes and villains in complex systems for the way investigations are conducted when something goes wrong inside of those systems.

4 Synthesis, Holism, and Emergence

Essentially, the properties of synthesis, holism, and emergence mean that one cannot hope to break down a complex system into its individual components to locate a broken part, unlike the Newtonian belief that such an act is possible [18]. Happenings within a complex system emerge due to the multiple interactions between various components and as such, a holistic view is needed to understand it. Some have proposed that individuals within complex systems are unable to fully understand every process involved in the system's functioning, due to its complexity [14, 20]. This means that individual effort is unlikely to play a huge role in steering a system both away from and towards disasters, as an individual's lack of understanding means that s/he would be unable to deliberately make changes that will have a positive impact throughout the system. In other words, heroes and villains do not exist.

Yet, Rochlin et al. [20] observed that anyone working on an aircraft carrier had the right to stop all flights if they think that such an action would avoid accidents, suggesting that perhaps heroes (and as a consequence, villains) can exist in complex systems. A second example of an individual who was able to steer a complex system away from potential dangers and towards improvements is perhaps Florence Nightingale.

4.1 *Case Study: Florence Nightingale*

In her lifetime, Florence Nightingale made numerous positive contributions to the British healthcare system and hence is regarded as the founder of modern nursing [21–23]. Her first major set of contributions took place during the Crimean War, when then Secretary of State at War Sidney Herbert requested Florence Nightingale to lead a group of nurses to the battlefield [24, 25]. Upon arrival, Florence Nightingale began to tap into her political connections to improve the state of affairs in various hospitals. For instance, she was able to increase the quality of the male nursing orderlies (they were responsible for majority of the work), and secure valuable equipment for her hospitals despite some difficulties.

After the Crimean War, Florence Nightingale returned to England and began an attempt to establish a royal commission for the study of: (a) sanitary conditions, (b) management and organisation of barracks and military hospitals, and (c) administration, training, and organisation of the Army Medical Department [24, 26]. Despite some obstacles, her attempt was a success and the formation of the royal commission led to several positive reforms to the British healthcare system.

Furthermore, Florence Nightingale's popularity led to the creation of the Nightingale Fund, used to establish the Nightingale Training School for nurses [27]. Although there were problems with the school (e.g. poor management), it did exert a profound influence in the nursing domain. For example, Lucy Osburn, the founder of modern nursing in Australia, received her training in the Nightingale Training School [28]. Likewise, Linda Richards, who was a prominent figure in American nursing (and perhaps Japanese nursing), was heavily influenced by Florence Nightingale's work [29, 30].

Besides the above contributions, Florence Nightingale also played a critical role in the conception of the Metropolitan Poor Law Act [31]. Back then, workhouses exist to provide the disabled and the poor a small amount of assistance in terms of food, shelter, warmth, and care [32]. The Metropolitan Poor Law Act led to the setting up of the Metropolitan Asylum District for the separation and healing of those with insanity, fever, or small-pox [31]. Well-built dispensaries and infirmaries were also established for the poor who were afflicted with non-infectious illnesses. Most importantly, the Act led to the launch of the Metropolitan Common Poor Fund which was used to financially support the various programs.

Lastly, Magnello [33] claims that Florence Nightingale was among the first to employ statistics in the analysis of healthcare data, thereby revolutionising the nursing domain and causing a decline in avoidable deaths. For example, Florence Nightingale made use of statistics to convince government officials of the urgent need for sanitary reforms. She also introduced standardised forms which were used in various major London hospitals and parts of Europe to facilitate the collection of medical data.

In summary, Florence Nightingale appears to have played an influential role in steering the healthcare systems of various countries away from disasters and towards improvements. In complex systems, as the truism goes, one actor cannot control much. But they can influence almost everything. This seems to indicate that the efforts of an individual can have enormous repercussions on a complex socio-technical system, meaning heroes and villains can potentially exist in these systems.

5 Foreseeability of Probabilities, not Certainties

According to Newtonian Science, the future of a component can be predicted with absolute certainty if its state at any time was known in complete details [18]. With sufficient knowledge of the initial conditions of a component, and the laws that govern the motion of the component, all subsequent events are foreseeable. In other words, if someone can be shown to have known (or should have known) the initial positions and momentum of the components of a system, and the forces acting on these components (which includes both external forces and those determined by the positions of these and other components), then this person could, in principle, have predicted the further evolution of the system with complete certainty and accuracy.

If complete and accurate knowledge is achievable, then harmful outcomes should be predictable [18]. Therefore, it makes sense (at least from a Newtonian Science perspective) to ask why someone failed to predict an accident when it occurs. Were they unaware of the laws governing their part of the universe (i.e. were they incompetent or unknowledgeable)? Or have they failed to plot out the possible effects of their actions? Indeed, legal reasoning follows this feature of the Newtonian model when determining if an individual was negligent (p. 6): "Where there is a duty to exercise care, reasonable care must be taken to avoid acts or omissions which can reasonably be foreseen to be likely to cause harm. If, as a result of a failure to act in this reasonably skilful way, harm is caused, the person whose action caused the harm, is negligent" [34].

In other words, someone can be considered as negligent or villainous if he/she did not avoid actions that could be foreseen to lead to negative effects [18]. Effects that would have been predictable and therefore avoidable if the person had made an effort to understand the laws governing his/her part of the universe and plot out the possible effects of his/her actions. Most road traffic legislation is based on this Newtonian idea of foreseeability. For example, a road traffic law in a typical Western country might specify how a motorist should adjust speed so as to be able to stop the vehicle before a crash occurs, and at the same time remaining aware of the circumstances that could influence speed selection (such as wet roads). Both the foreseeability of accidents and the awareness of circumstances are steeped in Newtonian Science. Both are also heavily susceptible to outcome bias. If something was not foreseen, then speed was surely too high. As a result, the system's user is always wrong [35].

Simply put, the opposing property from the perspective of Complexity Theory states that decision makers tend to make decisions that they think have the highest probability of producing a good outcome, based on their aims, limited focus and knowledge at a particular point in time [36]. Still, whether the good outcome occurs remain to be seen. Unfortunately, a system's complexity means that such actions might bc locally adaptive, but globally maladaptive [37]. In other words, a seemingly logical decision may end up leading to a negative outcome. If so, this would suggests that an individual may not have the ability to direct a system away from disasters, meaning it is unlikely for heroes and villains to have a place in complex socio-technical systems.

5.1 Case Study: Emergency Physician

In their paper on medical diagnostic error, Croskerry and Nimmo [38] presented a case of a 28 years old female patient who was sent by her addiction treatment facility to a neighbouring emergency department. She claimed to be suffering from anxiety and chest pains over the past week, and was afraid that she might have a heart problem. The physician who was attending to her was skeptical of her claims as addicts from the treatment facility have been known to fake their injuries to obtain drugs [39]. Moreover, the patient left a bad impression on the physician as he had a busy schedule but was made to wait while she took a smoke break.

When the physician later reviewed the patient's electrocardiogram results and found no abnormalities, he diagnosed her with anxiety and discharged her [38]. After returning to the addiction treatment facility, the patient complained that her chest hurts, but her griping was not taken seriously as she had been medically cleared. She later passed away from a cardiac arrest caused by blood clots in her lungs, a result of her smoking habits and being on birth control pills.

In this scenario, the doctor made a decision which he believed would lead to the best possible outcome. He thought that the patient was just like the other addicts that came before her, feigning injuries to get their hands on drugs and hogging valuable resources in an emergency department that was already busy [38]. As such, he quickly discharged her to avoid wasting resources on a fake patient. Unfortunately, his actions led to an unexpected negative outcome. This might indicate that an individual's ability is limited and as a consequence, it suggests that heroes and villains are unlikely to have a place in a complex socio-technical system as they effectively have little control over the outcome of their actions.

6 Time-Irreversibility

The trajectory of a Newtonian system is not only towards the future, but also towards the past [18]. In principle, we can reverse the evolution of a system to reconstruct any earlier state that the system has gone through. Such assumptions give accident investigators the confidence that an accident can be reconstructed by starting with the outcome and then tracing its causal chain back into time. Knowledge about past events is assumed to be obtained via uncovering a pre-existing order. The only thing between an investigator and a good reconstruction are the limits on the accuracy of the representation of what happened. It follows that accuracy can be improved by "better" methods of investigation.

Complex systems, in contrast, are constantly adapting and evolving, and as such, it would be an uphill task for investigators to reconstruct the interactions leading up to an adverse outcome [18]. Additionally, any retrospective analysis of an accident would require the investigator to make a subjective judgement on which piece of information is important, and which is not, further skewing the reconstruction process [18, 40, 41]. It should also be noted that accident investigations in some domains (e.g. healthcare) can be problematic as it is more about what activities took place rather than physical evidence [41]. Hence, there is likely to be a greater reliance on the memories and perceptions of those involved. Yet contemporary research on memory strongly suggests that it is highly unreliable [42–44].

For example, Loftus and Palmer [43] presented participants with video footage of vehicular accidents before asking them questions about the footage. They found that when the verb *smashed* was used, participants' speed estimates tended to be higher. Furthermore, these participants were also more likely to report seeing broken glasses in the video footage, despite the fact that there was none.

Similarly, Itsukushima et al. [45] showed participants coloured slides portraying the daily activities of a woman before giving them additional information which they falsely claimed were the impressions and thoughts of past participants. Some participants received information that were consistent with the slides, while others were given fake information. They found that participants had a tendency to incorporate the bogus information into their memory, even though they were told that the information were simply the impression and thoughts of previous participants and therefore may not necessarily be true. In sum, retrospective analysis of an accident tends to be skewed, and the poor reliability of one's memory makes retrospective analyses even more unreliable.

6.1 Case Study: Eric Cropp

In 2006, an infant passed away after being administered sodium chloride that was over twenty times stronger than the prescribed dosage [46]. Prosecutors held Eric Cropp, a pharmacist supervising the pharmacy technician who prepared the fatal solution, responsible for the accident and charged him with involuntary manslaughter. Eric Cropp ended up being given six months in jail, six months of home confinement, and three years of probation. Yet, should Eric Cropp really be blamed for this mishap?

According to information from newspapers and the Institute for Safe Medication Practices, the working conditions on the day of the accident were dreadful [47, 48]. Some examples being: (a) a build-up of orders as the pharmacy computer system was not functioning earlier that day, (b) staffing shortage, (c) unable to perform normal work or take breaks, due to high workload, (d) the pharmacy technician who prepared the mixture was thought to be highly distracted as she was planning her wedding during working hours, and (e) being rushed to dispense the medication for the infant, even though it was not required for several hours.

Current research strongly indicates that stress and time pressure can adversely affect one's performance [49–52]. For example, Thompson et al. [52] gave 241 experienced nurses 50 vignettes depicting actual clinical scenarios and had the nurses assess the situation to decide if an intervention was necessary. Their study revealed that time pressure had a debilitating impact on nurses' performance, such that they were less likely to identify the need for an intervention and were also less likely to intervene.

Despite the presence of multiple detrimental factors that could have easily affected the performance of anyone, the prosecutors chose to hold Eric Cropp solely responsible for the accident [46]. In other words, Eric Cropp is the villain in this story. As mentioned above, the reconstruction process of an accident may be skewed as investigators make subjective judgements on which pieces of information are important, and which are not [18, 40, 41]. This seems to be what is happening here, with all the disadvantageous factors being deemed as unimportant pieces of information. The ignorance of these information and the labelling of Eric Cropp as a villain seems to suggest that heroes and villains do not objectively exist in complex systems. Rather, they seem to be constructs formed by investigators based on the information they collect and evaluate during the reconstruction process.

7 Perpetual Incompleteness and Uncertainty of Knowledge

Typical accident investigations seem to hold the view that an objective truth exists, and that knowledge about the truth can be obtained with the right investigation method [18]. Hence, the more information an investigator gathers, the more it leads to a better investigation, a better representation of "what happened". In other words, the true story where there is no gap between external events and their internal representation [53]. Those equipped with better methods, and those who enjoy greater "objectivity" (i.e. those who are free of biases, and will consider *all* the facts) are thought to be in a better position to construct such a true story.

However, such a notion is unlikely to be valid, at least for complex systems. Instead, it is believed that objective knowledge in complex systems cannot be obtained [18]. Every individual in complex systems will form their own set of knowledge as they each experience different inputs and outputs, with no objective mean of assessing who is in possession of the "truer" knowledge [18, 54]. This is simply a result of the system being too complex, leading to each individual only being able to take into consideration a limited amount of inputs. Hence, their knowledge would always be incomplete and uncertain.

7.1 Case Study: Mid Staffordshire

In 2009, a scandal erupted at Stafford hospital when it was revealed by regulators that the hospital provided an appallingly low level of care [55]. Examples include patients being forced to soil their sheets and then being left in them, patients not being bathed for a long period of time, patients being given either the wrong food or none at all, and patients being given the wrong diagnoses [56]. The outrage was perhaps made worse by claims that hundreds of patients may have died due to these factors [57].

In response to this scandal, some have expressed a desire to find and punish the healthcare practitioners responsible [58, 59]. Additionally, there have been reports of nurses losing their sense of compassion towards patients, further cementing the idea that the cause of the whole debacle comes from those at the front end [60, 61]. In summary, there are those who advocate for the disciplining of healthcare practitioners based on the "input" of patient maltreatment and claims of nurses being callous.

The desire for retribution is understandable. However, evidence strongly indicates that these healthcare practitioners are working under incredibly harsh conditions. For one, staffing levels were cut to save money despite the high workload of the healthcare practitioners [57, 60, 62]. A BBC Radio interview with a midwife working in Stafford hospital revealed that she was chastised for caring due to time wasted [60]. Likewise, junior doctors were not properly supervised and were pressured to hastily meet and diagnose patients so that patients can be moved out from A&E within four hours [57].

Moreover, people were prevented from speaking out as they would be labelled as a troublemaker or risk losing their job [60]. Julie Bailey spoke out against the Mid Staffordshire NHS Foundation Trust when she felt that her mother was receiving an atrocious level of care during her eight weeks at the Stafford hospital, and was bullied into leaving Stafford [63]. Examples of bullying behaviours against her include her car tires being slashed and her mother's grave being vandalised. Helene Donnelly, a nurse who once worked at Stafford hospital and later spoke out against it, recounted being threatened and told to watch her back [64]. Lastly, a midwife reported feeling that there was a lack of support in the hospital for correcting bad behaviours, and that management would often belittle their suggestions for improvements [60].

Existing research in psychology strongly indicates that high workload and time pressure positively correlate with emotional exhaustion, one of the symptoms for burnout [65–67]. Additionally, emotional exhaustion correlates with depersonalisation, which is the development of distrust or unsympathetic attitudes toward others [68, 69]. Research also suggests that holding the belief that one has little control over events in life makes one susceptible to burnout, a likely occurrence in Stafford hospital, given the hostile atmosphere towards those who step out of line [70]. Lastly, as mentioned earlier, psychological studies suggest that stress and time pressure can have a detrimental effect on one's performance [49–52].

Taken together, it would suggest that the "input" of high workload and time pressure could have caused practitioners to experience emotional exhaustion, which led them to depersonalise their patients and thus producing the "output" of no compassion. Likewise, the "input" of having to quickly assess patients under pressure and without guidance might be the reason for the "output" of misdiagnoses.

To conclude, both the advocates for punishment and the healthcare practitioners at the hospital have formed their own ideas based on their different inputs and outputs. Any attempts to objectively determine who is right would be a difficult task (if not impossible). It is an undeniable fact that the healthcare practitioners at Stafford hospital have caused both suffering and a loss of dignity to patients. Therefore, the calls for them to be punished seem like a reasonable response. Yet, it hardly seems fair to label the practitioners as the "villains", given that the system was rigged against them.

8 What Does This Mean for Heroes and Villains in Complex Socio-technical Systems?

To recap, the case of Florence Nightingale exerting great influence over the healthcare systems in various countries seems to strongly indicate the possibility for heroes and villains to exist in complex socio-technical systems. In contrast, the case of the emergency physician suggests that heroes and villains are unlikely to exist in such systems as they have little control over the outcome of their actions.

Similarly, the case with Eric Cropp can be said to demonstrate that villains (and therefore heroes) are simply social constructs formed by investigators based on their subjective judgments. Lastly, the case with Mid Staffordshire suggests that an objective set of truth does not always exist. Instead, each individual will form their own set of knowledge grounded in their own experiences and one man's hero might be another man's villain. After all, while the actions of Julie Bailey and Helene Donnelly are arguably heroic, they have been vilified by others [63, 64].

Taking the above into consideration, this can mean that heroes and villains can exist in complex socio-technical systems. However, to call them as such might be highly inappropriate. As argued above, the healthcare practitioners in each case were normal people working to the best of their ability, making decisions that appeared rational to them at that point in time, and/or managing various external factors that were working against them. While some were successful in their endeavours (i.e. Florence Nightingale), others were not as the odds were stacked against them. Calling these people villains simply because they were unable to overcome systemic handicaps or because they do not fit our subjective judgments seems unfair.

Interestingly, Strachey [71] suggested that perhaps Florence Nightingale's achievements were possible partly due to systemic factors being in her favour. He suggested that the timing of the Crimean War was perfect for Florence Nightingale, since she would have been too inexperienced if the war had begun a few years earlier. Conversely, if the war started a few years later, Florence Nightingale might have been preoccupied with other works and may perhaps be too old to handle the demands that would have been imposed on her.

Furthermore, Florence Nightingale was born into a well-connected political family, which likely gave her an edge in some areas [24, 25, 71]. For one, then Secretary of State at War Sidney Herbert was a personal friend of Florence Nightingale. The decision to choose Florence Nightingale as the one to lead female nurses to the front line was said to be a result of her offering her services, and Sidney Herbert's belief that she was the right woman for the job based on his knowledge of her in a personal capacity. Moreover, as mentioned above, these connections also gave her some influence which she used to push for improvements in hospitals on the front line. Lastly, Florence Nightingale's fame from the Crimean War resulted in the establishment of the Nightingale Fund that was used to found the Nightingale Training School for nurses, which had a deep impact on the nursing domain [27].

In summary, heroes and villains can potentially exist in complex socio-technical systems. However, it would be illogical to call them as such. A myriad of external factors are at play, working for and/or against every individual in the system. To hail some as heroes simply because the odds were in their favour, while vilifying others because the odds were against them seems both absurd and unfair.

9 Conclusion

When success or failures are seen as complex, emergent phenomena, there is no longer an obvious relationship between the behaviour of components in the system (or their malfunctioning, e.g. "human errors" or their spectacular success, e.g. "heroes" or "heroic recoveries") and system-level outcomes [18]. Instead, system-level behaviours emerge from the multitude of relationships and interactions within the system. Investigations that embrace the complexity produced by such relationships and interactions, then, might stop looking for the "causes" of failure or success. Instead, such investigations gather multiple narratives from different perspectives, which give partially overlapping and partially contradictory accounts of how emergent outcomes arise. The complexity perspective dispenses with the notion that there are easy answers to events within a complex system, answers that supposedly can be achieved by those with the best method or most objective investigative viewpoint.

Perhaps the question about the existence of heroes or villains in a complex system is unanswerable. Reconstructing events in a complex system is impossible, primarily as a result of the characteristics of complexity [18]. The complex system that is subjected to an investigation after an incident is not the same system that produced the incident. It will have changed, partly because of the incident, partly because of the passage of time, and the nature of complexity. Psychological characteristics of retrospective investigations make it so too. Take, for instance, the idea that a sequence of events or chain of causes precede an accident. Who makes the selection of the "events" or "causes" and on what basis? And why only a single chain? The act of separating factors deemed important from unimportant ones is an act of imagination, of construction, of the creation of a story, not the reconstruction of a story that was already there, waiting to be uncovered. Such actions of imagination bring with it a bunch of selection mechanisms and criteria into the supposed "re"-construction. There is no objective way of doing this as the investigator's own background, preferences, experiences, biases, beliefs, and purposes all influence the investigation.

Not surprisingly, Complexity Theory suggests that the observer is not just a passive consumer, but the active creator of the observed [72]. In complexity, there is no method to decide which narrative is correct [18]. However, some narratives will deliver more interesting results than others, depending on the goals of the audience. The selection of "causes" (or "events" or "contributory factors") is always an act of construction by the investigators. If the sheer notion of heroes and villains is part of their narrative fore-structure, for example because of institutional or cultural needs, then the complex system, and the interactions between components and events in it, may well be construed so that heroes or villains become key players in the story.

References

1. Haq H (2013) How human error can cause a plane crash. BBC Travel [Internet]. UK. Available from: http://www.bbc.com/travel/story/20130521-how-human-error-can-cause-a-plane-crash
2. Holden RJ (2009) People or systems? To blame is human. The fix is to engineer. Prof Saf 54:34–41
3. Levitt P (2014) When medical errors kill: American hospitals have embraced a systems solution that doesn't solve the problem. Los Angeles Times [Internet]. Los Angeles. Available from: http://articles.latimes.com/2014/mar/15/opinion/la-oe-levitt-doctors-hospital-errors-20140316
4. Murty M (2014) MH370: sack those who 'slept on the job'. The Rakyat Post [Internet]. Kuala Lumpur. Available from: http://www.therakyatpost.com/news/2014/06/18/mh370-sack-slept-job/
5. National Transportation Safety Board (2015) Lack of consideration for human factors led to in-flight breakup of SpaceShipTwo. Available from: http://www.ntsb.gov/news/press-releases/Pages/PR20150728.aspx
6. National Transportation Safety Board (2015) Most wanted list of transportation safety improvements. Washington
7. Topham G (2013) Railway accidents happen because someone makes a mistake. The Guardian [Internet]. London. Available from: http://www.theguardian.com/uk-news/2013/jul/25/railway-accidents-human-error-warning-systems
8. Alper SJ, Karsh B-T (2009) A systematic review of safety violations in industry. Accid Anal Prev 41:739–754
9. de Brito G (1998) Study of the use of Airbus flight-deck procedures and perspectives for operational documentation. Int Conf Hum Comput Interact Aeronaut 195–201
10. de Saint Maurice G, Auroy Y, Vincent C, Amalberti R (2010) The natural lifespan of a safety policy: violations and system migration in anaesthesia. Qual Saf Health Care 19:327–331
11. Vicente KJ (1999) Preface. Cognitive work analysis: towards safe, productive, and healthy computer-based work. Lawrence Erlbaum Associates, New Jersey, p xiii–xix
12. Xiao-qian C, Hui-tao L (2013) Gray relational analysis on airline employees' pay satisfaction and violations. In: International conference on management science and engineering (20th). p 288–293
13. Dekker S (2014) Safety differently: human factors for a new era. CRC Press, Florida
14. Hollnagel E (2014) Safety-I and safety-II: the past and future of safety management. Ashgate Publishing Ltd, Farnham
15. Reason J (2008) The human contribution: unsafe acts, accidents and heroic recoveries. Ashgate Publishing Limited, Surrey
16. Clarke DM (2010) The human contribution: unsafe acts, accidents and heroic recoveries. J Risk Res 14:144–145
17. Leveson NG (2002) System safety engineering: back to the future. MIT Aeronautics and Astronautics, Boston
18. Dekker S, Cilliers P, Hofmeyr J-H (2011) The complexity of failure: implications of complexity theory for safety investigations. Saf Sci 49:939–945

19. Dekker S (2011) Drift into failure: from hunting broken components to understanding complex systems. Ashgate Publishing Limited, Surrey
20. Rochlin GI, La Porte TR, Roberts KH (1998) The self-designing high-reliability organisation: aircraft carrier flight operations at sea. Nav. War Coll. Rev 51:97–113
21. Arnstein MG (1956) Florence Nightingale's influence on nursing. Bull NY Acad Med 32:540–546
22. Horsley K (2010) Florence Nightingale. J Mil Veterans Health 18:4–5
23. Winkelstein W Jr (2009) Florence Nightingale: founder of modern nursing and hospital epidemiology. Epidemiology 20:311
24. Baly ME, Matthew HCG (2011) Nightingale, Florence. In: Goldman L (ed) Oxford dictionary of national biography. [Internet]. Oxford University Press. Available from: http://www.oxforddnb.com/view/article/35241?docPos=2
25. Helmstadter C (2010) Navigating the political straits in the Crimean War. In: Nelson S, Rafferty AM (eds) Notes on Nightingale: the influence and legacy of a nursing icon. Cornell University Press, New York, pp 28–54
26. Smith CW (1950) Florence Nightingale, 1820–1910. Constable and Company Ltd, London
27. Godden J (2010) The dream of nursing the empire. In: Nelson S, Rafferty AM (eds) Notes on Nightingale: the influence and legacy of a nursing icon. Cornell University Press, New York, pp 55–75
28. Godden J (2006) Lucy Osburn, a lady displaced. Sydney University Press, New South Wales
29. Fox JC (2014) Mass. General exhibit honors a pioneer of nursing education. Boston Globe [Internet]. Massachusetts. Available from: https://www.bostonglobe.com/lifestyle/health-wellness/2014/04/27/mass-general-portrait-exhibit-honor-pioneer-nursing-education/zwKAMEeWXycyXwToeBp03I/story.html
30. Snodgrass ME (1999) Historical encyclopedia of nursing. ABC-CLIO, California
31. Cook ET (1913) The life of Florence Nightingale vol. 2 of 2. Macmillan and Co., Limited, London
32. Thomson D (1983) Workhouse to nursing home: residential care of elderly people in England since 1840. Ageing Soc. 3:43–69
33. Magnello ME (2010) The passionate statistician. In: Nelson S, Rafferty AM (eds) Notes on Nightingale: the influence and legacy of a nursing icon. Cornell University Press, New York, pp 115–129
34. Global Aviation Information Network (2004) Roadmap to a just culture: enhancing the safety environment. Global Aviation Information Network Working Group E, p 46
35. Tingvall C, Lie A (2010) The concept of responsibility in road traffic (Ansvarsbegreppet i vägtrafiken). Transp. Linköping Swed
36. Woods DD, Dekker S, Cook R, Johannesen L, Sarter N (2010) Behind human error, 2nd edn. Ashgate Publishing Ltd, Surrey
37. Dekker S, Pruchnicki S (2014) Drifting into failure: theorising the dynamics of disaster incubation. Theor Issues Ergon Sci 15:534–544
38. Croskerry P, Nimmo GR (2011) Better clinical decision making and reducing diagnostic error. J R Coll Physicians Edinb 41:155–162
39. Anderssen E (2012) When doctors make bad calls. The Globe and Mail [Internet]. Toronto. Available from: http://www.theglobeandmail.com/life/health-and-fitness/when-doctors-make-bad-calls/article549084/?page=all
40. Lundberg J, Rollenhagen C, Hollnagel E (2009) What-you-look-for-is-what-you-find—the consequences of underlying accident models in eight accident investigation manuals. Saf Sci 47:1297–1311
41. Nemeth C, Cook RI, Patterson E, Donchin Y, Rogers M, Ebright P (2004) Afterwords: the quality of medical accident investigations and analyses. SAGE Publications, pp 2084–2088
42. Lindsay DS (1990) Misleading suggestions can impair eyewitnesses' ability to remember event details. J Exp Psychol Learn Mem Cogn 16:1077–1083
43. Loftus EF, Palmer JC (1974) Reconstruction of automobile destruction: an example of the interaction between language and memory. J Verbal Learn Verbal Behav 13:585–589

44. Ramirez S, Liu X, Lin P-A, Suh J, Pignatelli M, Redondo RL et al (2013) Creating a false memory in the hippocampus. Science 341:387–391
45. Itsukushima Y, Nishi M, Maruyama M, Takahashi M (2006) The effect of presentation medium of post-event information: impact of co-witness information. Appl Cogn Psychol 20:575–581
46. Carrabine N (2009) Ex-pharmacist sentenced to jail in death of Mentor girl. The News-Herald. [Internet]. Ohio. Available from: http://www.news-herald.com/general-news/20090814/ex-pharmacist-sentenced-to-jail-in-death-of-mentor-girl
47. Institute for Safe Medication Practices (2009) Ohio government plays whack-a-mole with pharmacist. ISMP Newsletters. [Internet]. Available from: https://www.ismp.org/newsletters/acutecare/articles/20090827.asp
48. McCoy K, Brady E (2009) Rx for errors: drug error killed their little girl. USA TODAY [Internet]. Virginia. Available from: http://usatoday30.usatoday.com/money/industries/health/2008-02-24-emily_N.htm
49. Gimmig D, Huguet P, Caverni J-P, Cury F (2006) Choking under pressure and working memory capacity: when performance pressure reduces fluid intelligence. Psychon Bull Rev 13:1005–1010
50. Kellogg JS, Hopko DR, Ashcraft MH (1999) The effects of time pressure on arithmetic performance. J Anxiety Disord 13:591–600
51. Slobounov SM, Fukada K, Simon R, Rearick M, Ray W (2000) Neurophysiological and behavioural indices of time pressure effects on visuomotor task performance. Cogn Brain Res 9:287–298
52. Thompson C, Dalgleish L, Bucknall T, Estabrooks C, Hutchinson AM, Fraser K et al (2008) The effects of time pressure and experience on nurses' risk assessment decisions: a signal detection analysis. Nurs Res 57:302–311
53. Heylighen F (1989) Causality as distinction conservation. a theory of predictability, reversibility, and time order. Cybern Syst Int J 20:361–384
54. Heylighen F, Cilliers P, Gershenson C (2007) Complexity and philosophy. In: Bogg J, Geyer R (eds) Complexity, science and society. Radcliffe Publishing, Oxford
55. Telegraph Reporters (2013) Mid Staffordshire Trust inquiry: how the care scandal unfolded. The Telegraph [Internet]. London. Available from: http://www.telegraph.co.uk/news/health/news/9851763/Mid-Staffordshire-Trust-inquiry-how-the-care-scandal-unfolded.html
56. Francis R (2010) Independent inquiry into care provided by Mid Staffordshire NHS Foundation Trust January 2005–March 2009 volume I. London. The Stationery Office, London, p 452
57. Commission Healthcare (2009) Investigation into Mid Staffordshire NHS foundation trust. Healthcare Commission, London
58. Pearson A (2013) Forget hacking hacks, let's jail some callous Mid Staffs NHS staff. The Telegraph [Internet]. London. Available from: http://www.telegraph.co.uk/news/health/news/9867843/Forget-hacking-hacks-lets-jail-some-callous-Mid-Staffs-NHS-staff.html
59. Triggle N (2013) Stafford Hospital: hiding mistakes 'should be criminal offence'. BBC News [Internet]. London. Available from: http://www.bbc.co.uk/news/health-21341766
60. BBC News (2013) NHS nurses 'penalised for caring' [Internet]. Available from: http://www.bbc.com/news/uk-21348797
61. Campbell D (2013) Mid Staffs hospital scandal: the essential guide. The Guardian [Internet]. London. Available from: http://www.theguardian.com/society/2013/feb/06/mid-staffs-hospital-scandal-guide
62. Francis R (2013) Report of the Mid Staffordshire NHS Foundation Trust public inquiry: executive summary. The Stationery Office, London
63. Adams T (2013) Mid Staffs whistleblower Julie Bailey: 'I don't go out here on my own any more'. The Guardian [Internet]. London. Available from: http://www.theguardian.com/society/2013/oct/27/julie-bailey-mid-staffordshire-nhs-whistleblower

64. BBC News (2011) Stafford Hospital inquiry: final witness's tears at bullying. BBC News [Internet]. London. Available from: http://www.bbc.com/news/uk-england-stoke-staffordshire-15210964
65. Crawford ER, LePine JA, Rich BL (2010) Linking job demands and resources to employee engagement and burnout: a theoretical extension and meta-analytic test. J Appl Psychol 95:834–848
66. Janssen PPM, De Jonge J, Bakker AB (1999) Specific determinants of intrinsic work motivation, burnout and turnover intentions: a study among nurses. J Adv Nurs 29:1360–1369
67. Maslach C, Jackson SE (1981) The measurement of experienced burnout. J Occup Behav 2:99–113
68. Glasberg A-L, Eriksson S, Norberg A (2007) Burnout and 'stress of conscience' among healthcare personnel. J Adv Nurs 57:392–403
69. Laschinger HKS, Leiter MP (2006) The impact of nursing work environments on patient safety outcomes: the mediating role of burnout engagement. J Nurs Adm 36:259–267
70. Schmitz N, Neumann W, Oppermann R (2000) Stress, burnout and locus of control in German nurses. Int J Nurs Stud 37:95–99
71. Strachey L (1977) Eminent victorians. Penguin Books, Middlesex
72. Wallerstein I (1997) Open the social sciences: report of the Gulbenkian Commission on the restructuring of the social sciences. Stanford University Press, California

Patient Safety and Disaster Forensics: Understanding Complex Causality Through Actor Network Ethnography

Anthony J. Masys

Abstract In 1999, the Institute of Medicine (IOM) estimated that 48,000–98,000 Americans die each year in the hospital stemming from mistakes and oversights in medical care. Human error is often cited as a major contributing factor or cause of incidents and accidents. For example, in the aviation domain, accident surveys have attributed 70 % of incidents to crew error citing pilot error as the root cause of an aviation accident (Woods et al. in Behind human error: cognitive systems, computers, and hindsight, 1994 [82: 2], Helmreich in Br Med J 320:781–785, 2000 [39], Shappell and Wiegmann in Hum Factors Aerosp Saf 1(1):59–86, 2001 [70: 60]). Similarly in the healthcare domain, medical errors are reported to be a major cause of morbidity and mortality. Aggarwal et al. (Qual Saf Health Care 19(2): i3–i8, 2010 [3]) argues that '… an increasing awareness of medical injuries has been paralleled by the rise in technology, and the increasing complexity it causes'. According to Woods et al. (Behind human error: cognitive systems, computers, and hindsight, 1994 [82]), human error can be characterized either as a cause of failure or as a symptom of a failure. The label 'human error' as reported by Woods et al. (Behind human error: cognitive systems, computers, and hindsight, 1994 [82]) is considered prejudicial and unspecific. They argue that the label 'human error' retards rather than advances our understanding of how complex systems fail and the role of the human in both successful and unsuccessful system operations. This is contextualized and further supported in the healthcare domain (clinical encounters) by Artino et al. (Clin Pharmacol Ther 91(2), 2012 [4]). A systems view of the problem space regards human error as a symptom of '…contradictions, pressures and resource limitations deeper inside the system' (Dekker in The re-invention of human error, 2002a [24: 2]). Actor Network Theory provides a conceptual foundation and lens to facilitate a systems thinking based analysis (Masys in Fratricide in air operations: opening the black box—revealing the social, 2010 [56], Wickramasinghe et al. in Int J Network Virtual Organ 4(3):257–279, 2007 [80]) to examine the key dynamics that reside in the black box of human error pertaining to patient safety. This chapter frames patient safety and human error through the lens of Actor Network Theory by leveraging insights from accidents and disasters such as the

A.J. Masys (✉)
University of Leicester, Leicester, UK
e-mail: Anthony.masys@gmail.com

A.J. Masys (ed.), *Disaster Forensics*, Advanced Sciences and Technologies for Security Applications, DOI 10.1007/978-3-319-41849-0_4

2003 US/Canada Blackout (Masys in Networks and network analysis for defence and security. Springer Publishing, 2014 [58]) and the Uberlingen mid-air crash (Masys in Int J Disaster Prev Manag 14(4):548–557, 2005 [54]). An example of the medical errors associated with the use of an Emergency Department Information Systems (EDIS) in a clinical situation is given. This ANT facilitated 'disaster forensic analysis' reveals a complex causality that is rhizomal rather than linear thereby challenging our notion of human error and highlights where intervention strategies can be focused to support patient safety.

Keywords Patient safety · Actor network theory · Human error

1 Introduction

According to Woods et al. [82], human error can be characterized either as a cause of failure or as a symptom of a failure. Patterson et al. [63: 35] argue that the frequent attribution of human error as a 'root cause' often serves as a stopping point for an investigation. However, in the new view of human error, it becomes a starting point thereby revealing how multiple interacting factors combine in a complex socio-technical system. A systems view of the problem space regards human error as a symptom of '…contradictions, pressures and resource limitations deeper inside the system' [24: 2]. This systems view supports a systems perspective, whereby the attribution of human error is seen as an oversimplification of a complex aetiology resulting from a number of causes [18, 70].

Described in Masys [56] and reported by Woods et al. [82: 4], the question surrounding thc attribution of human error is a complex matter highlighting that human performance is a multidimensional issue that is a function of the context in which an incident takes place; that technology shapes human performance thereby creating new pathways and forms of error and failure; that human performance involves a set of interacting people; that competing goals within the organizational context creates dilemmas shaping accident aetiology; and that the attribution of error is a social judgment rather than an objective conclusion.

Dekker [25] differentiates between Old and New thinking regarding human factors, system safety and accident aetiology as shown in Table 1.

Table 1 Old view/new view of human error [25: vii]

The old view of human error	The new view of human error
• Human error is a cause of accidents • To explain failure you must seek failure • You must find people's: inaccurate assessments, wrong decisions, bad judgments	• Human error is a symptom of trouble deeper inside a system • To explain failure, do not try to find where people went wrong. Instead, find how people's assessments and actions made sense at the time, given the circumstances that surrounded them

Dekker [26: 7] argues that the old view paradigm is rooted in a Cartesian-Newtonian worldview of linear causality. This perspective traces the failure of a system to a chain of events within a system that linearly define the path towards an accident. The value of articulating the sequence of events is not disputed. However, it has been cited by Leveson [53] that event-based models provide a poor representation of systemic accident factors and focus primarily on proximate events. According to Leveson [53: 9]:

> Viewing accidents as chains of events may limit understanding and learning from the loss. Event chains developed to explain an accident usually concentrate on the proximate events immediately preceding the loss. But the foundation for an accident is often laid years before.

As a consequence de Almeida and Johnson [22: 1] remark that:

'It is becoming increasingly difficult to identify the causes of incidents and accidents back through the complex interactions that lead up to an adverse event. At the same time, there is a growing appreciation of the need to consider a broad range of contextual factors in the aftermath of any mishap'. Hence the requirement for a methodology that challenges the old model of human error by opening the blackbox.

Systems Thinking [59] represents a theoretical framework, a perspective and a set of methodological tools that has been applied across various domains including health care [18]. The systems perspective reveals properties of the whole that are not evident with an examination of the components revealing inherent nonlinear dynamic interactions and emergent behaviour. Systems theory as discussed in Senge [67] emphasizes interconnectedness and causal complexity thereby challenging traditional linear thinking and simple causal explanations. As a worldview, systems thinking recognizes that systems cannot be addressed through a reductionist approach that reduces the systems to their components.

In 1999, the Institute of Medicine (IOM) estimated that 48,000–98,000 Americans die each year in the hospital from mistakes and oversights in medical care [42]. Wears [79: 374] argues that 'Efforts to improve the safety, cost, or quality of health care are difficult because the clinical work that produces care is hard to capture. It involves multiple activities, complex artifacts, shifting and often conflicting goals, cognitive work, coordinated activity, and cooperative dynamics within a complex socio-technical system'. Artino et al. [4: 167] describe how clinical success within the health domain are often framed in terms of physician quality. This tends to '…underestimate the influence of situational factors' in successful and unsuccessful clinical encounters'. Aggarwal et al. [3] highlights the influence of technology on patient safety. Aggarwal et al. [3: i3] argues:

> It is important to note that an increasing awareness of medical injuries has been paralleled by the rise in technology, and the increasing complexity it causes. The number of different medications for blood pressure, the various approaches and techniques for hip surgery and the multitude of blood tests for diagnosis of infections, all provide infinitely more opportunities for things to go wrong. This is not to say that the patient safety problem is caused by technology, but rather that technology has led to a complex and confusing environment within medicine

The question becomes, how can we open the blackbox of human error within the complex landscape of work and risk to analyze this entangled 'social' space?

Actor Network Theory is a lens to facilitate a systems thinking based analysis [56, 80] to examine the key dynamics that reside in the black box of human error pertaining to patient safety. Dekker [23: 3] argues that 'human error is not an explanation but is something to be explained'. Challenging the traditional view of human error, this chapter recognizes that '...accidents are seen as emerging phenomena in complex systems and as the result of an aggregation of conditions rather than the inevitable effect of a chain of courses' [40: xv].

1.1 Actor Network Theory

Detailed in Masys [56], Actor Network Theory (ANT) emerged from the sociological studies of science and technology through the contributions of Serres and Latour [68], Callon and Law [15] and influenced by the work of Foucault [35, 36], Deleuze and Guattari [27] with a focus on the socio-technical domain. Masys [54, 56, 58] and Bennett [7, 8] apply the systems perspective of ANT to examine the inter-connectedness of the heterogeneous elements within a disaster context characterized by the technological and non-technological (human, social, organizational) elements. Yeung [81] notes that much of the work that draws on actor network theory places its analytical focus on unearthing the complex web of relations between humans and non-humans. Latour [46: 806] argues that '...it is impossible even to conceive of an artifact that does not incorporate social relations, or to define a social structure without the integration of non-humans into it. Every human interaction is socio-technical'. The 'social' is thereby described as 'materially heterogeneous' [16: 166].

Developed to analyse situations where separation of the social and technical elements is difficult [14], ANT provides a methodological approach to analysing the socio-technical domain, through 'radical ethnography'. As described in Masys [56], Latour [48: 29] argues that '...the choice is thus clear: either we follow social theorists and begin our travel by setting up at the start which kind of group and level of analysis we will focus on, or we 'follow the actors' own ways and begin our travels by the traces left behind by their activity of forming and dismantling groups'. 'Following the actors' [12, 13, 47] lies at the foundation of the ANT methodology. 'Following the actors' allows the researcher to investigate those actors that have been 'silenced or deleted'. With this in mind Latour [48: 82] argues that '...if objects are not studied it is not due to a lack of data, but rather to a lack of will'. What ANT facilitates is an examination of networks of relations in terms of composition, emergence, maintenance, influence and volatility. It is this very approach that lies at the heart of disaster forensic analysis.

2 Case Studies

Drawing upon select case studies [54, 58] and leveraging the methodological approach described in Masys [56], the application of ANT to support Forensic analysis is explained. The insights will then be applied to patient safety.

2.1 2003 US/CAN Blackout

As described in Masys [58], Hurricane Katrina in 2005 devastated New Orleans thereby revealing inherent vulnerabilities that resided in the socio/political/ecological/technical infrastructure (system) of the city and the nation. These 'unseen' vulnerabilities that emerged at the 'seams' of interconnection and interdependencies can be characterized as 'resident pathogens', in that the hurricane as a 'triggering mechanism' interacted with the '…city's fragile physical environment, aging infrastructure, and declining economic and social structure' [21: 501] as well as policies, regulations and politics.

Comfort [21: 502] asks the question '…Was the damage in New Orleans due to Hurricane Katrina, or was it some combination of human and technical factors that failed under the stress of the hurricane?' This questions opens up the space of possibilities to unearth the silenced and deleted voices. Similarly, this question resonates with other events such as: Fukushima in 2011 whereby an earthquake and resulting tsunami had a devastating effect on the Fukushima Daiichi nuclear power plant [41]; Hurricane Sandy in 2012 that resulted in significant disruption to New Jersey and New York [33]; Ash Cloud in 2010 stemming from the eruption of Eyjafjallajökull and resulting in significant disruptions to air travel and trade in Europe [38]. All these cases highlight the interconnectedness and interdependencies that characterize how we live and what we depend upon.

This helps to frame the case study in the US/Canada Power Outage Task Force Final Report [76: 1], 'on August 14, 2003, large portions of the Midwest and Northeast United States and Ontario, Canada, experienced an electric power blackout. The outage affected an area with an estimated 50 million people and 61,800 megawatts (MW) of electric load in the states of Ohio, Michigan, Pennsylvania, New York, Vermont, Massachusetts, Connecticut, New Jersey and the Canadian province of Ontario'.

2.2 Uberlingen Mid Air Collision

As described in Masys [54], on 1 July 2002, a Russian Federation-registered Tupolev 154M (Tu-154M) of Bashkirian Airlines, carrying a party of schoolchildren, collided with a Bahrain-registered DHL Boeing 757 cargo aircraft. The

Tu-154M passenger aircraft (with Russian crew) was heading west. The DHL Boeing 757 freighter (with a British Captain and Canadian First Officer) was heading north. Both aircraft were approaching the same waypoint at an altitude of 36,000 ft. Both were under control of Skyguide's Zurich air traffic control centre (ATCC). As discussed in detail in Bennett [5, 6], "the disaster originated in the two crews' different understandings of the primacy of their on-board TCAS systems. The western crew obeyed their TCAS 'descend' resolution, putting their aircraft into a dive. The Russian crew ignored their concurrent TCAS 'climb' RA, choosing instead to obey Zurich ATCC's 'descend' instructions. They dived their aircraft into the DHL freighter". Seventy-one died in the air.

3 Discussion

The domains of critical infrastructure vulnerability analysis [58], aviation safety [56] [Masys 55], nuclear safety [Masys et al. 60], oil and gas safety [57, 58] can be characterized as complex socio-technical systems. The analysis of socio-technical systems is not new to the social sciences [9, 51]. Coakes [20: 2] described Socio-technical thinking as: '...holistic in its essence; it is not the dichotomy implied by the name; it is an intertwining of human, organizational, technical and other facets'. This systems or holistic approach is further expounded by Senge [67] who argues that the world exhibits qualities of wholeness, the relevance of systemic thinking is captured within its paradigm of interdependency, complexity and wholeness. Although events can be considered to be discrete occurrences in time and space '...they are all interconnected. Events can be understood only by contemplating thc wholc' [34: 13].

In the health domain, Carey et al. [18: 7] argue that '...public health problems are already deemed complex, and system-based approaches can contribute to changing the language, methods and methodologies for conceptualizing and acting within this complexity'. The complex socio-technical system is represented by Aggarwal et al. [3: i3] in that 'Every clinical process is supported by numerous other systems and processes, variously dependent on technology'.

ANT is an approach that is interested in the tensions between actor, network and technology, and how they manifest in practice [47, 49]. With this understanding of the complexity of socio-technical systems as manifested in the health domain and pertaining to patient safety, the holistic perspective of ANT makes it well suited to facilitate a Forensic analysis of this network of heterogeneous elements.

3.1 Actor Network Theory

What makes ANT such a powerful approach is that it treats both human and machine (non-human) elements in a symmetrical manner, thereby facilitating the examination of the situation (such as an accident) where Callon [14: 183] argues, '…it is difficult to separate humans and non-humans, and in which the actors have variable forms and competencies'.

The 'follow the actors' approach is rooted in an understanding of the actor and the inherent relationality that characterizes the network. An actor-network as described by Latour [43] and Callon [12, 13] is characterized as a network that is inherently heterogeneous, where the relations between the actors are important, rather than their essential or inherent features.

The actor, whether technical or non-technical, is examined within the context of a heterogeneous network. In fact the actor is a network in itself '…in the same way, elements in a network are not defined only by their "internal" aspects, but rather by their relationships to other elements, i.e., as a network' [2: 360]. The actors or actants of ANT can be humans, organizations, cultures, ideas, animals, plants or inanimate objects and are described in terms of the alliances and exchanges they exhibit in the interconnected network of relations.

Examination of actors such as those characterized traditionally as technologies, facilitates an exploration of how these 'actors' mediate action and how they are entangled in local techno-social configurations. This becomes particularly relevant in the healthcare domain as patient/healthcare provider is more and more mediated by technology. For example Doyle [29] presents an interesting observation regarding technology mediated surgery:

> The surgeon's immersion in the 3D visual field of robotic surgery requires different visuospatial skills than open surgery, although training programmes are still in their infancy (Hance et al.). Simulations, however, are already used for the training of laparoscopic surgery (Hance et al.). The skills required in virtual laparoscopic surgery simulators are compared to those used for computer gaming, with students who play computer games "11 % more efficient than those who did not play computer games" (Enochsson et al. 878).

Harbers [37: 10] argues that '…we are confronted here with a hybrid situation in which human beings and technology are tightly interwoven—a mixture, a muddle of man and machine'. We address these questions and arguments through the concept of the 'hybrid collectif' [15, 56, 57] that emerges from the analysis. As described in Masys [56], action rather takes place in 'hybrid collectifs' [15] that entangle human actors as well as non-human actants in multiple ways. Tools, for example, are not just things that are used to achieve certain ends: 'They contribute to the making of the universe of possibilities that make action itself' [17: 18]. In Actor Network Theory the network is not purely social, but is constructed by hybrids of social (human) and non-social (technological, natural, material) elements simultaneously.

3.2 Actor

As described in Masys [56], an actor-network, as cited in Aanestad [1: 6–7], '…is *a* heterogeneous network of human and non-human actors... where the relations between them are important, rather than their essential or inherent features [12, 13, 47]'. The actor, whether technical or non-technical, is examined within the context of a heterogeneous network. In fact the actor is a network in itself '…in the same way, elements in a network are not defined only by their "internal" aspects, but rather by their relationships to other elements, i.e., as a network' [2: 360]. In this sense, the Actor Network becomes a network of aligned interests formed by the heterogeneous actors, characterised as full of hybrid entities [45].

3.3 ANT Processes

Fundamental processes within ANT are inscription and translation. These processes have been described in application to accident aetiology in Masys [54–58]. Inscription is about how technical artifacts embody patterns of use and thereby shape perception [54, 56] and action [54, 56, 58].

The process of translation has been described as pivotal in any analysis of how different elements in an actor network interact [Somerville 71]. Masys [56: 53] captures this element:

> Translation rests on the idea that actors within a network will try to enroll (manipulate or force) the other actors into positions that suit their purposes. When an actor's strategy is successful and it has organized other actors for its own benefit, it can be said to have translated them. Translation as argued by Callon [13: 143] '…are embodied in texts, machines, bodily skills [which] become their support, their more or less faithful executive'.

With this understanding of ANT as a systems lens to explore socio-technical systems, the insights derived from its application are captured in the analysis of the 2003 US/Canada Blackout [58] and the Uberlingen midair collision [54].

3.4 Resident Pathogens (US/Canada Blackout)

Masys [58] describes the US/Canada Blackout in terms of vulnerabilities that emerge from the spatially, temporally and characteristically heterogeneous actors. In a sense, these vulnerabilities can be likened to 'resident pathogens' [65: 198] that reside 'inscripted' within the system and through the process of translation can precipitate cascading failures.

As described in the final report (U.S.-Canada Power System Outage Task Force Final Report), August 14, 2003, blackout was caused by deficiencies in specific practices, equipment, and human decisions by various organizations that affected conditions and outcomes that afternoon. Deficiencies in corporate policies, lack of adherence to industry policies and regulatory framework inscripted into the system are symptoms of translation process that facilitated inadequate management of reactive power and voltage which at the sharp end caused the blackout. This

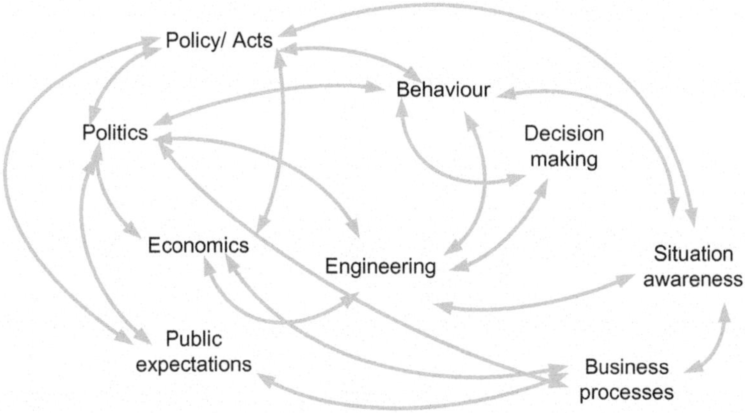

Fig. 1 Actor network relational mapping [58]

realization shows the spatial, temporal and physical heterogeneity that characterizes the actor network. This is described in Masys [58].

Following the actors reveals a macro-relational mapping of the actors as depicted in Fig. 1.

Drawing upon ANT, Dolwick [28] argues that: '…if one were to try to draw a map of all of the actors present in any interaction, at any particular moment in time, instead of a well-demarcated frame, one would produce a highly convoluted network with a multiplicity of diverse dates, places and people'. The interdependencies and interconnectivity within the system contributes to its vulnerability [64]. Comfort [21: 9] argues that '…the vulnerability of technical systems that support basic operations in a city cannot be calculated separately, but rather must be based upon careful estimates of the degree of interdependence or dependence across the entire socio-technical system that provides services to an urban region'.

Expanding upon the relational mapping, through thematic analysis [10] and memoing [19], insights into the aetiology emerge revealing how decisions taken spatially and temporally displaced become relevant and present in the accident.

Figure 2 depicts an expanded actor network representation highlighting key actors and relations that emerged from the thematic analysis [10, 19].

3.5 *Hardwired Politics: Uberlingen Case*

Understanding the Uberlingen mid air collision requires opening the blackbox of situation awareness (SA) and the intersection of the human and nonhuman actors. Shared devices such as common displays, a common environment or communication are essential in the development of effective SA as noted by Endsley [30].

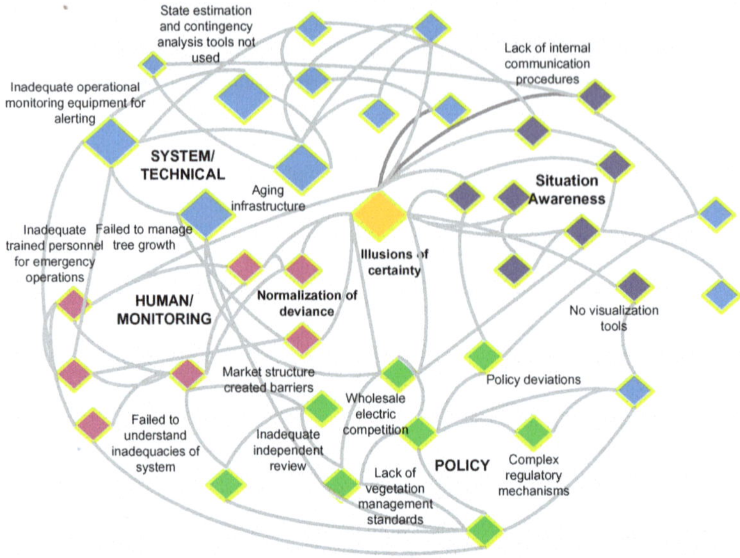

Fig. 2 Resident pathogens [58]

The formation of shared SA in relation to team processes requires additional activities such as coordination and information exchange [Entin and Entin 31].

Masys [54] situates the SA through the lens of the hybrid collectif thereby framing the TCAS and ATC within the actor network. As highlighted in Bennett [5, 6]: "inconsistencies in the matter of the primacy afforded TCAS advisories with respect to conflicting air traffic controller (ATC) instructions" arc factors within the aetiology of this accident. The notion of an inscription may be used to describe how concrete anticipations and restrictions of future patterns of use are involved in the development and use of a technology [61]. To make the technology work, all these elements must be aligned, i.e. cooperating toward a common goal [2]. The inscribed patterns of use may not succeed because the actual use deviates from it. Rather than following its assigned program of action, a user may use the system in an unanticipated way; he/she may follow an anti-program [44, 61].

As described in Masys [54], "The different understandings regarding the primacy of the aircraft-borne collision warning system (known as the TCAS) originated in national variations in the operationalisation of the system; in deference to a flight-deck cultural practice known to the UK's Civil Aviation Authority as 'commander's discretion'; and in the failure of the world community to establish a single, authoritative rule-making body for commercial aviation" [5, 6]. It highlights the hegemony of the technical solution to socio-technical issues. It is important to recognize that the operational role of the TCAS, ATC and aircrew must be assigned such that the proper structure facilitates the construct of a high level of SA. However this is not the case.

Described as a network of heterogeneous elements [56], the degraded SA in this case study is rooted within the relationality resident within the network. These relations include the beliefs, culture, policies, SOPs and politics of the aviation industry.

The shared SA resident within this network of heterogeneous elements was shaped by the socio-technical and socio-political characteristics of the system. These elements coalesced to produce a degraded SA that precipitated the convergence of circumstances that resulted in the accident [56]. Through the application of ANT methodology, we opened the black box to reveal the network space, replete with actors and relations and examine the processes of inscription and translation in the construction of the black box, thereby 'undeleting' the 'silenced voices' that reside within.

3.6 Socio-technical Systems: EDIS and Patient Safety

Parallels can be drawn between the US/Canada Blackout, Uberlingen case study and quality and patient safety illustrations. In the complex socio-technical systems characterized by the health domain, studies have emerged examining the benefits and unintended consequences stemming from the introduction of new technology and new processes. For example, Aggarwal et al. [3: i4] notes that '…The use of minimally invasive therapies to treat disease has radically altered the patient experience for common procedures such as gallbladder removal and tubal sterilisation. This technology improved safety in many respects but also introduced new hazards; in some centres, bile duct and aortic injuries increased initially owing to the absence of adequate preparation, training and assessment prior to the routine use of the new techniques'.

The complex and dynamic environment of the Emergency Departments (ED) makes it a high risk area. ED are characterized by '…rapid turnover, frequent transitions of care, constant interruptions, large variation in patient volumes, and unfamiliar patients, make the ED environment particularly error prone' [32: 399]. To manage such complexity, Emergency Department Information Systems (EDIS) have migrated into the environment to support improving quality and outcomes with electronic health records. The EDIS is the primary tool used in emergency departments to track and document on patients. Farley et al. [32: 399] report that '…variation in EDIS functionality affects physician decision making, clinician workflow, communication and ultimately, the overall quality of care and patient safety'.

Within this complex socio-technical system, we 'follow the actors' Callon [12], Latour [47, 48] and 'let them set the framework and limits of the study themselves' [73: 80]. Through this approach we search out the interactions, negotiations, alliances and networks that characterize the network space. Latour [43: 175–176] remarks:

Fig. 3 **a** Actors represented as nodes. **b** Actors and relational connectivity

> ...we have to be as undecided as possible on which elements will be tied together, on when they will start to have a common fate, on which interest will eventually win out over which. In other words, we have to be as undecided as the actors we follow...The question for us, as well as those we follow, is only this: which of these links will hold and which will break apart.

In the analysis of patient safety in technology mediated environments, Fig. 3a illustrates a number of actors (represented by nodes) that have emerged from the analysis. One then pursues each one of these actors (follow the actor) to see what they are connected with and how relationally they are connected. It is the relations that shape action and inaction and therefore become the key focus of an Actor Network analysis: it is about influence, dynamics and relations.

Figure 3b shows the notional connectivity between actors.

The steps associated with the ANT approach involve following the actor and then determining the relational significance of their role in the actor network.

In step 1, identifying and tracing the network begins with 'follow the actor' [47] in order to investigate the relevant 'leads' each actor suggests. This is significant in that it is essentially the actor themselves and not the researcher that determines the direction of the investigation. For example a policy or standard operating procedure (SOP) may lead (influence) directly to a physical (technical) or informational actor. These interrelations mark the direction of analysis. In Step 2 the goal is to 'interview' the actors. This is accomplished through a relational mapping of the influences across both human and non human actors. The aim of this step is to see how these actors relate to each other and the associations they create—to identify how they interact, how they negotiate, and how they form alliances and networks with each other. The relational mapping described resonates with the propositional networks described in detail in Stanton et al. [72].

Take for example the complex and dynamic environment of ED. Patient safety and health care quality is about ensuring high reliability and consistency in practice. Farley et al. [32: 400] argues that the '...EDIS are intended to decrease practice variability and improve system reliability by ensuring legible communication, facilitated retrieval of past information (e.g. Physician notes, diagnosis studies), and access to computerized physician order entry to aid in clinical decision support'.

As argued by Law and Callon [52: 285] '...we are concerned to map the way in which they [actors] define and distribute roles, and mobilize or invent others to play these roles'. Essentially ANT is likened to ethnography extended to nonhumans.

Aggarwal et al. [3: i5] argues that it is '...important to recognise that technologies can pose many hazards. Some of these relate to improper or careless use; for instance, if a technology is introduced too quickly and without thought for the culture and clinical setting or without appropriate training, it can become a danger rather than a benefit'. Contextually with regards to EDIS and the benefits they afford, '…EDISs, may also lead to medical errors and cause patient safety and quality concerns. A recent report released by the IOM, Health IT and Patient Safety: Building Safe Systems for Better Care, states that…[p]oorly designed, implemented, or applied, health IT can create new hazards in the already complex delivery of health care…As health IT products have become more intimately involved in the delivery of care, the potential for health IT-induced medical error, harm or death has increased significantly' [32: 400].

Wears [79: 374] argues that 'Making invisible work visible is both highly important and highly dangerous. In efforts to redesign work or introduce complex technology, one must understand the activities that actually occur during clinical work and not be limited to the nicely ordered tasks that are envisioned in managers', supervisors', or even the workers' minds so that valuable activities are not inadvertently degraded or eliminated'.

New kinds of error (unintended consequences) emerged from the introduction of EDIS. These included: '…juxtaposition errors, in which users select an item next to the intended choice, such as wrong patient being selected; desensitization to alerts or alert overload; confusing order option presentations; and system design issues related to poor data organization and display…new forms of communication failure can be introduced if users are not sensitive to the limitations and pitfalls of an overreliance on the EDIS. …overreliance on the EDIS as a primary source of communication can degrade the quality of communication, leaving providers with the dangerous task of decoding generic messages' [32: 401].

Figure 4 illustrates an actor network representation that shows that accident aetiology associated with human error is decentered and recognizes the influence of

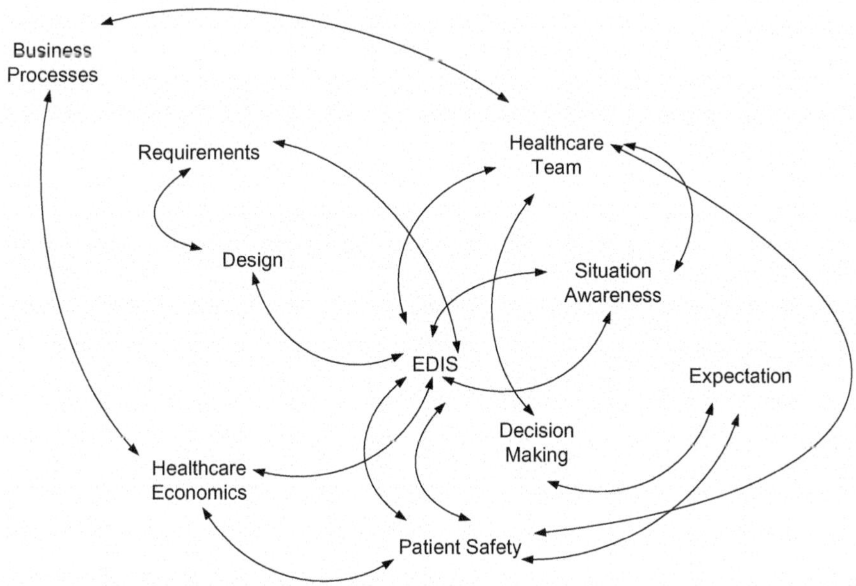

Fig. 4 High-level actor network relational mapping

other factors that shape action/inaction. EDIS does not necessarily reduce errors, but rather changes the way work is performed.

With examination of EDIS, the Actor Network Theory does not focus on these domain areas in the analysis a priori but rather focuses on the intersection of the **P**hysical, **H**uman, and **In**formational which is interpreted and represents the actor network relational space. This space represents what we term the 'hybrid collectif' within which the dichotomy associated with the human and the technological is dissolved. The relational effects of EDIS emerge from the analysis and reinforce as argued in Van der Duim [77: 88], '…It does not make sense to ignore materials and to treat them separately, as though they were different in kind: the characterization of materials is just another relational effect'.

What we discover is the healthcare team essentially 'patient safety' is a heterogeneous network comprised of:

'person' + supporting team and technicians + orders, SOPs + policy + EDIS + sensors + virtual team members + communications + training + doctrine + Professional and work culture + work of engineers + design + politics of healthcare.

It is this very relational network of actors that creates the possibility for action. In defining the actor network, patient safety emerges as this entangled network of heterogeneous elements, a hybrid collectif. What this reveals is that action, as seen in the case studies, takes place in a 'hybrid collectif' that is comprised of entangled human actors as well as non-human actors in multiple ways. Viewed from this perspective, tools (such as the hardware and software) that are embedded in the actor network are as Callon and Caliskan [17: 18] remark not just things that are used to achieve certain ends: 'They contribute to the making of the universe of possibilities that make action itself'.

As argued by Verbeek [78: 130], the concept of delegation indicates that programs of action can be 'inscribed' into artefact's. Within the context of the case studies (Uberlingen, US/Canada Blackout and patient safety), technical mediation expands our notions of action and experience. Actors within the network, such as a technical artefact, co-shapes the human world relations by giving shape not only to people's actions but also to people's experiences. The experience of healthcare providers was mediated by their participation within this hybrid of human and non-human actors that relationally transcend linear temporal and spatial conceptualizations.

Considering the EDIS example for patient safety, the seduction of technology often obscures the fact that new computerized and automated devices also create new burdens and complexities for the individuals and teams of practitioners responsible for operating, troubleshooting, and managing high-consequence systems [82]. Law [50: 9] argues that 'Adding complexity to the relations which make up a system in order to strengthen those relations may actually dissolve those relations in practice'. Failure to understand the reverberations of technological or process change on the operational system and the socio-political framework behind them hinder the understanding of important issues surrounding the evolution of human error within a system and how breakdowns occur. An artefact's capacity for influence (whether physical or informational) is thus dynamic and not static [1].

4 Conclusion

Action rather takes place in 'hybrid collectifs' that entangle human actors as well as non-human actants in multiple ways. Tools, for example are not just things that are used to achieve certain ends: 'They contribute to the making of the universe of possibilities that make action itself' [17: 18].

Through the hybrid collectif, the event-based 'domino' perspective disappears revealing a complex temporal and spatial heterogeneity. What emerges from the analysis of the actor network is the notion that time and space are folded thereby recasting the concept of latent effect/errors as purported by Turner [75] and Reason [65], in terms of a network schema.

Olson et al. [62: 220] argue that 'Expectancies form the basis for virtually all deliberate actions because expectancies about how the world operates serve as implicit assumptions that guide behavioral choices'. Within the context of accident aetiology, Reason [66: 32] argues that the 'path to adverse incidents is paved with false assumptions'.

The Actor Network perspective reveals a de-centered aetiology that is reflected by the distribution of relational network of heterogeneous elements that participate and shape action and inaction. What becomes apparent is that the attribution of blame is re-examined. Opening the black box of human error we realize that the accident aetiology resides as a property of the associations within the hybrid collectif rather than human agents.

As described in Masys [56], we recognize from the analysis that:

1. Non-humans have significance and are not simply resources or constraints.
   (a) Non humans intervene actively to push action in unexpected directions.
2. Entities are interactive effects.
   (a) They are networks of associations of human and non-human.
3. Action results from the complex interactions resident within the actor network that is dynamically shaped by inscription and translation processes.
4. The actor network lens reveals that action cannot be explained in a reductionist manner, as a firm consequence of any particular previous action [16: 172].

It demonstrates as William-Jones and Graham [81: 275] argue that 'Entities whether people or technologies, are not fixed and do not have significance in and of themselves. Instead, they achieve significance through relations with other entities'.

As noted in Brey [11: 76], '...that any fact about the competencies and performances of a particular technical artefact is the product of a network of actants that jointly work to "produce" this fact.' Thrift [74: 1468] makes a supporting observation: ...no technology is ever found working in splendid isolation as though it is the central node in the social universe. It is linked-by the social purposes to which it is put-to humans and other technologies of different kinds. It is linked to a chain of different activities involving other technologies. And it is heavily contextualized.

For patient safety in this techno-mediated healthcare system we draw upon Verbeek [78: 131] who remarks that 'Artifacts influence the way in which people do things, and this influence could be deliberately inscribed into them'. This is supported by the Actor Network analysis. What became apparent is 'technical mediation, whereby artifacts co-shape the relational world (network space) by influencing or '...giving shape not only to people's actions but also to people's experiences' [78: 139]. As noted in Shadrick et al. [69: 4] and demonstrated from an actor network perspective 'introducing new technology is not manipulating a single variable, but a change that reverberates throughout a system transforming judgments, roles, relationships, and weightings on different goals'.

Through the lens of ANT what emerges from the analysis is a network characterized by actors that are neither purely technical nor purely social, but rather what Callon and Law [15] terms 'a hybrid collectif'. This actor network comprised of 'heterogeneous' elements/relations erases the dichotomy that traditionally exists between the human and nonhuman, and thereby challenges the attribution of blame associated with 'human error'.

Acknowledgments The theoretical foundation of this work is derived from Masys [56].

References

1. Aanestad M (2003) The camera as an actor design-in-use of telemedicine infrastructure in surgery. Comput Support Coop Work 12(1):1–20
2. Aanestad M, Hanseth O (2000) Implementing open network technologies in complex work practices: a case from telemedicine. In: Organizational and social perspectives on information technology. Kluwer Academic Publishers, Aalborg, Denmark, pp 355–369 (Proceedings from the IFIP WG 8.2 conference IS2000, 10–12 June)
3. Aggarwal R, Mytton OT, Greaves F, Vincent C (2010) Technology as applied to patient safety: an overview. Qual Saf Health Care 19(2):i3–i8
4. Artino AR, Durning SJ, Waechter DM, Leary KL, Gilliland WR (2012) Broadening our understanding of clinical quality: from attribution error to situated cognition. Clin Pharmacol Ther 91(2)
5. Bennett SA (2004a) The 1st July 2002 mid-air collision over Überlingen, Germany: a holistic analysis. Risk Manage Int J 6(1):31–49
6. Bennett SA (2004b) The 2002 Überlingen mid-air collision. Crew Safety, No. 7
7. Bennett SA (2014) A conspiracy of bastards? In: Masys AJ (ed) Networks and network analysis for defence and security. Springer International Publishing, Cham, pp 217–245

8. Bennett SA (2016) The benefits of a systems-thinking approach to accident investigation. In: Masys AJ (ed) Applications of systems thinking and soft operations research in managing complexity. Springer International Publishing, Cham
9. Bijker WE, Pinch TJ (1984) The social construction of facts and artefacts: or how the sociology of science and the sociology of technology might benefit each other. Soc Stud Sci 14:399–441
10. Boyatzis RE (1998) Transforming qualitative information: thematic analysis and code development. Sage Publications, Thousand Oaks, CA
11. Brey P (2005) Artifacts as social agents. In: Harbers H (ed) Inside the politics of technology: agency and normativity in the co-production of technology and society. Amsterdam University Press, Amsterdam, pp 61–84
12. Callon M (1986) The sociology of an actor-network: the case of the electric vehicle. In: Callon M, Law J, Rip A (eds) Mapping the dynamics of science and technology. MacMillan Press, London, pp 19–34
13. Callon M (1991) Techno-economic networks and irreversibility. In: Law J (ed) A sociology of monsters? Essays on power, technology and domination, sociological review monograph. Routledge, London, pp 132–161
14. Callon M (1999) Actor-network theory: the market test. In: Law J, Hassard J (eds) Actor network and after. Blackwell and the Sociological Review, Oxford and Keele, pp 181–195
15. Callon M, Law J (1995) Agency and the hybrid collectif. South Atlantic Q 94(2):481–507
16. Callon M, Law J (1997) After the individual in society: lessons on collectivity from science, technology and society. Can J Sociol 22(2):165–182
17. Callon M, Caliskan K (2005) New and old directions in the anthropology of markets. Paper presented at the conference new directions in the anthropology of markets, New York University, 5 Apr 2005
18. Carey G, Malbon E, Carey N, Joyce A, Crammond B, Carey A (2015) Systems science and systems thinking for public health: a systematic review of the field. BMJ 5:1–9
19. Charmaz K (2005) Grounded theory in the 21st century. In: Denzin NK, Lincoln YS (eds) The SAGE handbook of qualitative research. Sage Publications, London, pp 507–536
20. Coakes E (2003) Socio-technical thinking—an holistic viewpoint. In: Clarke S, Coakes E, Hunter MG, Wenn A (eds) Socio-technical and human cognition elements of information systems. Information Science Publishing, Hershey, pp 1–4
21. Comfort LK (2006) Cities at risk: Hurricane Katrina and the drowning of New Orleans. Urban Aff Rev 41(4):501–516
22. de Almeida IM, Johnson CW (2008) Extending the borders of accident investigation: applying novel analysis techniques to the loss of the Brazilian space programme's launch vehicle VLS-1 V03. Saf Sci 46(1):38–53
23. Dekker S (2001) Reconstructing human contributions to accidents: the new view on error and performance. Technical Report 2001-01, Lund University School of Aviation
24. Dekker S (2002a) The re-invention of human error. Technical Report 2002-01, Lund University School of Aviation
25. Dekker S (2002) The field guide to human error investigations. Ashgate Publishing Ltd., Aldershot
26. Dekker S (2005) Ten Questions about human error: a new view of human factors and system safety.Lawrence Erlbaum Associates, Inc., New Jersey, PA
27. Deleuze G, Guattari F (1987) A thousand plateaus: capitalism & schizophrenia. University of Minnesota Press, Minneapolis
28. Dolwick JS (2009) 'The Social' and beyond: introducing actor-network theory. J Marit Archaeol 4(1):21–49
29. Doyle J (2007) Cybersurgery and surgical (Dis)embodiment: technology, science, art and the body. Transformations 15
30. Endsley MR (2000) Theoretical underpinnings of situation awareness: a critical review. In: Endsley MR, Garland DJ (eds) Situation awareness analysis and measurement. Lawrence Erlbaum Associates, Mahwah, NJ, pp 3–28

31. Entin EB, Entin EE (2000) Assessing team situation awareness in simulated military missions. In: Proceedings of the Annual Meeting of the Human Factors and Ergonomics Society, vol 1. Santa Monica, CA
32. Farley HL, Baumlin KM, Hamedani AG, Cheung DS, Edwards MR, Fuller DC, Genes N, Griffey RT, Kelly JJ, McClay JC, Nielson J, Phelan MP, Shapiro JS, Stone-Griffith S, Pines JM (2013) Quality and safety implications of emergency department information systems. Ann Emerg Med 62(4):399–407
33. FEMA (2013) Hurricane Sandy After Action Report. 1 July 2013. Accessed 14 Aug 2013. https://s3-us-gov-west-1.amazonaws.com/dam-production/uploads/20130726-1923-25045-7442/sandy_fema_aar.pdf
34. Flood RL (1999) Rethinking the fifth discipline: learning within the unknowable. Routledge Publishing, London
35. Foucault M (1980) Power/knowledge: selected interviews & other writings 1972–1977. Pantheon Books, New York
36. Foucault M (1986) Of other spaces. Diacritics 16:22–27
37. Harbers H (2005) Inside the politics of technology: agency and normativity in the co-production of technology and society. Amsterdam University Press, Amsterdam
38. Harris AJL, Gurioli L, Hughes EE, Lagreulet S (2012) Impact of the Eyjafjallajökull ash cloud: a newspaper perspective. J Geophys Res 117(B00C08):1–35
39. Helmreich RL (2000) On error management: lessons from aviation. Br Med J 320:781–785
40. Hollnagel E (2004) Barriers and accident prevention. Ashgate Publishing, Hampshire, England
41. IAEA (2012) Protection against extreme earthquakes and Tsunamis in the light of the accident at the Fukushima Daiichi Nuclear Power Plant. International Experts meeting 4–7 Sept 2012, Vienna, Austria. http://www.iaea.org/newscenter/focus/actionplan/reports/protection040912.pdf
42. Kohn LT, Corrigan JM, Donaldson MS (eds) (1999) To Err is human: building a safer system. National Academy Press, Washington, DC
43. Latour B (1987) Science in action: how to follow scientists and engineers through society. Open University Press, Milton Keynes
44. Latour B (1991) Technology is society made durable. In: Law J (ed) A sociology of monsters—essays on power, technology and domination. Routledge, London, pp 103–131
45. Latour B (1993) We have never been modern. Harvester Wheatsheaf, Hemel Hempstead
46. Latour B (1994) Pragmatogonies: a mythical account of how humans and nonhumans swap properties. Behav Sci 37(6):791–808
47. Latour B (1996) On actor-network theory. A few clarifications. Soziale Welt 47:369–381
48. Latour B (2005) Reassembling the social: an introduction to actor network theory. Oxford University Press, Oxford
49. Law J (1999) After ANT: complexity, naming and topology. In: Law J, Hassard J (eds) Actor network theory and after. Blackwell, Oxford, pp 1–14
50. Law J (2000) Ladbroke grove, or how to think about failing systems, published by the Centre for Science Studies, Lancaster University, Lancaster LA1 4YN, UK. http://www.comp.lancs.ac.uk/sociology/papers/Law-Ladbroke-Grove-Failing-Systems.pdf
51. Law J (2002) Aircraft stories: decentering the object in technoscience. Duke University Press, Durham and London
52. Law J, Callon M (1988) Engineering and sociology in a military aircraft project: a network analysis of technological change. Soc Probl 35(3):284–297
53. Leveson NG (2002) System safety engineering: back to the future. Aeronautics and Astronautics Department, Massachusetts Institute of Technology, MIT, Cambridge, MA. http://sunnyday.mit.edu/book2.pdf
54. Masys AJ (2005) A systemic perspective of situation awareness: an analysis of the 2002 mid-air collision over Uberlingen, Germany. Int J Disaster Prev Manag 14(4):548–557
55. Masys AJ (2008) Pilot Error: dispelling the hegemony of blamism. A case of decentered causality and hardwired politics. Int J Disaster Prev Manag 17(2):221–231

56. Masys AJ (2010) Fratricide in air operations: opening the black box—revealing the social. Ph. D. dissertation, June 2010, University of Leicester, UK
57. Masys AJ (2012) The emergent nature of risk as a product of 'heterogeneous engineering—a relational analysis of the oil and gas industry safety culture. In: Bennett S (ed) Innovative thinking in risk, crisis and disaster management. Gower Publishing, UK
58. Masys AJ (2014) Critical infrastructure and vulnerability: a relational analysis through actor network theory. In Masys AJ (ed) Networks and network analysis for defence and security. Springer Publishing
59. Masys AJ (ed) (2015) Applications of systems thinking and soft operations research in managing complexity. Springer Publishing
60. Masys AJ, Ray-Bennett N, Shiroshita H, Jackson P (2014) High impact/low frequency extreme events: enabling reflection and resilience in a hyper-connected world. In: 4th International Conference on BuildingResilience, 8–11 September 2014, vol 18. Procedia Economics and Finance, Salford Quays, United Kingdom, pp 772–779
61. Monteiro E (2000) Actor network theory and information infrastructure. In: Ciborra C (ed) From control to drift. The dynamics of corporate information infrastructures. Oxford University Press, Oxford, pp 71–83. www.idi.ntnu.no/~ericm/ant.FINAL.htm. Accessed 11 July 2005
62. Olson JM, Rose NJ, Zanna MP (1996) Expectancies. In: Higgins ET, Kruglanski AW (eds) Social psychology: handbook of basic principles. Guilford Press, New York, pp 218–238
63. Patterson ES, Woods DD, Roth EM, Cook RI, Wears RL, Render ML (2006) Three key levers for achieving resilience in medication delivery with information technology. J Patient Saf 2 (1):33–38
64. Perrow C (1999) Normal accidents: living with high-risk technologies, 2nd edn. Princeton University Press, Princeton, NJ
65. Reason J (1990) Human error. Cambridge University Press, New York
66. Reason J (2004) Beyond the organisational accident: the need for "error wisdom" on the frontline. Qual Saf Healthc 13(2):28–33
67. Senge P (1990) The fifth discipline: the art and practice of the learning organization. Doubleday Currency, New York
68. Serres M, Latour B (1995) Conversations on science, culture and time. Michigan University Press, Anne Arbor
69. Shadrick SB, Lussier JW, Hinkle R (2005) Concept development for future domains: a new method of knowledge elicitation. Technical Report 1167: Army Project number 622785A790
70. Shappell SA, Wiegmann DA (2001) Applying reason: the human factors analysis and classification system (HFACS). Hum Factors Aerosp Saf 1(1):59–86
71. Somerville I (1997) Actor-Network Theory: a useful paradigm for the analysis of the UK cable/on-line socio-technical ensemble? http://hsb/baylor.edu/eamsower/ais.ac.97/papers/somervil.html
72. Stanton NA, Salmon PM, Walker GH, Jenkins DP (2009) Genotype and phenotype schemata and their role in distributed situation awareness in collaborative systems. Theor Issues Ergon Sci 10(1):43–68
73. Tatnall A (2000) Innovation and change in the information systems curriculum of an Australian University: a socio-technical perspective. Ph.D. dissertation, Faculty of Education. Central Queensland University
74. Thrift N (1996) New urban eras and old technological fears: reconfiguring the goodwill of electronic things. Urban Stud 33(8):1463–1494
75. Turner BA (1978) Man-made disasters. Wykeham, London
76. U.S.-Canada Power System Outage Task Force Final Report on the 14 Aug 2003 Blackout in the United States and Canada: Causes and Recommendations April 2004. Accessed on 1 June 2013. http://www.nrcan.gc.ca/energy/sources/electricity/1378
77. Van der Duim VR (2005) Tourismscapes, an actor-network perspective on sustainable tourism development. Dissertation Wageningen University, 29 June 2005

78. Verbeek P (2005) Artifacts and attachment: a post-script philosophy of mediation. In: Harbers H (ed) Inside the politics of technology: agency and normativity in the co-production of technology and society. Amsterdam University Press, Amsterdam, pp 125–146
79. Wears RL (2012) Work, visible and invisible. Ann Emerg Med 59(5):374–375
80. Wickramasinghe N, Tumu S, Bali RK, Tatnall A (2007) Using actor network theory (ANT) as an analytic tool in order to effect superior PACS implementation. Int J Network Virtual Organ 4(3):257–279
81. William-Jones B, Graham JE (2003) Actor-network theory: a tool to support ethical analysis of commercial genetic testing. New Genet Soc 22(3):271–296
82. Woods D, Johannesen L, Cook R, Sarter N (1994) Behind human error: cognitive systems, computers, and hindsight. Crew Systems Ergonomics Information Analysis Center (CSERIAC), Wright-Patterson Air Force Base, Ohio, December 1994
83. Yeung HWC (2002) Towards a relational economic geography: old wine in new bottles? Paper presented at the 98th Annual meeting of the Association of American Geographers, Los Angeles, CA. 19–23 Mar 2003. http://courses.nus.edu.sg/course/geoywc/publication/Yeung_AAG.pdf

The Fog of Battle in Risk and Crisis Communication: Towards the Goal of Interoperability in the Digital Age

Allan Bonner

Abstract It is axiomatic that clear communication which produces an attitudinal or behavioural change is crucial in business, government, and politics. Selling products, providing entitlements, and obtaining consent from voters, relies on communication. This is more crucial when communicating about a risk or during a crisis—recalling product, providing health information, or preparing an urban area for an emergency, for example. The risk communication literature provides guidance in these crucial cases. The author samples this literature, and finds gaps and contradictions. Another field in which clear communication is crucial is in military operations. Precision in communication, as well as clear command and control of subordinates is required, and is said to save lives and achieve objectives. Western military analogies are often borrowed by those in general administration. But day to day operations are not actually the life and death struggles of the battlefield. Moreover, command and control was beginning to be questioned 100 years ago, even in the military. It is thus doubly ironic that military references exist in the civilian context, which may not resonate with some, as a result of age, gender, egalitarianism, pacifism, or other matters. It may be that in crisis management, military metaphors are used even more frequently, but are less apt. The author contends that the one military analogy rarely, if ever, cited in public and private administration—the fog of war (Von Clausewitz and Graham in On war. Project Gutenberg, London, 1909 [1])—may provide the best guidance on how to deal with those affected by a crisis. This fog may be akin to the gaps the author has found in the risk communication literature, and needs to be acknowledged in the same way fog is acknowledged in the military. With regard to military terminology, it may be that some eastern Martial Arts aphorisms are more appropriate than their western counterparts. The practical application of risk and crisis communication may be in the summarizing of findings of a survey of thousands of pages of urban emergency plans from the top 100 English speaking cities in the world, and related literature. These reveal contradictions in communication practices when the stakes may be life, death, and serious injury. These emergency communication tasks include

A. Bonner (✉)
Toronto, Canada
e-mail: allan@allanbonner.com

A.J. Masys (ed.), *Disaster Forensics*, Advanced Sciences and Technologies for Security Applications, DOI 10.1007/978-3-319-41849-0_5

notifying residents of the need to prepare, how to prepare, the fact that they are in danger, that they must shelter in place, or leave a danger zone. Most plans do not seem to be capitalizing on the risk communication literature, or acknowledging the fog of war. Ironically, many plans seem to create both fog and friction, which may cost lives. The author repurposes the term interoperability to include differing terminology, jargon, and extraneous information, which delay or cloud communication, and may thus impede action (Fessenden-Raden and colleagues 1987 [2], note that "Receivers of risk information are not just empty receptacles..." 'p. 101, Drucker (Harvard business review on knowledge management. Harvard Business School Press, Boston, 1998 [3]) notes that' "...command and control...is outmoded", p. 3, and the generic difficulties in organizational communication, p. 7, Carey (Bus Week 24:104–107, 1991 [4]) points out the danger in pilot-crew communication in aircraft operations, and Bowling and Hoffman note the costs of miscommunication the author speculates on why urban emergency plans seem so ineffective, and offers some approaches to improving plans in certain jurisdictions.

Keywords Risk communication · Risk perception · Emergency planning

1 Introduction

In 1882, the time of the author's great-grandfather, there was limited sensitivity to the public interest or even to communicating with the public. W.H. Vanderbilt, President of the New York Central Railroad reportedly quipped "the public be damned" (1882)[1] to a journalist asking about the public benefit to be achieved with a potential service. In 1925, the author's grandfather's time, US President Calvin Coolidge noted that "the chief business of the American people is business." When the author's father was in his late 20s (1952) Charles Erwin Wilson, CEO of General Motors noted that "what was good for our country was good for General Motors and vice versa." These simple aphorisms, lacking empathy with the general public, seemed to capture the times.

In the author's time, there have emerged multiple attempts to implement a range of egalitarian initiatives which are the antithesis of the approaches implied by the quotes above. Corporate social responsibility programmes, public consultations, empowering legislation, the rise of environmentalism, and scientific methods have enriched these initiatives.[2] The attempts to inject science into advertising, polling, and public policy could justify terming the 20th Century the 'Century of the

[1]There is debate over who made this statement, Commodore or William H. Vanderbilt. It is more widely accepted that the phrase should be attributed to William H. Vanderbilt, who may have denied making it.

[2]For a concise history of the rise of environmentalism, see Jamieson in Lash et al. p. 224.

Scientist.' And yet, the great promise of that century is largely unrealized in the area of communication.

When the author was in his late 20s, academic literature promised proven, scientific methods with which to communicate with audiences in distress and crises. Abe [5] examined what sources of information are trustworthy to audiences. Even physicians were thought to benefit from risk communication, in part, defined as presenting information about human health to a lay audience.[3] In fact, early practitioner Vince Covello, taught, in part in the medical faculty to help physicians communicate with patients. Some of his articles date to the early 1980s. The work of Tversky and others date to the 1970s, making interest in this discipline almost 50 years old.

Risk communication included "low trust, high concern" situations in which the audience was worried or even fearful, and the communicator had low credibility [6].[4] Some practitioners saw more general communication from government about budgets, health care, defence, and other matters within this framework. The risk communication literature appeared to be the best available advice on how to affect risk perception and influence risk decisions. Former journalist, Peter Sandman, took risk communication theories into a more general sphere, focusing on how news media report on risk (1985), [7, 8]. Practitioners and researchers also owe debts to Slovic [9, 10] and his colleagues who worked on methods to educate and inform audiences, and probed the value and dangers of making risk comparisons. Roth [11] seems to have differed, calling "…multiple perspectives … a meaningless gesture unless recipients can put them into context" (pp. 381–382). Pidgeon [12] brought a psychological perspective to his studies. The former head of the US Environmental Protection Agency, Ruckelshaus [13, 14], cautioned that our inability to discuss ever more minute quantities of risk with the same precision of the new instruments which measure that risk, can actually undermine democracy through an information deficit. Douglas and Wildavsky [15] located risk perception in a social process. They also questioned the perceived wisdom of having multiple sources of information, which might actually increase concern, (p. 49). Clarke [16] probed how and why organizations misperceive risk.[5] One aspect of risk perception is the tendency to over-and under-estimate risk (Fischhoff, citing colleagues and others 1985 [19], p. 88). Wilson ([20], p. 43) noted how large infrequent accidents draw more

---

[3]Covello has used this definition in speeches. The author has spoken at the same conference of the International Association of Business Communicators in Los Angeles, and had a dinner conversation with Vince Covello in New York on the occasion of the author's finishing his M.Sc. in Risk, Crisis and Disaster management.

[4]"Low trust, high concern" is a term often associated with Covello (2001).

[5]See also Clarke [16], Slovic [17] and Kahneman et al. [18] on the mismatch between risks and perceptions due to heuristics, p. 299. This summary is not submitted as an exhaustive literature search, but the highlights of articles used by the author in practice with government and industrial clients and in academic research and writing. The author has also spoken at conferences at which both Peter Sandman (Niagara-on-the-Lake, Ontario) and Vince Covello (IABC International Conference, Los Angeles, CA) and is drawing on these lectures as well.

attention than frequent small events which may cumulatively have more impact, and (citing Tversky and Kahneman) noted the illogic of our estimates of two events happening at the same time (pp. 17–18).

Risk communication principles seem to have a new life in the post-9/11 age of terrorism, and perhaps after SARS, mad cow disease, Ebola, and fears of a new flu pandemic.[6]

The rules of risk communication have always had much promise, and seemed intuitively correct. If a risk were known, a source of a benefit, voluntary, fair, understood, controllable, human in origin, or natural (a weather event), then risks are more acceptable. But, if the risk is unknown, involves high technology, not understood, involves children, is a dread or catastrophic risk (nuclear, chemical, terrorism), dramatic, with dramatic pictures, high media attention, inequitably distributed through society, complex, fatal, emanating from technology perceived to be a luxury, and one about which experts disagree, then concern will go up, and acceptance of the risk will go down [23, p. 289] Slovic [24], pp. 404 and 408, [9, 19, p. 92].[7] Morgan et al. ([17], p. 140) cited "dread" risks, and whether an event is a "signal" that the risk is not being handled well. Ironic aphorisms include noting that people reserve the right to smoke, but will protest less harmful emissions from an industrial plant.[8] How concerned one is with the asbestos in the neighbourhood school will be dependent on whether you have children in that school. On the way to a risk-taking event such as sky-diving, a practitioner may bemoan dangerous drivers—exhibiting risk-aversion.[9] Axiomatically, someone who is not agitated need not be calmed down, and thus the dosage of risk communication techniques can be altered for the person, the topic, and the context.[10]

Risk communication was becoming a popular field of research in the decades after all of science and perhaps all authority figures had come under scrutiny. The related and apparently harder discipline of risk assessment was deemed "…not a science per se…" (Cumming [28], p. 1). It had perhaps "…trans-scientific elements … questions which can be asked within the framework of science, but which are

[6]For further discussion on these topics see columns by the author: Ottawa Citizen [21, 22] and a CBC Radio Interview on October 21 2014.

[7]Note also Roth's [11] critical analysis of Covello, Sandman and Slovic's findings about the acceptability of risk, p. 375. Roth does not find the correlation that the other authors cite. Many of the researchers use versions of these terms. The author is also drawing on the speaking notes Sandman and Covello used at conferences at which the author also spoke.

[8]The example used by Douglas and Wildavsky [15] is "skin cancer caused by leisure-time sunburn…" While equally harmful as cancer caused by pollution, it is hard to mobilize the public against voluntary, leisure activities.

[9]See Covello et al. [25, pp. 55–56]. A protracted case of risk-taking is "nuclear jumpers" who are highly paid to perform quick repairs in nuclear plants, during which time they receive intense exposure to radiation in Raymond [26, p. 100].

[10]Fischhoff [27] has noted the irony of risk information increasing alarm and causing more harm than the risk itself, p. 13. Covello has spoken of the value of four empathetic statements to generate receptivity, followed by four substantial statements to change perception.

beyond the capacity of science to answer" (Cumming [28], pp. 1–2).[11] The craving to know and ability to measure, did not necessarily translate into knowledge and productive action. So, it was fitting that in 1987, when risk communication was very popular in industry, Baruch Fischhoff pointed out that the discipline is not in fact a science, but uses scientific techniques to attempt to measure communication effectiveness when the topic is risk. He also made medical analogies, as if risk information were a prescription to be used as a public health measure [27, p. 14], requiring appropriate ethical concerns by practitioners—akin to launching a new drug. To continue Fischhoff's analogy, imagine the medical ethics of administering the same dosage of a drug, even an over the counter variety, to an auditorium filled with a diverse audience worried about exposure to the flu, SARS, or asbestos, but without obtaining medical condition, medical history, or informed consent.

Other researchers noted gaps in the risk communication promise, even as the field was at the height of its popularity. Kasperson [29] pointed out that we know more about how individuals react to risk information than we do about how social groups react.[12] Kasperson [6, 29, p. 279] also raised the compelling notion that a variety of individuals and groups who speak about risk, affect risk perception.[13] What is said by politicians, lawyers, clergy, university professors, and others in both public and private must have an effect—not just what's said on the evening news.[14] This was congruent with the British Royal Society's [31] study indicating that "[r]isk perception in inherently multi dimensional and personalistic, with a particular risk or hazard meaning different things to different people and different things in different contexts" (p. 7). Nelkin and colleagues [32] added information about the credibility of spokespeople and the order in which they speak, during an evolving event. These two elements affected both news coverage of events and perception by the recipients of risk information.

This complexity became an inconvenient truth muddying the promising risk communication waters. A practitioner had to ponder who was saying what to whom, via what medium, and at what point during a crisis. This would take highly sophisticated media monitoring, consuming valuable time during a turbulent crisis, and might require knowing, in advance, what stories news media were about to run—an impossibility. Moreover, a practitioner had little information about how demographic groups—women, First Nation/Aboriginals, the elderly and so on—would react.[15]

[11]Note also Pidgeon [12] and his contention that "…risk assessment will always be conditional on, among other things, a range of modeling assumptions, with the choice of assumption often heavily dependent on the skills and judgment of the risk analyst" p. 132.

[12]See also Covello [23, p. 286] on our lack of knowledge about how "structural variables" or membership in a demographic group affects risk perception.

[13]See also Mazur [30] on coverage and consciousness of a variety of events, especially the lack of coverage of a PBB poisoning episode in Michigan, p. 49.

[14]See also Covello [23, p. 287] on the variables of frequency and impact.

[15]See Aptekar's question on "age, gender, social class, and ethnicity…" as variables, p. 75, and Anderson and Gardenio's similar question on gender, p. 725. Aptekar also adds the variable of human caused versus natural disasters, p. 75.

Perhaps Disraeli was wrong to contend that all crises are the same,[16] although the concept of isomorphism would indicate that crises have similar attributes [33].

Amongst the complexity were also some apparent contradictions. Covello [34] noted the power of positive speaking in his 1N = 3P equation, meaning that a negative is equal to three positives, and more memorable. Those who remember "I am not a crook" from President Nixon [35] or "No new taxes" from Bush [36] could hardly disagree. Yet, Sandman [37] and Stanton ([38], p. 16) counselled spokespeople to raise the negative themselves, knowing that negatives were in the minds of a public audience. It is hard to imagine people feeling fearful about a risk they are not aware of, and yet this advice was to raise the issue of risk and, perhaps paradoxically, reduce fears. Practitioners might have received the impression that an 'out of sight, out of mind', could become a top of mind, yet acceptable risk. Experience dating to the founding of Greenpeace in the late 1960s shows anecdotally that many protesters will indeed be passionately opposed to matters even if they understand the technology, don't have children exposed to the risk, receive a benefit, have some control, and so on.[17] Sometimes the goal is the protest, fellowship among protesters, negative assertion, and such, perhaps not adequately explored in the risk communication literature.

There is another contradiction worth raising. The perceived wisdom among public commentators is that it is not the crisis, but the reaction to it that helps determine success or failure. While not quantifiable, verifiable data, few could argue that US President Richard Nixon might have survived the Watergate crisis had he burned the tapes of his White House conversations and implemented remedies right after the 1972 American election. Few could argue that Lee Iacocca's fast and forthright handling of an odometer spinning controversy served Chrysler well [40].[18] Few might question that President Clinton would have done better by admitting to his sexual relationship with a White House intern much earlier, rather than deny.[19] The complexity remains in these and other cases because of the variables of the times, the person, the organization, and public attitudes. So, taking the advice above about making a risk known, voluntary, fair, understood, controllable, human in origin, and natural ignores the competing news of the day, the credibility of the spokesperson, other commentators, the order in which they

---

[16]Usually attributed, but unverifiable.

[17]The author was living in Burnaby, a suburb of Vancouver before and during the founding of Greenpeace, and participated in several events, including a protest against the Amchitka nuclear test and the invasion of Cambodia at the Peace Arch on the US-Canadian border, which ended in the 'invasion' of Blaine, Washington, noted as the first time the American border had been breached since the war of 1812. A previously published article (The Hill Times) on this topic appeared in the compilation entitled Political Conventions, Bonner [39, pp. 233–235].

[18]See Barton, pp. 75–76.

[19]Clinton's inappropriate relationship with the intern came to light during a case wherein a former Arkansas State Employee, Paula Jones, attempted to sue the then-president for sexual harassment. During the sexual harassment case the president was asked about his relationship with the intern to which he replied, "there is not a sexual relationship, an improper sexual relationship or any other kind of improper relationship."

speak, the length of time before response, and so on. Remedies appear to be like a seasonable special in a fruit store—subject to time, chance, entropy, contamination, and thus only useful within an ever-diminishing window of opportunity. Conner [41] catches the same idea: "[w]hat has changed about change is its magnitude, the approach it requires, the increasing seriousness of its implications, and the diminishing shelf life of the effectiveness of our responses to it." (pp. 37–38)

Much has happened in the decades since this research became compelling in academic circles and among practitioners. In Canada, for example, there are two slightly different Supreme Court rulings on the "duty to consult" First Nations/Aboriginals [42].[20] Ontario's Municipal [43] and Planning Acts [44] require public consultation. More self-regulated professionals are involved in such consultations, of late. Lawyers are duty-bound by the concept of 'zealous advocacy' to "carry on the practice of law and discharge all responsibilities to clients, tribunals, the public and other members of the profession honourably and with integrity" for their clients [45, p. 8]. "Architects must respect man and man's environment, and take into account the impact of their research, work and interventions on the life, health and property of every person" [46, I.7]. Planners "have a primary responsibility to define and serve the interests of the public" [47, s. 1.0]. While the Professional Engineers' Act states that: "the safeguarding of life, health, property, economic interests, the public welfare or the environment" are innate to professional practice [48, s. 1]. These laws and codes of conduct make effective consultations a requirement, not a frill or option. It is not reasonable to expect that techniques which may have worked decades ago in a general population, will work among First Nations/Aboriginals, or new Canadians during consultations. It is also not reasonable to expect that what may work, or be possible in the Canadian federation, with a weak mayor system, in which municipalities are creations of the provinces [49],[21] might work in the United States in which many cities have constitutional protection and strong mayors [51] operating in a political system featuring 'states' rights'. Thinking that anything that works in either of these systems might work in a traditional unitary state[22] such as Holland or England, may also be an error. Overlaying culture, or different decision making process (Japan versus America perhaps) may only show that rules are not only made to be broken, but that rules may not actually exist. As in archetypes in psychology (obsessive, compulsive, introvert, extrovert, etc.), rules may point to, or illuminate a way, more than define a hard and fast reality.

Social science researchers have long grappled with how to conduct accurate surveys. All scientists have grappled with methodology, post Heisenberg and Kuhn.[23] What we choose to research, what is funded, the gender and background of

[20]Haida Nation versus British Columbia (Minister of Forests), and Tlingit First Nation versus British Columbia (Project Assessment Director). Both cases dated 2004.

[21]See also Makuch et al. [50].

[22]A unitary state is a region governed by a central government that has ultimate control.

[23]See Merton's standards for science in Sellnow [52, pp. 5–7].

the editors, preconceptions, and even the scientists' presence are all factors which can skew data.[24] Some classic errors in social science research have resulted from reliance on magazine subscription lists and telephone surveys, which, by definition only survey people who subscribe to that magazine, or who have telephones. Decades ago we were cautioned that research may feature "…an over-representation of people with telephones, time on their hands, or who crave human contact…" [54]. In recent decades, new challenges have emerged. More than 10 % of the population have only hand-held telephones [55], and 35 % ignore the home phone for fear it will be a sales call [56]. Call display alerts people that the call may be a survey and thus 'refuse rates' approach 70 % in some cases. Human behaviour regularly defies the statistical norm. People sometimes tell researchers one thing and do another, as seen in recent elections in both the UK and Canada in 2015. This may also have been the case with some risk communication research.

Three of the most critical comments about risk communication did not seem to deter use of the data, and continued study. Far from contradictions or gaps, Ruckelshaus noted in 1984 that "…risk assessment data can be like the captured spy; if you torture it long enough, it will tell you anything you want to know".[25] Albeit with reference to a specific risk, Sharlin [57] noted that there "…is nothing in the scientific method that can be used to reassure the public once fears have been aroused over the threat of cancer in food containing parts per billion of EDB[26]" contaminants.[27] The Royal Society noted in 1992 that "…few definitive empirical studies of risk communication, and in particular of its consequences and effectiveness are available (p. 7)."

So, there has always been an apparent certainty to the risk research and advice that should be unsettling for the sceptical scientist. At least the contradictions and gaps are unsettling. More perplexing is how modern crisis managers and risk communicators should incorporate theory into practice. Psychologist Kurt Lewin has noted that there is nothing so practical as good theory [58]. But what is good theory today?

2 Social Media as a Communication Tool

In the social media era, it is interesting to ponder Kasperson's social amplification theory of risk (1988), Sandman's interest in media coverage [59], and Nelkin's study of the order and timing of interviewing spokespeople and experts during an

[24]See Bonner [53] What Constitutes Professional, Legal, Ethnographic Research?, unpublished paper.

[25]Ruckelshaus [14, pp. 157–158].

[26]EDB stands for ethylene dibromide. This chemical is believed to be cancer causing, and can be found in some food items commonly bought in grocery stores. EDB is also a common chemical found in many non-food items, and formerly was used as an insecticide in the US.

[27]Sharlin, p. 196.

environmental incident [60, 61]. It is also interesting to ponder the notion that risk is amplified by dramatic news coverage, and experts disagreeing now that all-news channels predominate along with social media.

It seems many of the terms we have historically used to describe experiences with media are no longer applicable. In living memory, the term mass media referred to the fact that 80 % of television viewers in the United States watched comedian Milton Berle on Tuesday nights [62].[28] Fifty million people listened to Bing Crosby's radio shows [63]. Richard Nixon drew 60 million viewers [64] for his famous 'Checkers' speech in 1952, when 20 % of Americans didn't own TV sets. In 1980, more than 52 million Americans watched the three network supper hour newscasts.[29] News anchor Walter Cronkite and others referred to this as a shared experience. Today, supper hour newscasts draw just over 20 million, mainly older viewers, and thus are not a widely shared experience. We referred to the 'electronic hearth' or TV as a metaphorical gathering place, and the "public sphere" (Daley and O'Neill, referencing Habermas, pp. 52–53). These terms seem anachronistic when the hearth has 500 fires (channels), and families are separated via their individual media experiences on different devices, in different rooms. We now use the questionable term, 'join the conversation' to describe what is often one-way communication.

So how shall a practitioner view the evolution of crises and communication about them in this new environment? The author has argued elsewhere that the Exxon Valdez oil spill took on larger than warranted significance as a result of certain types of amplification [65, 66].[30] Exxon was a large, well-respected company. Alaska was perceived as America's last physical, pristine frontier, and CNN was a new player in news coverage.[31] The Valdez spill was about the 30th largest oil spill to that date in the world, albeit the worst in America.[32] Prince William Sound, while appearing to be pristine in postcards, already had traces of oil from shipping, and several spills over many decades. Oil is the product of dead organisms, a significant percentage of which is biodegradable [67]. A large amount of spilled oil disappears within about six hours as a result of the light ends (toluene, benzene, etc.) evaporating. The spill happened at a time when it was known that leaving oil on beeches to allow bioremediation to occur was most likely the best course of action [67]. However, Exxon officials, probably rightly, felt that communicating this in an emotionally charged atmosphere was futile. They felt they needed a photo-opportunity for CNN and other networks, and that involved hosing beaches with high pressure water, and blotting rocks with absorbent towel. Hosing probably harmed living organisms beneath the surface of the beach, and blotting

[28]Comedian Milton Berle was the host of 'Texaco Star Theatre'.

[29]Statistical analysis provided by Nielsen Media Research, used under licence by Pew Research.

[30]See Keeble [67, p. 23] on the various previous spills in the area, pp. 55 and 57 on dispersants and p. 205 on bioremediation.

[31]See Peaks [68, p. 14].

[32]McAlpine [69, p. 17].

accomplished little, or just relocated the problem to waste facilities.[33] Is it credible to assert that certain other spokespeople, alternative ordering and timing of messages, and other aspects of risk communication theory and practice would have caused a better result for Exxon in Alaska?[34] Moreover, what might have been the result had social media existed during the oil spill? If CNN skewed and magnified risk, what might social media have done?

It is beyond the scope of this examination to arbitrate whether there are larger or smaller challenges when the needs of not only CNN are in responders' minds, but also those of several other cable networks and a half-dozen social media platforms. There are two obvious approaches though. One is that the fragmentation of media markets means that a news story on the major broadcast networks lasting less than two minutes, may be missed by much of the population. People may be watching specialty cable. Moreover, in the US, one may be watching a left or right-leaning cable network, the electronic equivalent of only reading a particular tabloid newspaper in the UK. If one is consuming news on Twitter, one may see a good sampling of mainstream media coverage of an event, reposted by fellow users. But it may also be that the consumer skips a story that appears uninteresting, or too complex—the digital equivalent of turning the page in a newspaper. Certainly the data about the timing and ordering of spokespeople's statements during an evolving event seems less relevant when social media and specialty TV channels can be accessed by consumers on demand. Timing and ordering is also affected by news aggregators who choose and assemble stories. The progenitor of this selective consumption was during the era of partisan newspapers and pamphleteering—one could refuse to buy a newspaper or take a pamphlet on the street. Today, one's Facebook and Linkedin feeds and Youtube channel consumption may only reveal what the consumer's compatriots view as newsworthy—strengthening existing views. The fragmentation of the news marketplace may mean that a major story simply does not take on the importance it did before social media and the all-news networks.

[33]See also Busch [70, p. 773], Clarke [16, p. 296] for other spills in Alaska, Daley [71, p. 45] on the apparent slow reaction, Davidson [72, pp. 44–52] on burning and dispersants, The Economist ([73], p. 100) on impact and cost.

[34]Exxon's Manager of Communication at the time, Ken Kansas, when interviewed for the author's 1999 dissertation on the topic, was adamant that these contextualizing messages would not have been heard or understood in the frenzy of negative news stories about the spill. Mr. personally signed checks in the hundreds of thousands of dollars to media consulting firms during the spill and has an overall annual budget of US$60 million and thus could have attempted other solutions, one supposes. Mr. Kansas tried to involve the Disney characters and organization, having a long personal association with Walt Disney and his company as a graphic artist, and a professional association, having funded the energy pavilion at EPCOT Center, Disneyworld. Mr. Kansas reported to the author that his Disney contacts said any association with Exxon at the time of the spill might have "sunk" the Disney organization. Mr. Kansas was well-educated, with a Masters in Communication, well-funded, and thoughtful. The author spent many dozens of hours discussing this topic with Mr. Kansas both socially and as they worked together in Toronto, Ottawa, New York, New Jersey, Rhode Island and Washington.

Conversely, it is hard to argue that more communication, instant communication, and communication vehicles that are portable and constantly accessible have no implications. They must. The mass and instant blast of information from social media must be heard by many, if only in a two-step hypothesis in which small numbers of recipients communicate with larger numbers [74], or social media influence the content of legacy media. The modern version of the two-step hypothesis might not only include face to face communication, but also mechanized re-postings via several social media platforms—a powerful new development. Legacy media also re-transmit the content of social media. But is this more powerful on a daily basis than the reach of the old broadcast supper hour evening newscasts?

As in the pamphleteering and partisan press era, misinformation exists with social media. The names of victims in the Virginia Tech shootings were Tweeted before official confirmation [75]. Re-tweeting calls for help in the 2011 earthquake, tsunami, and Fukushima nuclear event surely were a diversion, as was the dissemination of incorrect information about a suspect in the Boston marathon bombing [76].[35] But this may just be the technological and modern equivalent of a person trapped in a building in the San Francisco earthquake of 1906 shouting "I'm here" and the reverberation and echo confusing responders. We may not need protracted analysis of this phenomenon, other than that it is mechanization of a problem. We may also see terrorists use social media more, and during an event to cause more damage—a further mechanization of a problem.

Social media is easy to dismiss. Much content seems trivial. Yet when one describes content as angry denunciations, pictures of kittens, and error filled coverage of the trivial, one is also describing much newspaper, radio, and TV content. As George Bernard Shaw may have noted, "…newspapers are unable, seemingly to distinguish between a bicycle accident and the collapse of civilization."[36] In the competition for the lead news story on a given day, it often appears that none actually win. Langer [77] describes this as "…non-news about non-events…". All media contain trivial features, backgrounders, and fillers. In risk or crisis coverage, these elements dwindle, and the coverage of the serious issue fills the space and time. But traditional print and electronic coverage is still general, speaking in broad terms about the number of city blocks affected, for example. Social media gets serious, but also becomes more specific with users Tweeting and geo-tagging exact addresses on those affected blocks. This specificity holds promise for emergency responders.

Recent academic study of social media confirms that it may have much to offer in risk and crisis situations. Public health officials may be able to track the outbreak of regular flu, SARS or a pandemic via Tweets from users indicating they don't feel well or are taking the day off work. Large numbers of certain key words will be a good situation analysis. Tweets may work faster in alerting a population to a

---

[35]Sunil Tripathi was incorrectly named by users of social media as the suspect for the Boston bombing in 2013.

[36]Usually attributed but unverifiable.

seismic event than scientific instruments. Instead of hours, 75 % of Tweets are sent within two minutes of a perceived event. A Northeastern University study indicates social media users are potential suppliers of "massive data sets" which can be used for study [78].[37]

Newspapers and TV have a finite space and time for the dramatic pictures that risk theory says increase our trepidation. But Instagram and Twitter have an infinite capacity to accept such pictures from citizen journalists. Research has not yet been definitive on how these new forms of communication fit into the social amplification of risk, and whether they have diminished the role of mainstream media in this process. Intuitively, media that is not consumed cannot have an effect on those who are not consumers, but media which is consumed may have an effect.

What is not in doubt when it comes to social media is its prevalence and power among users. The Pew Research Center found in 2012 that 33 % of adults under age 30 obtain their news from social media [79]. Ninety per cent of 18–29 year olds say they sleep with their phones in or right next to their beds [80]. According to Google [81], 56 % of Canadian adults were using smart phones in 2013, up from 33 % the previous year. During the 2013 Calgary, Alberta flood, Twitter logged more than 850,000 tweets [82]. Official Tweets from the city and police seemed to garner the most attention, meaning they were sought out by the general public, and credible. The Canadian Red Cross indicates that 33 % of the population would call for help using social media [83]. Palen [75] confirms high usage of social media, especially during brush fires. These numbers are out of date when cited because of the speed with which social media is gaining acceptance, and are probably now larger than reported above.[38]

Given the high usage of social media, and even misuse as a method of calling for help, it is likely that social media will become a more official medium for emergency communication. The Canadian Red Cross reports that 33 % of respondents would use social media to call for help, and 63 % think that emergency response officials should respond to these calls [83]. The US Federal Communications Commission may consider using social media for emergency alerts, much like existing reverse 911 calls.

2.1 The Fog of Communication

The term 'fog of war' or 'fog of battle' is often attributed to Von Clausewitz. Whether he actually used either phrase is debated. But he did write about "fog and

[37] Alan Mislove is an Associate Professor in the College of Computer and Information Science at Northeastern University. Currently the University is undertaking in-depth studies in social networking data mining.

[38] The author's research into social media was originally conducted for essays submitted to the M.Sc. (Pl) programme at the University of Toronto and later more widely published in syndicated columns through Troy media and in the book Safer Cities of the Future, pp. 106–109.

friction".[39] The notion of fog speaks to what is unknown at the beginning of an event, and that which becomes unknown as a result of changes in the environment during an event. In military history, fog was not only water droplets in the air, but also the "opacity of the black powder battlefield".[40] Less literally, it is "poor intelligence".[41] The military find this concept pervasive in that "three quarters of the factors on which action is based are wrapped in a fog of greater or lesser uncertainty".[42]

The military also focus on the changing nature of fog and the short life-expectancy of plans—'battle plans go out the window at the first contact with the enemy'. There is an additional temporal element to this fog:

> The further ahead we think, the less our actual influence becomes. Therefore, the further ahead we consider, the less precision we should attempt to impose. Looking ahead thus becomes less a matter of influence and more a matter of laying the groundwork for possible future actions. As events approach and our ability to influence them grows, we have already developed an appreciation for the situation and how we want to shape it [85, pp. 83–84]

Friction refers to impediments to action, even if there is no fog. With friction, one may have good knowledge and a good plan, but an obstacle, known or unknown, creates friction, which slows or prevents action. To complicate matters, this slowing of action or progress may result in new conditions—more friction, fog, or perhaps opportunity.

There are many things that both crisis planners and military people know. The military know the speed of a tank, ship, or aircraft. They know the trajectory of a projectile. They train in war games to gain the best knowledge possible of people and materiel. A military appreciation of a situation and intelligence gathering helps predict future events. But there is still fog and friction from the unpredictable—the enemy, weather, human behaviour, topography, the convergence of elements, and so on.

Similarly, crisis planners know climate, the built environment including roads, topography, resources available, legal cases, past events, and so on. But there may still be fog and friction—a new court ruling, unforeseen weather events, terrorism, and more. As systems theory tells us, knowing the bending moments of the components of a scaffold may tell us little about how that assembled scaffolding will perform ([86]; see also von Bertalanffy [87]).

The military use rigorous training methods. These methods are virtually continuous for enlisted members who drive, fly, shoot, march, and perform other functions. The military have a well-educated officer corps, whose professional development also appears to be nearly continuous. One doesn't hear that a skill—flying a plane, driving a tank, parachuting, firing a shell or gun—is forgotten in a regiment or on a base. Rusty perhaps, but not forgotten. The military do not seem to

[39] Kiesling [84, p. 85].

[40] Kiesling, p. 85.

[41] Kiesling, p. 87.

[42] Kiesling, p. 86.

speak about their field manuals gathering dust on shelves.[43] Yet one hears industry and government managers speaking of crisis plans gathering dust, and of skills disappearing with an individual who retires.

If the military can embrace fog, so can public and private sector crisis managers and risk communicators. One should not embrace just the catchy military phrase, while ignoring the core value of acknowledging that one may have to work through fog.

One can see why the term "fog" is not embraced by crisis managers and risk communicators.[44] Crisis planning is often about both predicting the future and taking actions to have an influence on future outcomes. One builds a levee to contain high water because multiple past events and data tell emergency responders how high the water may rise. One surrounds spilled oil with a floating fence called a boom, tells citizens when to evacuate neighbourhoods, and where to get help. Many emergency management documents transmit a certainty that these actions will be required, and followed.

But this feeling of engineering and scientific certainty has been eroded by numerous events over the years. Who would have thought that workers at the Flixborough Petro-Chemical plant would have run toward the evacuation alarm [89], rather than leave the building? We may never know because this was fatal for many [90]. Were some workers more curious as to the cause of the evacuation order than they were fearful of their lives? Where risks so familiar that risk perception diminished? Who would have thought that police and fire fighters, arriving at the same scene at much the same time, would have the opposite approach? Yet, at the King's Cross underground fire, police evacuated subway cars and pushed commuters up and out of the station, while firefighters pushed others down and out onto the open air platform. An inquiry noted that many people in the cars would be alive had they been left in the cars [91]. Who would have thought that response to spilled gasoline on a street during a car accident, would also feature a cardiac event with the driver, explosion risk if jaws of life or a defibrillator were used, explosion risk if fire-retardant foam were used at the wrong time, water-based response when gasoline migrated into a nearby river and finally, an explosion in a pub cellar, caused when fumes were ignited by an electric timer [92]. But these are some of the myriad events in the Walton Town Centre petrol tanker overturning—events constituting fog and friction.

Events may sometimes unfold in a predictable way. Much human behaviour may be more orderly than evidenced above. But the military concept of fog may serve as a warning that when an event is complex, large, unexpected, and with multiple

[43] The author is drawing on both his general readings on military matters, and on personal experience as a consultant to Canada's Department of National Defence between 1987 and 2002. This work was done on most Canadian military bases, headquarters, and in Cyprus.

[44] The author has used this theoretical concept and reference in urban planning (2015), business administration (2005) and the subsequent leadership book Wounded Leaders [88] based on the author's DBA research and dissertation.

players, fog may become greater. Fog may be greater among a multi-cultural population with multiple frames of reference or diverse understandings of danger. The prediction of outcomes may become more difficult for a number of reasons.

2.2 *Interoperability as Fog*

Much was made after the 9/11 attack on New York's World Trade Center about 'interoperability.' This was defined as the ability of emergency responders to speak with each other over various communication devices. It is not clear that this objective of compatibility among hardware and software has been widely achieved even at this date. Lack of interoperability impedes communication during an emergency. But there may be another type of interoperability challenges, evidenced in the lack of clear policies and communication in urban emergency plans.

Observers of the aftermath of some disasters note the victims who busy "themselves in unimportant, but manageable tasks" [93].[45] What might be termed activity traps or busy work involves picking up "soda cans" on a site devastated by an earthquake which provided a "sense of mastery over a world that had abruptly lost its predictability" [93]. What may be a manifestation of shock is understandable after a trauma, but why there would be busy work codified in emergency plans written long before an event is a mystery. Such unproductive and distracting content falls into the category of friction, as defined by Clausewitz.

After reading the emergency plans for the top 100 cities in the English speaking world, this new definition of interoperability presents itself in sharp relief. If what is being said is not understood by recipients, this is another version of fog and friction. This is as pronounced as if radios and radio frequencies were not compatible among responders, and transmissions could not be received by citizens. A Presidential directive notes that plans should be consistent, in plain language, and thus understood by users.[46] This does not appear to be the case. Nor is it the case in Ontario where Guelph's emergency plan notes that "…there has been a variance of approaches … a lack of standardised tools to manage incidents, and hence no single province-wide system to ensure effective coordination" [94, p. 11].

The first aspect of interoperability involves access. About 20 % of the population may not have access to the Internet [95], even in industrialized jurisdictions. Alerts and information sent via the Internet are thus not received by some. Notices of power outages via devices which require power may also not get through, especially when many power outages last longer than the battery life of hand-held devices. The Canadian Red Cross notes that almost half of Canadians have lost electricity for more than 72 hours—well beyond battery life [83]. When the power

[45] Aptekar, p. 99

[46] Presidential Directive on Homeland Security, February, 2003, cited in Jacksonville/Duval County Emergency Plan.

is on, and for those who do access the Internet, the emergency plans with blank pages, redacted text, 401 errors, maps without street names, drop down and pop ups which obscure text, with small font, and which print off the page are also a form of fog, friction, and miscommunication. These impediments are prevalent in most plans. Dozens of pages containing the names or titles of contributors to city emergency plans, slow a reader down, at best. Pages of glossaries of terms, unnecessary definitions, lists of vulnerable facilities, and other features have the ring of padding in a college essay and don't seem to add to the safety and security of citizens. It is hard to imagine how many of the features would help emergency responders. Rather, they would act as friction, or a slowing down of their life-saving activities. Little may have changed in the 25 or so years since a study of earthquake emergency plans noted that "[s]ome problems have no apparent solutions, and some solutions are suggested in the absence of a matching problem" [96, pp. 28–29]. This also describes many of the urban emergency plans studied.
Consider these definitions and explanations:
Baltimore:

> Winter storms are large storms occurring during the cold, winter months… [97, p. 17]
>
> Flooding occurs when rivers, creeks, streams, ditches, or other water bodies receive too much water from rain or snowmelt. (Ibid., p. 7)
>
> Hail is a large frozen raindrop… (Ibid., p. 16)
>
> Freezing rain is "[r]ain that falls and freezes to a cold surface such as a road or tree causing a glaze of ice…" (Ibid., p. 18)

Brisbane:

> Self Evacuation: Evacuation from an area under a person's own arrangements. [98, p. 11]

Philadelphia:

> Precipitation deficits are the earliest indicators of a potential drought… [99, p. 76]

Pittsburgh:

> Notification – "To make known or inform" [100, p. 37]
>
> Notify—"To inform about a condition, event or situation." (Ibid.)

Kitchener:

> Chief of Police: "The Chief or alternate of the Waterloo Regional Police Service." [101, p. iv]

There are similar tautological definitions in Kitchener's plan for the City Solicitor, Commissioner of Public Health, Medial Officer of Health, Commissioner of Social Services, Fire Chief and many others.

In Baltimore, SPW stands for shelter in place, EVE means evacuation, and TOE signifies that there is a 911 outage. The need for a three letter substitute for a term well known by three numbers is not evident. If used with members of the public, few would know the meaning.

Some plans have advice for citizens that also constitutes friction. Sherbrooke, Quebec notes that one should "[e]nsure that the water from the city's system is drinkable…" after a flood [102, p. 2]. But there is no guidance on how to do this. This is in contrast to Richmond, BC's link to detailed instructions on how much bleach, and of what kind to use to purify water.

Among the most diversionary advice comes from Barrie, Ontario site of a tornado in 1985 which killed 8 people in that city, five elsewhere, and injured hundreds [25]. About 800 people were left homeless, hundreds of buildings were damaged or destroyed and the cost was estimated to be about $100 million [103]. In Barrie's written plan of 2007 there is the direction that a Federal Ministry, Human Resources Development Canada (HRDC) could help:

> The local office has an obligation to support the communities it serves. Each local office has plans to offer assistance when called upon to do so. Managers of local offices at all points of service are authorized to take immediate action at the municipal level when so requested by an authorized representative of local government. …[Assistance includes:]
>
> - Reception and registration of available paid workers and volunteers…
> - Selection and dispatch of paid workers and volunteers to the emergency scene as requested…
> - The early post emergency assessment of the capability of employers in the afflicted area to resume operation, re-deployment of former workers, and assessment of the capability of employers in the area to absorb displaced workers from the disaster area… [104, p. 45].

This kind of inventory can be invaluable and save lives in an emergency. Kansas City recognizes this with its "Plan Bulldozer" (2015)—an inventory of private sector heavy equipment that may be needed. But in Barrie's case, it turns out that the service is probably not actually available. Officials at the federal ministry in Ottawa and its regional representatives in Toronto indicate that they don't actually provide such services and would refer requests for such help back to the city—full circle,[47] with a delay which might cost lives.

"The medium is the message" said playful media theorist Marshall McLuhan [105]. But is the message sent also received and understood? Detroit's plan calls for citizens to leave their porch lights on if they've evacuated their homes [106]. Communicating this directive in normal times would be a challenge. Assuming distraught citizens will remember to turn lights on as they leave is a larger assumption. Moreover, if the emergency is a power outage, high winds which damage the light bulb, or if the resident can't replace a burned out bulb, the directive is pointless. In any case, one assumes emergency responders will find that this directive is not universally known or followed, and will thus take more time double checking many doors, regardless of the status of the porch light. This is a further example of a lack of interoperability.

[47]This summarizes several phone calls lasting more than 1 h over two days with HRDC/Service Canada officials in Ottawa and Toronto.

One does not need to look far afield to find that good communication can saves lives and poor communication can endanger.[48] A lack of information and misinformation is the norm in most events. In fact, "…a natural disaster is usually also an information disaster. …[W]hen information flows are needed the most, they are least able to function".[49] Sood and colleagues specifically note the chronic difficulty that diverse responders, and more responders than expected have in finding and using resources, and communicating with each other and with the chain of command.[50] Stories of offers of help being ignored, rescue dogs being held up in quarantine, and rescue workers not knowing which ruins have been searched illustrate forms of miscommunication and friction as well [112, p. 34].

3 Communicating with Creative Clarity

The lack of clarity in many plans, termed a lack of interoperability in this writing, is in contrast to some best practices which are clear and creative. Creativity, or perhaps just hard work and thought, is sometimes required because concepts are hard to communicate, and target audiences may be hard to reach. Perhaps the most obvious case of the need for more clarity in the terroristic age is the admonition—"If you see something, say something." It is obvious that this leaves out what one might see, and to whom one might speak. But leaving aside that semantic game, it is worth noting that Leeds, UK has included vivid pictures of the equipment that terrorists might use and their behaviour—multiple communication devices, trouble at banking machines and so on [113]. Plan writers in Leeds are not content just asking readers to beware of ill-defined "suspicious" packages. They provide a picture of one (pp. 2–3), leaking oil, tied up asymmetrically, with a hand-lettered address, and too much postage.

Louisville plan writers have noted the need to be vigilant, but also when to be vigilant. They list the anniversary of the US Supreme Court ruling on abortion in Roe v Wade, the storming of the Waco, Texas religious facility by federal agents, the Oklahoma City bombing, and 9/11 [114, pp. 243–245]. This appears to be more effective and practical than asking for unspecified vigilance 365 days per year.

Several cities are not content with the policy of notifying citizens that their power is off, via devices which require power—ironic at best. Birmingham [115] has plans to use door to door leaflet drops and temporary street signage. Halifax [116] plans to use door to door visits, in addition to leaflets. Leeds, UK [117] plans to use "sky shout"—loud speakers on helicopters, and can send messages, segmented by the type of

[48]Bowling and Hoffman [107] have approached this notion from the perspective of mediation and negotiation, for example. The Economist [108], citing a seismologist, notes that perception can be "dangerous", p. 4. While not a matter of life and death, Giulliani [109] notes that effective communication is a competitive advantage. See also Richardson [110], on responders not having proper directions to the Hillsborough soccer crush incident.

[49]See Sood et al. ([111], p. 28).

[50]The author is paraphrasing Sood et al. ([111], p. 28).

organization—pub, club, hotel. Albuquerque, NM [118] assumes that cell phones won't work in an emergency, and Fort Worth, TX [119] has 149 sirens. Houston cites "AlertFM…a strobe light or bed-shaker alerting device to alert persons sleeping…" or who have disabilities (Annex 1 Public Information, p. 2).

Several cities provide reality checks about matters we ignore or misinterpret. Milwaukee's Office of Emergency Management website [120] tries to dispel myths about lightning—rubber tires will protect you, and lie flat on the ground. Those not fearful of hail will be chastened by the picture of a hailstone about as large as a baseball in Oklahoma City's plan [121, pp. 3–22]. One can easily imagine the damage to a person, home or vehicle that is hit with such an object. Auckland, New Zealand [122] notes that a volcanic event there could result in a reduction of their GDP with twice the impact of the reduction during the Great Depression. Bradford, UK on its website [123] notes that six inches of water can knock a person off her feet, and that a car will float in two feet of water. Jacksonville/Duval County, Florida's plan vividly describes a hurricane as akin to "…a bulldozer clearing everything in its path…" and flying debris as "…a battering ram destroying objects in its way…" [124, p. 25] Boston's climate change study notes that Massachusetts "could have a climate similar to Maryland's" or the Carolinas by the end of the century, and the implications of increased heat days and poor air quality, especially for vulnerable populations [125, p. 26]. Boston may motivate people to be more self-sufficient by noting that the city only has enough food on hand for 3–5 days.

There are other examples of creative clarity and thus interoperability:

- Oshawa, Ontario [126] lists hidden water sources in homes, including ice cubes, the hot water tank, in the pipes, and in the toilet reserve tank
- Guelph, Ontario's [94] emergency kits for cars, first aid kits, and food for self-sufficiency
- New Orleans [127] now has 14 foot street art as evacuation pick up points
- Albuquerque, NM [118] uses a lightening detector alarm at their soccer complex
- Houston, TX [128] and Long Beach, CA have good advice on dealing with reporters
- Burnaby [129] and Surrey [130], BC provide or link to a 26 steps to preparedness break down preparation for residents who may be overwhelmed by the tasks
- Long Beach, CA provides an historical example of how routine official communication spun out of control with tragic consequences. This city plan points out that the Watts riot of 1965 started out as a traffic stop by police. A crowd gathered, tensions rose and a result was 32 deaths, 874 injured and up to $100 million dollars in damages (City of Long Beach Hazard Mitigation Plan [131]).[51] Several American cities appear to have re-lived a version of these tragic events.

[51]See City of Long Beach 2014 Hazard Mitigation Plan, p. 143.

Evacuation—A Communications Imperative

Kansas City's evacuation plan begins with an analysis of events in New Orleans during and after Hurricanes Katrina in August 2005, and Rita in September of 2005 [132].[52] This is not an academic critiquing an operational plan from a theoretical perspective. It is not a government inquiry where politics may play a role. It appears to be one city's emergency planners trying to do best for local residents, and trying to learn from previous tragedies in another city. It is a credible and powerful source.

Kansas City's plan states that New Orleans:

> failed to provide adequate transportation for citizens without their own vehicular transportation. …The citizens of New Orleans have always known of the danger, but many failed to provision food, water and medical supplies, and failed to make adequate plans for their own safety. …[T]raffic quickly overwhelmed the limited evacuation routes. Then, because of the slow movement of vehicular traffic, some evacuating vehicles ran out of gasoline and others were forced to 'turn back' before reaching a safe location. [132, p. 1] (see Footnote 52)

This analysis is congruent with the academic literature on the topic. During Katrina there were conflicting orders, sometimes counter-manning each other, and a lack of coordination among 500 response organizations [133]. City officials knew there was a significant risk, and assumed that risk on behalf of residents. What city officials knew was that 60 % of citizens couldn't or wouldn't leave during an emergency [134]. This is probably true for many urban residents. Hess and Gotham note that between 25 and 56 % or urbanites do not own cars [135]. For those who do, Kendra et al. refer to "fantasy" plans relying on busses, stadia and the availability of fuel [136]. They note that it might take 50 hours to evacuate one million people since those in two car families may take both cars, tow trailers and drive either extra cautiously or erratically [136]. They point out that there were more deaths as a result of the evacuation during Hurricane Rita than because of the direct results of the hurricane. Moreover, New Orleans is the 40th largest population centre in the US. In a larger centre, there could have been larger problems.

Worse, New Orleans had an earlier event during the 1998 storm season in which evacuation didn't work well, and still didn't after an expenditure of USD $7 million on retrofitting roadways to accommodate contra flow [137], in which vehicles use all lanes to try to escape danger. Houston, which received many of the New Orleans' evacuees and then had to call its own evacuation, now maintains two-way traffic when at all possible to accommodate emergency vehicles.[53] Wolshon [137] terms contra flow 'potentially life threatening'. It may only increase capacity by 70 %, not double, and prevent the movement of emergency responders and supplies into the affected area. He argues that the drawbacks almost outweigh the benefits [137].

[52]See Kansas City, Missouri Evacuation and Transportation Services Plan Public Version [132, p. 1].

[53]See Houston's emergency plan and Wolshon who terms contra flow "potentially life threatening."

Kansas City's plan writers appear to have found similar frustrations as found during the author's study of 100 emergency plans:

> There is a lack of competent guidance on how to prepare a mass evacuation plan. Consequently we were unable to find a single large metropolitan area that had a plan which they considered good enough to model. The federal government's planning guide, entitled Guide for All-Hazard Emergency Operations Planning (SLG-101) provides virtually no information on how to prepare a mass evacuation plan. Similarly, state documents provide only checklists of what should appear in a mass evacuation plan. The most useful source of evacuation was a 1984 EEMA document entitled Transportation Planning Guidelines for the evacuation of Large Populations (CPG 2-15). [132, p. 1][54]

And yet, Kansas City's planners make more progress than evident in most other plans studied. Perhaps writers felt they had to after research found little guidance on the topic. Kansas City's plan enumerates the transit and school bus capacity, number of drivers, paratransit capacity and drivers, and the estimated population during the day and at night. Destinations are not enumerated, but there seems little doubt about how many people might need to be moved, and how many could be moved—surely a basic emergency planning tool. Sacramento, CA [138] adds estimates of the number of people who can stand in light rail cars and the height of water the cars can pass through. Most urban emergency plans would benefit from this level of specificity.

Evacuation should be an especially rich area of study when it comes to communication. What is being communicated may be a matter of life and death. One assumes both senders of messages and the recipients are keenly attuned to the task. Moreover, the behaviour requested is clear—leave the affected area. Yet, evacuation is not clearly dealt with in the plans studied. Many plans list the multiple authorities by which evacuation orders can occur, and the several managers and committees which will have input into the decision. Still others note how difficult a decision it will be. But that is not advice for citizens hoping to prepare, or who are already in danger.

This seems akin to parents instructing children on safety around the home. Parents who began with a discussion of family law, the physics of combustion, tort and trespass laws, and the strength of door bolts would be achieving little. One surely begins with directions to drop and roll in case of fire, keep doors locked, how to dial 911 and so on.

Extraneous matters, in family discussions about safety, or in urban emergency plans, constitute fog and friction. So does ineffective communication. It is beyond the scope of this discourse to explain the barriers to clear risk communication. Prisms exist—heuristics, familiarity, risk taking behaviour,[55] denial, complacency, thrill seeking, and so on. There are also barriers to general communication

[54]This is congruent with other jurisdictions, Guelph, Ontario's plan notes: "…there has been a variance of approaches within Ontario, a lack of standardised tools to manage incidents, and hence no single province-wide system to ensure effective coordination." [94, p. 11].

[55]References to voluntary sky-diving, Karate and other activities might be explained as also bestowing a benefit (thrills, fitness) on the participant. But the risk imposed on train travellers documented by Klasson [139] in which passengers applauded an engineer's efforts to go faster. The 1855 incident resulted in derailment, but is explained by Klasson as support for progress.

including listening skills, memory and more [140]. These theories cannot be arbitrated here. What can be stated with clarity though, is that a barrier, or a variety of barriers stand between those issuing an order to evacuate and the intended recipients. One may not be able to see or explain the barrier, but one sees the manifestation of the barrier—non-compliance. For decades, evacuation protocols have been termed "unrealistic" [141].

A document available to those searching for Charlotte, North Carolina's plan puts the challenges of evacuation into focus. The document is not an emergency plan, or perhaps even part of Charlotte's plan. But it is a good research paper on the issue, called an Interactive Emergency Evacuation Guidebook [142] and written by John Sorensen and Barbara Vogt, and summarized below.[56]

The evacuation documents provide observations about human behaviour. Families may try to reunite, before evacuating as a group. Office workers may evacuate in groups. People in motion have a tendency to stay in motion and thus, if engaged in going to or from school or work may evacuate more quickly than those who are not in motion.[57] Households with a plan tend to comply with evacuation orders more than those without a plan. People are reluctant to abandon pets, will return to evacuation zones to tend to their pets and livestock, and are anxious to return home, regardless. A longer warning seems to be needed for some events such as flooding, and a shorter warning for nuclear power or chemical plant releases. There are "shadow" evacuations of those near the danger zone of between 12 and 59 % of the population, with 26 % being average. There are spontaneous evacuations before official evacuation orders. Many refuse to comply with evacuation orders.[58] People have died after taking down barriers in order to drive on unsafe or closed roads. Drivers have used logging roads to by-pass road closure notices.

Into this mostly unexplained mix of behavior and event, comes advice on how communication may cause compliant and predictable action by recipients. The advice seems congruent with communication literature stressing the need for clarity, simplicity, specificity, and repetition. It may be that multiple sources of warnings, such as "door-to-door … coupled with emergency vehicle public address systems and TV and radio announcements" may reinforce each other (p. 6). The public often

[56]This summary of the challenges of evacuation the paper cited in text, which may form part of Charlotte, NC's emergency plan. It is entitled "Interactive Emergency Evacuation Guidebook" and was "Prepared for the Protective Action IPT—Chemical Stockpile Emergency Preparedness Program—Februa…" The interactive quality is not apparent and the exact date prints off the page. The document is 35 pages long and is dated 07/09/2013 at the bottom of pages, which may be the date of Charlotte's posting. Corroborating citations will be credited to appropriate other authors.

[57]Takuma (undated) has found behaviour which may be culturally specific in which people may go from home to office and vice versa, and those who wait at home for other family members before evacuating, p. 163.

[58]See also Kendra who notes that between 32 and 98 % evacuate.

wants confirmation of the evacuation order from multiple, official sources (Mayor, emergency manager, Red Cross, local scientists).

Congruent with the general business literature on communication is the notion of the need to be specific [143]. How is a lay person expected to know the meaning of emergency technical jargon? In fact, how is a lay person expected to know where high ground is, or even what shelter or evacuation means? The advice in the Charlotte document is to define such terms by number of feet or topographical features "the colored post upriver from the bridge…higher than the top of City Hall", (pp. 21 and 23). Evacuation orders should specify "now, not 10 min from now, but now" (p. 24). Further advice is to say to "stay away from brick buildings because bricks might fall down" (p. 24) on you, or that "[w]e don't know nor can it be known which buildings in the city will be safe and which will not be safe…but we do know that most people will be safer if they go home now" (pp. 25 and 26). In order to cause people to shelter in place it may be necessary to state that "the risk is in the air outside" your building (p. 24). Without this specificity, people will invent their own meanings, and construct different responses to ambiguous terms.

Most concerning, is the notion that emergency preparedness programmes may not be effective at all. Decades of formal and informal commentary have cited the notion that education is the key to a variety of ills. Just what constitutes 'education' is often ill-defined. In the social media era, it is often blogging, Tweeting, and posting, with little thought about whether such transmissions are received, understood, or acted upon. In communications theory, this is sending, not necessarily receiving. There is little new here. A newspaper, radio station, or TV broadcast which is not read or heard, is not received. This is also true of new media. It is oddly both profound and obvious that there is:

> …no conclusive evidence regarding whether or not preparedness programs, public education or information program actually makes a significant difference in increasing human response to warnings. … [A] good pre-emergency information program will increase response although the amount cannot be estimated. …a poor program will not likely make a great overall difference. …effectiveness will drop off over time… (Sorenson and Vogt, p. 9)[59]

What then are emergency plans really expected to achieve? After reading them, and the academic study of them, one cannot not be sure.

[59]This is consistent with 25 years old data about earthquake preparedness by Dooley et al. [144] who found that "…insistent publicity campaigns warning of earthquake predictions have not proven successful over long periods…", p. 467. The authors also studied one's demographic group as a determinant of propensity to prepare and had to leave this "unexplained", p. 454. On the difficulty of predicting earthquakes, see also The Economist [108, p. 79].

3.1 *Quantitative Risk Assessment as a Guide to Action*

Quantitative risk assessment (QRA) may be akin to risk communication—a discipline which uses scientific techniques, but is not a science itself.[60] Individuals may trump QRA with their own assessments as well. Those who drive because they perceive driving to be safer than flying when the reverse is the case may prove this point. Thus, the fear of risk may be as dangerous as the actual, apparently quantifiable risk. Even advocates Slovic et al. [10][61] have referred to the "…impotency of quantitative risk assessment…".

Yet, many emergency planners are adding quantitative risk assessment to their plans. One assumes this is to prioritize mitigation efforts or predict response requirements. But does QRA achieve this? Does QRA transmit valuable information to the general public? Nashville's plan notes that "…the probability of dam failure is low and not predictable" [145, s. 4-1, p. 12]. Yet within a few sentences, plan writers also note that there have been 55 past failures and 21 partial failures. This seems to show that this risk is actually somewhat predictable and higher than one might think. There is no advice for citizens who face this risk. Perhaps QRA is best used as a comparative tool, thus requiring a definition of "low" compared to other risks in order to add meaning.

Risk is defined as equaling impact plus vulnerability times likelihood in Nashville [145, s. 4.0, p. 1]. What will the average citizen do with a calculation that has risk equaling two dead plus falling buildings multiplied by every three years (during high winds)? This appears to violate the grade school dictum against adding apples to oranges and multiplying by bananas.

Surely the "100 year flood" term is misunderstood by the general public:

> The terms '10-year,' '50-year,' '100-year,' and '500-year floods are used to describe the estimated probability of a flood event happening in any given year. A 10-year flood has a 10 percent probability of occurring in any given year, a 50-year event a 2 percent probability, a 100-year event a 1 percent probability, and a 500-year event a 0.2 percent probability. While unlikely, it is possible to have two 100-or even 500-year floods within years or months of each other. [145, s. 4.1, p. 12]

Long Beach, CA [146, p. 49] uses FEMA's Calculated Priority Risk Index (CPRI). This involves four categories for hazards and a weighting scheme. This is the formula for an earthquake:

$$\text{CPRI} = [(3 \times 0.45) + (3 \times 0.30) + (4 \times 0.15) + (1 \times 0.10)] = 2.95 \text{ (Ibid., p. 51)}$$

Other hazards (Tsunami, windstorms, flooding all have what appear to be similar ratings, even for those which have never occurred (2.55, 2.90, 2.45) (Ibid., p. 53).

---

[60] See Griffiths [90, pp. 13–14].

[61] See Slovic et al. [10, p. 391].

In the end, perhaps advice on protecting people and property from these hazards would be a better use of space in an emergency plan.[62]

Kansas City [147] provides a simple, prose analysis of the risks it faces. There is little commentary, but mainly the noting of the proximity of Fort Leavenworth, the 35th infantry division, Army Command and General Staff College, an Air Force base, America's only B-2 stealth bomber base, and 442nd Fighter Wing. The Federal government is the largest single employer with the Federal Reserve Bank, a courthouse, office building, and an IRS processing center.

The targets are obvious but the calculation to decimal points is spurious.

3.2 *The Shape of Emergency Response—Augmenting Alpha-Numerics?*

Many emergency plans describe escalating crises with some form of non-textual aid. The severity of a crisis progresses from 1 to 3 in most cases, with three being the worst—a numerical aid. London, Ontario [148], and Houston, Texas [149] appear alone with the opposite approach—number one is the most severe crisis. Some cities have four stages and others seven. Houston has four readiness levels [149]. Some overlay colour coding [149]—a visual aid. Columbus [150] features the number three, but it is the admonition to be "ready in three." The definition of the three things or stages is not evident. Differing and even opposite terminology may impede the execution of mutual aid agreements with responders from other jurisdictions.

A further effort to describe risk and response may be the diagrams found in many emergency plans. Business schools have long decided how to represent an organizations' hierarchy. Organizational charts often feature pyramids or step-pyramids with one or a few people at the top and those who are subordinate below. There are dotted lines to signify a relationship, but not necessarily formal reporting. A box with a title beside another box may signify a support function—those reporting to the boss don't also report to the boss's assistant. Some organizations speak about having a 'flat pyramid' meaning fewer than average layers of management. These few lines are an oversimplification, but may illustrate a somewhat common approach to the organizational chart.

Conversely, the young, increasingly professionalizing, emergency response community may not have achieved a consensus on how to use a diagram to represent either responders or the stages of response. Many use what is akin to an organizational chart which would be familiar to the human resources department. Some American cities feature charts with the voters or citizens at the top—an appropriate egalitarian commentary in the American context, but of little significance in the management of an urban emergency [151, p. BP 42].

---

[62]The author also noted this example in his book, *Safer Cities of the Future*, 2015.

The diversity of visual approaches may be a potential distraction to new hires in the jurisdiction. If mutual aid agreements are invoked, these different visuals may confuse incoming responders. Worst of all, these diverse visuals are of little help to citizens seeking guidance on preparing for emergencies. It is not evident just what issues are being addressed and purpose served with many diagrams. There is limited text to describe or support them in most plans. One wonders why they are made public.

When a reader encounters one such diagram in an emergency plan, it is easy to skim over it. A reader might note the creative effort to illustrate a technical point. With an unclear diagram, that reader can rely on what is understandable in the plan's text. But if one studies several of these remarkably diverse diagrams, a reader is compelled to investigate what is behind these apparent diversions. Some diagrams are ovals showing the incorporation of lessons from training and real events [152, p. 21]. Others contain boxes, akin to an organizational chart with the names of managers, starting with the Mayor at the top of the chart [147]. Other, similar charts show a downward flow of authorities to policies and to the Community Emergency Management Plan (CEMP). Elsewhere the plan shows, a guide and a framework flow in from the sides, and finally down to field guides and other matters, which also flow back up [153, pp. i, BP I-6]. The hierarchical relationship of people or titles to each other makes sense at a glance, but the relationship of plans, guides, and frameworks flowing to and through each other must take some study to be of value. Those familiar with HR charts may find that guides, frameworks and plans seemingly reporting to each other gives pause. Elsewhere, there are pyramids, organizational charts whose boxes show people [154, p. 54], organizational charts with boxes showing the progression of an incident, [101, p. 6], donuts describing the emergency site [152, Annex: Risk Specific Plan: Flood, p. 15], and a bisected rectangle with a flowchart therein [153, p. BP III-17].

Jacksonville, Florida [124, p. 122] has a "planning P" which may be easy to remember. The "P" has the incident at the bottom, and response, briefing, meetings and so on rising up the stem of the P to Preparing for the planning meeting and circling around the top of the P to Operations briefing, executing of plan, and assessing progress. This "P" seems unique to Jacksonville, and may not be known to newcomers or those assisting Jacksonville when it asks for help from neighbouring cities. A short-hand indication of where responders are in the "P" may just constitute friction and delay for those unfamiliar with the schematic. Houston (Annex D, p. 33) represents "Victim Handling and Emergency Scene Flow" with a dotted circle labelled "Safety perimeter." Inside, ironically, there is an "Outer cordoned area" and a "Safety perimeter". Also inside is a hierarchical chart describing the condition of victims. Houston also uses FEMA's "Integrated Public Alert and Warning System" (IPAWS), which is not found in other plans (Annex A—Warning, Appendix 1-B, p. 27). Messages flow to, through and from IPAWS to government authorities and through various communication vehicles to the public. Most are "CAP" messages, a term not defined on the diagram.

Calgary has one of the most difficult diagrams to comprehend with squares, rectangles, dotted lines, and hoodoo-like peaks and undulating foothills. Taking the diagram at face value, it appears to be indicating that "Hazard Identification and Risk Assessment" followed by "Prevention Mitigation Response" didn't work because an event occurred. This is probably reasonable, but at variance with Sally Leivesley's notion of "risk engineering" [155] in which risks are engineered out of a situation or artefact. This is hard to justify, but a more positive approach. Then, in Calgary's diagram, "Community Restoration and Rehabilitation" and "Local Authority Recovery" seem to lag behind generic "Response", are not as intense and take longer to implement [156]. One wonders many things, including who is conducting the "Response" if it is not the "Local Authority", and why "Community Restoration and Rehabilitation" is not part of response. This may just be a semantic exercise, unfamiliar to many readers. One doubts whether local residents would be familiar with the meaning of this diagram, and why it is in a public plan. But what isn't in doubt in the diagram is the assumption that there will be business interruption. This seems to have a more prominent place than the restoration and rehabilitation of the community. Calgary emergency responders may have made that value judgement, but they need not concede the notion that businesses must be interrupted. Many emergency planners have long used the term "Business Continuity Planning" to note that the goal is to keep the business operating, not assume it will be interrupted and needs to be resumed. The California brewery which engaged in earthquake proofing its facilities is a case in point. Mitigation efforts were repaid when the brewery continued operating and didn't require repairs [108, p. 10].

4 Martial Metaphors as Communication Guides

Military concepts appear frequently in the business literature. They are usually from the western traditions. For example, acquisitions are described "…as a direct-march-up-the-hill kind of exercise" [157].[63] There are profiles of swash-buckling and successful executives, often in personal accounts, written by the executive [158]. Perhaps the most severe account, prophetic in its prediction of the demise of G.E. Capital, is one of the tough and abrasive Jack Welch [159]. Collins [160] notes that the focus on larger than life personalities who "…make headlines and become celebrities" (p. 68).[64] Something as prosaic as a sales meeting features "the killer karate school of combat sales meeting moves…" [161]. In one organization "due diligence teams were called 'commando squads'; its members got 18-in. Bowie knives with their names and that of the acquired business engraved on them" (Cliffe, p. 105).

[63]See Aiello and Watkins [157]. "The Fine Art of Friendly Acquisition." Harvard Business Review, p. 25.

[64]See also the uncritical history of business, as evidenced in Coolidge's notion that "the business of America is business" (1925).

Perhaps some eastern martial concepts are more valuable. These concepts from the several forms of Karate (open hand) are mostly codified in the oral traditions [162], handed down in schools or in Dojos by teachers called Sensei's (one who has gone before).[65] In the 1920s Japanese practitioners added the suffix "do" to Karate-do, signifying that the Art is also a way of life involving meditation, spirituality, compassion, and other elements. This helps explain the interest in the meditative art of calligraphy, Haiku poems, and other paths to self-knowledge.[66]

Some aphorisms from Karate are unambiguous. The first rule of Karate is said to be "Don't be there."[67] This speaks to crisis prevention. An industry analogy might be that it is better to be an unknown engineer who invents a flange which prevents oil well fires, than to be swashbuckling Red Adair who fought oil well fires and was the subject of a Hollywood movie [164] starring John Wayne. The second rule of Karate is "Run." This speaks to crisis avoidance. The industrial analogy might be to allow the oil to burn until there's no more oil, spray dispersants as the best of a bad lot of choices, or otherwise protect human health and safety by not dangerously engaging with the risky situation. The third rule of Karate is contained in the aphorism, "When a lion and a tiger fight, one is bound to be injured and the other killed."[68] The industrial application may be that one cannot be fearful of legislative, regulatory, or legal repercussions to the point of inaction. One must take action to address the crisis. There is no such thing as no risk.

At least one apparently contradictory phrase from Martial Arts lessons makes good sense upon examination. This phrase states that "Your block is a punch and your punch is a block."[69] Upon examination, a very hard block may indeed repel an opponent, or at least chasten one into worrying what a subsequent punch might be like after a powerful block. Similarly, one's punch is a block, because a well-placed, hard punch will block further action by the opponent. This

[65]For information on the ethnographic tradition, see Becker, p. 70 and Faraday and Plummer [163, p. 183].

[66]The author recognizes the challenge in citing data handed down through oral tradition. As in all ethnography, there is no claim that this information was scientifically obtained or is reproducible. However, it is authentic and was obtained under strict conditions. The author codified many of the Karate Dojo traditions and practices in which he participated for a DBA dissertation in 2005, and subsequently produced a more popularized version in the book Wounded Leaders. The author's Sensei regularly read drafts of this work. Essays were also assigned and read by the Sensei, and the Sensei's Sensei on the occasion of 3 Black Belt gradings.

[67]Thanks to Sensei Patrice Williams for many private discussions and his Dojo instruction on this topic. Thanks also to a more recent reading by fellow student Dr. Bheeshma Ravi, MD, Ph.D.—a physician, medical researcher and 4th Dan Karate practitioner. Dr. Ravi read this draft and concurred with the author's recounting of these Martial concepts.

[68]See also, Kim [165, p. 146].

[69]As in most of the observations about Dojo practice, this was recounted orally by Sensei Patrice Williams. See also Kim, p. 145. For a further example of the aphorisms from Karate, see Funakoshi, for example, p. 66 about lessons from watching a tug of war. For a popular attempt to import eastern philosophical tenants into western management, see Low, p. 2 (problem solving), 5 (growth and profit), 16–17 (governance) and 52 (sales).

examination of "bunkai" (application or meaning) of a technique is continuous and layered in Karate. It is continuous for at least two reasons. First, a practitioner wants to probe the meaning of a physical move in order to add power, effect, and meaning to that move. Secondly, for secrecy, many set moves in Karate are performed in reverse, partially, smaller, in a different order, or larger than in real application, as a metaphor for the actual move. This may be in contrast to the matters in the policy and procedures binder in a western corporation—codified, filed, and forgotten.

There may be a manifestation of this Martial Arts aphorism about blocks being punches and vice versa in risk communication. The hope is to communicate the right kind of powerful messages that not only deflect anger, but produce either acquiescence or support from the recipient. A Martial Artist might train for 1500 hours to achieve a Black Belt and the hopes of accomplishing this. Reading the risk communication literature is not comparable training.

The Martial Arts notion of embracing damage if unavoidable might serve crisis responders well. This may be true of the physical response as well as the risk communication explaining why a particular action has been taken. It is a moot point now, but one wonders the result had Exxon sprayed and/or burned the Valdez oil spill without approvals. The company was going to suffer damages regardless.

The three rules of Karate are augmented with temporal elements. These elements are more complex than the western military dictum 'speed equals power.' While true, the reason speed may equal a certain amount of power involves the element of surprise, and the notion that more power will be required the more an opponent can dig in and fortify, which takes time. A military rule of thumb is that one needs an attacking force three times the size of the defending force [166]. This is especially true if defenders are in fortified urban areas.

To the Martial Artist, "Late" means you deployed your technique (block, perhaps) too late to be effective. A result might be death, but certainly injury and a tough job catching up. "Late arrives ahead," involves starting out behind the proverbial "8" ball, but acting with enough speed and acumen to get ahead of the danger. In the Deepwater Horizon spill in the Gulf of Mexico of 2010, this might have involved deploying floating fences (booms) quickly and equipment to suck up the oil in order to contain it, after putting out the fire. "Ahead" speaks to knowing you're in danger and acting before that danger has an opportunity to befall you. In the Martial Arts, this might be deploying a technique before your opponent initiates one or before the opponent's move comes to fruition. In the Gulf of Mexico, it might have been automated responses to leaking oil. "Ahead of ahead" is much the same as "don't be there." Ahead of ahead involves seeing future potential danger and conducting oneself in such a way as to avoid it completely. In the Gulf of Mexico, this might have involved automated and redundant systems that suppressed failure prior to an oil leak, or perhaps new technology and energy obviating the need to drill for fossil fuels.

A Martial Artist does a reality check by noting the 'fallacy of expanded time.' It may be satisfying to think that one will duck, block, punch, and run, perhaps as in a Hollywood movie. But this is hard to do on the way down to the floor. Time did not allow all these judicious moves to be made. In crisis communication, a spokesperson may think s/he has compelling messages for an angry crowd, but can they be deployed? In a public meeting, held to discuss the merits of a municipal dump in Simcoe County Ontario (known as Site 41), the spokespeople were just able to introduce themselves and get out a sentence or two of background before the crowd began loud complaints. The public knew all the background, didn't want any more, and simply wanted the long-approved and studied dump cancelled [167]. They got their way over top of sound science and clear communication. The empathetic and skilled moderator, an Osgoode Hall Law School LLM in Alternate Dispute Resolution found that no one wanted to get to "Yes" or to "Maybe". They were content at "No."[70]

The fallacy of expanded time can be seen in the urban crisis plans studied. Several cities mandate a "windshield" survey of conditions or the use of an airplane for such a survey, for example Kansas City [147]. These techniques are time-consuming and out of date, since the widespread and easy use of drones. At least two cities mandate issuing identification to reporters—time consuming and unworkable at the scene of a crisis [168, 169]. Mississauga along with several other cities, have a workload for the communications person (media monitoring, advice, statement writing, media tours, photo-opportunities) which cannot be accomplished in days, let alone on media deadlines [170].

5 Ambiguity Will Clearly Prevail

Some perceived wisdom in crisis management includes that the CEO must go to the event quickly. Montreal Mayor Jean Doré was criticized for staying in the resort area of the Laurentian Mountains, north of his city, during a flood [171]. Exxon Chair, the late Larry Rawl was criticized for not going to Valdez, Alaska sooner [172], when he had no operational role to play on the beaches.[71] Union Carbide's Warren Anderson went to Bhopal, India after the chemical release there, was arrested, and put in jail [173]. President George W. Bush was criticized for flying over, but not stopping in New Orleans during Hurricane Katrina [174]. He may have been hampered by the American convention and constitution that promotes local response first and discourages the federal government from acting at the state or city level. One wonders if stranded residents in New Orleans were familiar with constitutional conventions.

There is a quip among emergency responders that senior politicians are a net loss to the response. Their photo-opportunity and remarks stop the work of nearby responders, they eat a sandwich or two, perhaps promise out of pocket settlements,

[70]The author was a consultant to Simcoe County on this dump and attended the public meeting cited. The author recommended colleague Paul Williams, LLM to moderate.

[71]McAlpine [69, p. 17].

and leave. It appears the 'rule' about going to the site is more of a guide, and complex at that.

Risk perception explains some ironies and ambiguities. Highway deaths in North America have hovered around 50,000 [175] for many years. This is not front page news. But a crash in Northern New Brunswick that killed seven high school students [176] did make the news and still does occasionally. We can visualize and understand the plight of seven young students in the prime of life more clearly than a large group of 50,000 unnamed people dispersed throughout the year across North America. Flying is safer than car travel, but it's estimated that nearly 1600 more people lost their lives in the year following 9/11 because they chose to drive rather than fly [177]. People are fearful of flying because it involves high technology, the principles of which are not widely understood. Moreover, 9/11 was horrific and created indelible memories. Drivers of their own cars felt in-control and safer, even if they weren't. The most expert risk communication may have done little good in these years. Statistics on how safe flying is, and how airlines were suffering financially would probably have gone unheeded.

On the issue of fast and clear communication in a crisis, this has not always achieved the results desired. Dow Beer was linked with mysterious deaths among habitual beer drinkers in Quebec City in 1966 [178]. The company acted quickly by reversing the brewing process and dumping beer down the sewers of the city to the point where the streets flowed with Dow's product. Although there was never a definitive link made between the product and the health effects on habitual drinkers, the brewery lost a large percentage of its market share. Who would have thought a brewery would pour good beer down the sewer? Consumers erred on the site of caution, given the visual evidence they had. Similarly, Ball Park Hot Dogs quickly withdrew their product and tacitly blamed their workers after a tampering case, later found to be a hoax [179]. Pepsi resisted withdrawing their product, strongly and rightly suspecting that the finding syringes in their soft drink cans were a result of a hoax and copycat crimes [180].

Much has been made of the Tylenol case study in the academic and popular literature, and in presentations at professional development conferences. The event in which the famous pain reliever was contaminated with cyanide may be fading memory, but it is still taught in many communication classes. The event seems to support the value of clear and fast communication. Executives acted quickly, withdrew the product, were open with reporters, installed an 800 number for concerned customers, and part of the product re-launch strategy was to use an excellent communicator, medical director Dr. Thomas Gates [181, 182]. What appears to have been lost, however, is that seven relatively young people died horrible deaths and no one has ever been charged or convicted of the tampering crime. Effective risk and crisis communication should not blind researchers to the human toll in an event.

6 Discussion

A Martial Artist deals in the same measurements as a crisis manager or risk communicator—inches. A Martial Artist may point to her solar plexus and note that the difference between death and injury is inches—the distance between the solar plexus and the ribs. Then, the difference between injury and vanquishing one's opponent is another few inches from the ribs to outside the silhouette of the body.

Everything is a trade-off. As ecologist Barry Commoner said in 1971, "There's no such thing as a free lunch" [183]. Jumping and spinning (as in the Martial Arts movies) is a wonderful diversionary technique, and the opponent doesn't know whether a foot, fist, or chop is coming out of the spin. But, the spinner is unstable and can be injured on landing. A kick is stronger than a punch, but kicking means you're balanced on one leg for a moment. A long lead with the arms means the opponent has a hard time getting through your defense. But, with a long lead, you lack the power in your punch that momentum adds, because you cannot harness the force of a twisting, rising, or falling body. Few can match Bruce Lee's famous "one inch punch" and need a longer driving force to develop power.

In risk communication there are also trade-offs, that are not prominent enough in the academic literature. While talking, even with great skill, one cannot also be listening. While explaining technology and making risks understood, some audience members may feel manipulated. Those who don't feel manipulated may simply not understand. How realistic is it for busy people, those with below average intelligence, or without a scientific education to understand heavier than air flight, radon gas in their basements, bioremediation in an oil spill, or other such matters? Has research determined what an ideal human might do in ideal circumstances, should they ever exist?

6.1 There Are Few Easy Answers

Fifty years after interest in risk and crisis communication, there is still an unrealized promise from serious academic study into the field. There remain contradictions and gaps. This may be tolerable in normal times, but in risk and crisis communication, when lives are at stake, we must surely rationalize the contradictions and close the gaps. This is evident in the failings of many aspects of the one hundred emergency plans studied.

Certainly best practices exist in urban emergency planning. Whether this is a result of a thoughtful practitioner, experience with tragedy, or happenstance cannot be decided here. But the practices cited above in Leeds, Kansas City, Sacramento, Guelph, Auckland, NZ, Oshawa, and elsewhere are cases in point.

But why is the language in most city emergency plans relatively inaccessible to average citizens—demonstrating inoperability? Are they written only to comply with legislation? If so, they still could be in plain language and standardized formats. Are they mainly written as guidance for emergency responders? If so, they need not be public. Are they written by committees, over time, with differing goals in a changing chain of command?

If these plans are written for emergency responders, there is at least one school of thought that indicates that many of the pages may not serve a useful purpose. These pages contain definitions, roles, responsibilities, the legal regime under which emergencies will be managed, glossaries, job descriptions, and the other extraneous matters discussed above. On first reading of these elements, it is hard to believe that police, fire, EMS, public works officials, and others need information about the jobs they do every day or how they might be required in a crisis. In fact, they may not need much information of any kind.

Researchers who have examined the decision making process of emergency responders note a process termed "naturalistic" decision making or NDM.[72] As one might imagine, firefighters work diligently to put out a fire, rather than engage in a discussion of the exigencies of combustion, or approaches from which to choose. Emergency room doctors take action until no more action is possible, rather than engage in a classroom-style discussion of what might cause or prevent death. NDM is best used in situations which feature fluid and changing conditions, real-time reactions, ill-defined goals, ill-structured tasks, and knowledgeable people (Klein [184], p. viii). The related term "recognition-primed decision-making" describes an emergency responder's actions on the scene of an event which are driven by recognition of situations and patterns. Those decisions are in a constant state of re-evaluation based on new environmental information. New information yields new action:

> Rarely did the fireground commanders contrast even two options. We could see no way in which the concept of optimal choice might be applied. Moreover, it appeared that a search for an optimal choice could stall the fireground commanders long enough to lose control of the operation altogether. (Klein [184], cited in Flin (1996) pp. 144–145).

Experience and training usually result in the responder picking a workable option on first attempt. The goal is to satisfy, not optimize. One might argue that decisions are made—where to stand, dousing flames, retreating and so on. But the terms "naturalistic" and "recognition-primed" contrasts with the usual decision-making process of selecting options, pondering outcomes, picking one option, examining the actual outcome, and so on. The contents of urban emergency plans as found in this study do not appear to be useful to responders. Klein, quoted above, implies that these distractions can cause a loss of control, and then even death.

Urban emergency plans suffer from a number of failings listed above, and the miscalculation of what responders or the public need. This perhaps should not be a surprise. There is no reason to believe that there is less miscommunication in offices and professional lives than in our personal lives. We know anecdotally that personal

[72]The author wrote about this phenomenon in his unpublished doctoral dissertation in 2007.

miscommunication is prevalent among loved ones at home. We should not be surprised to find miscommunication in organizational life:

> Much of the information in organizations has been gradually accumulated over a long period of time, without any broader plan or strategy guiding its management. The result is a very large body of content, much of which is out-of-date, inaccurate or poorly-structured (Robertson, p. 2).

This observation is not specifically about emergency planning documents, but applies. So does management expert Peter Drucker's observation that: "layers of management neither make decisions nor lead. Instead, their main, if not their only, function is to serve as relays—human boosters for the faint, unfocused signals that pass for communication in the traditional pre-information organization" (p. 7). Eric Kierans' position [185] that the modern organization is so complex that it is beyond even senior management to understand and manage operations. Why should we expect more from emergency plan writers, perhaps in the middle of the hierarchy in cities? These general observations may explain the miscommunication within emergency plans—no more mystery than the miscommunication in other offices and departments in any organization. With regard to the overall emergency plans themselves, at least one other researcher has noted deficiencies, and focused more on why they may be present:

> How does management proceed? Usually piecemeal. First, a security officer (probably unqualified) is hired, followed by an alarm company together, perhaps, with guards and gatekeepers. No attempt is made to make an independent assessment of the risk–present and future—and then construct an appropriate plan based on that assessment, available resources and vulnerability… Security, however, has even greater intellectual needs.

This author is describing general failings of the evolution of organizational security systems. However, it may describe the evolution of an urban emergency plan, as well.

In recent experiences in graduate school, the author has noticed behaviours which may be relevant. The first is what Schon [186] might call the 'professionalizing of everything.' Several disciplines refer to themselves as "professional" and try to exhibit professionalism. Some have self-regulating bodies, as in Social Work, Law, Planning, Education, and others. But there are degrees of authority, autonomy, and self-regulation. Architects and Engineers have stamps required before certain projects can be built; physicians can write prescriptions; and lawyers are officers of the court. But what is the authority of a city planner or emergency planner? A project can go forward without a city planner's permission and there is no planning stamp. Emergency management and emergency planning is not yet a profession in this sense and has even less authority. In many cities, the writing of plans may be personality driven, with a strong or charismatic person in the police, fire service, or, mayor's office leading. Or there may be someone with special experience or skills from previous experience (military, bigger city, major event). These matters may contribute to the asymmetrical approach found.

Another notion observed is that some who practice in a discipline seem intent on working partly in another field.[73] Planners study gender issues, settlement of refugees, urban agriculture, urban murals, and other diverse issues, when there are programs in gender, sociology, social work, agronomy, art, and forestry within other university faculties and city departments.[74] Whether those who appear to be encroaching on another discipline think theirs is the superior field and should subsume another, whether they are not sufficiently stimulated by their field, or if there is another explanation, cannot be decided upon here.

Finally, there seems to be a desire to use the technical and scientific terminology of what may be harder disciplines. The author has heard many civilians in the Department of National Defence in Canada adopting military terms ('in country… chain of command…alongside…haul onboard'). In the oil and mining industries, non-engineers speak of 'flanging up' something to complete a task. In a recent urban planning class dealing with bicycles as transport, there were references to epidemiological studies of injuries among non-medical practitioners. Few non-lawyers attempt to render a legal opinion, and fewer lawyers and law societies tolerate this. Yet there are examples in the urban emergency plans studied which seem to long for, and imitate the certitude of a legal ruling, military precision, scientific methods, the clarity of architectural renderings, and so on.

These and other matters constitute fog and friction. Asymmetry and adopting other disciplines' methods and terms may bespeak a lack of confidence in the emerging emergency planning field among its members. It may also be the written version of the traumatized earthquake victim picking up soda cans in order to inject meaning into a chaotic event. Jargon, diagrams, lists, glossaries and such may cause writers to feel that there has been a comprehensive job, well done, employing the semantic and schematic resources of several fields. If the document appears to incorporate military precision, scientific certainty, a designer's visualization, and a lawyer's citations, perhaps the writers and even readers will feel they've brought order to a future chaotic situation.

An often cited impediment to progress is a lack of funding. This is cited in many emergency plans. Yet, a summer student or intern could find the best practices cited above, and incorporate these into a city's emergency plan at little or no cost. The incuriosity in the emergency planning field[75] gives pause, as does the lack of communication among the response hierarchies in the cities studied.

More specific, clear, and standardized approaches would likely be safer.

---

[73]This seems particularly the case as of late, with the emergence of "Interdisciplinary Studies" as its own specialized field of study. Universities such as Royal Roads and Lakehead have developed well respected studies in this discipline.

[74]The author has witnessed this occurrence over the course of his Urban Planning studies.

[75]As part of the research for *Safer Cities* (2011), the author and his colleague Ashley McIntosh sent a survey to 87 cities covered by the study of plans. Of these, 14 replied and one partially replied.

7 Areas for Further Research and Action

Risk Communication still holds promise. In fact, it is surely still the best available data on how to influence people's risk perceptions and reactions to risk. And yet, not enough progress has been made in almost five decades of research into the topic. If it can be determined just what information at what stage of an event will affect which segment of the population, this would be valuable information for industry and government. This data needs updating in the context of social media.

One size will not fit all in a unitary state, a federation with weak mayors, and the American system of strong city governments and home rule. Research that cites experiences in multiple jurisdictions may be weak. Asymmetry is expected and may produce effective results. That asymmetry needs to be clearly stated and explored. But the diversity found in the 100 urban emergency plans studied is surely far too broad and varied. The animosity to federal aid in American is noteworthy and may date to the revolution and Civil War. Perhaps it's no surprise that the Presidential order specifying clear and standard communication has not borne fruit. Other than city by city, how one could make emergency plans more standard in the US is beyond this research to address.

In Canada, we have a similar problem, but it is more technical than constitutional. We don't have the history and culture which support the notion that local government is favoured. The technical issue is a dichotomy perhaps. Our constitution only mentions local matters but not local government [50]. One might speculate that most founders, being from a unitary state, saw little need to make provision for more than two levels of government. This is only speculation. What did occur though, is that local government grew out of local needs for fire protection, police, and other services. Provinces then enacted legislation creating municipalities, subject to provincial oversight.

The dichotomy occurs upon reading the Canadian Supreme Court's ruling on so-called subsidiarity. The term embraces the legal notion that the level of government which is closest to citizens is most able to provide services [50, 187]. The Court has stated that "law-making and implementation are often best achieved at a level of government that is not only effective, but also closest to the citizens affected and thus most responsive to their needs to local distinctiveness and population diversity." The Court has only said that this is "often" the case, and "often" may well not include the writing of emergency plans. It may not always be accurate to describe city governments as "effective" either. This case could be argued, perhaps, but it is not immediately obvious that city governments should be described as effective, or more effective than provincial, or the national governments. Do they achieve a value for dollar spent or benefit to the taxpayer that exceeds other levels of government? One suspects this research has not been done. Nor might many city governments be "closest to the citizens." Perhaps many voters feel closer to their provincial or federal representatives. Media coverage may focus more on senior governments. Moreover, are cities more responsive to the needs of constituents than other levels of government? This is a case that needs to be proven, not merely asserted.

In terms of "local distinctiveness and … population diversity", these are popular terms in contemporary discourse. They once were code words for 'states' rights' in the US, even as recently as the 1964 Barry Goldwater campaign, when the Presidential candidate referred to "diversity" as just such a code word [39, pp. 245–248]. Yet, notwithstanding legal quips to the contrary, people do not drown differently during floods in the Fraser Valley versus the Saint John River Valley.[76] Burn victims suffer in the same ways, regardless of city of residence. So do those impacted by weather events, terrorism, and so on. There are dietary needs, local culture, religion, and languages that may affect response, but it is hard to imagine diversity being so pronounced that the vastly different approaches to emergency response found are in fact required.

There is another reason for standardization. Many Canadian cities are close to or straddle a provincial border. Lloydminster, Saskatchewan, Ottawa, Ontario, and Sackville, New Brunswick come to mind. Many Canadians commute across a provincial border for work or to recreation properties. The notion that emergency plans should be consistent for these and other citizens is a case that appears to be made by its own assertion.

Thus, at least some provincial action seems required. There are forums such as Question Period in the legislature, estimates committees which scrutinize the Municipal Affairs Ministries' spending, as well as standing and select committees of legislatures. Officials in the appropriate ministries, from political staff to the Director level in the civil service, might rightly inquire into the state of municipal emergency plans. While not as newsworthy as a Council of the Federation or First Ministers' conference, there could be discussions among the appropriate ministers in a dedicated national forum, or as an agenda item in existing bi-and multi-lateral fora.

For those who rightly agree that emergency response lacks a political constituency [19], one needs a strategy other than suggesting that politicians campaign for government on a platform to improve resiliency, evacuation times, and such.[77] Perhaps Albuquerque, NM holds a key to success. Their city plan states that:

> The benefits of mitigation planning go beyond solely reducing hazard vulnerability. Related measures emanating from a mitigation plan such as preserving open space, protecting vital infrastructure, designing sustainable buildings, maintaining environmental health, and protecting critical facilities meet other important community objectives including public safety, natural resource protection, and business development.[78] [118, pp. 2–3]

There are thus multiple benefits from, and surrogate ways to approach urban resiliency and effective emergency planning. These city building matters do have broad constituencies which can and should be mobilized.

---

[76]Makuch [50] entertainingly expresses a similar thought: "What's good for Goose Bay is not necessarily good for Gander." p. 168.

[77]See Berke and Beatley, p. 6.

[78]See the City of Albuquerque, NM, Hazard Mitigation Plan, page 2.

8 Conclusion

That more work needs to be done is a simplistic and hoary prescription. While there are gaps and inconsistencies in the risk communication literature, it is still a good guide, especially in the cautionary note from the military that there will be fog and friction. There is no reason to abandon good, but incomplete data. There is also no reason to refrain from vigorous improvement.

The main prescription advocated here is for senior governments to take the lead to ensure adequate and standard emergency plans. This may be difficult in the United States, but possible in other countries. Governments have numerous fora in which to make urban preparedness issues known. If there isn't the will, law or tradition to impose improvements, these senior governments could publish a list of large urban areas which are in or out of compliance with existing laws. This might engage news media and citizen groups, and embarrass cities into taking action.

Governments also have a regulatory role. If emergency response organizations fail to evolve as self-regulating professions, others could take action. Colleges and universities could embark on a more rigorous diploma and degree-granting program. Governments could encourage or enforce standards, qualifications, and outcomes of the work of emergency responders.

Otherwise we may repeat history. The loss of life during the 2003 Great White concert in Rhode Island [188] brings to mind the hundreds of lives lost in similar fires over one hundred years. The Laurier Palace Theatre Fire of 1927, Iroquois Theater fire of 1903, Triangle Shirtwaist Factory fire of 1911 [189] and many others are testament to our not having learned sufficiently from these tragedies.

There will be power outages in the future. There will be ice storms in Canada, tornadoes and hurricanes in the southern US, and floods across North America. These may increase as a result of climate change, as indicated in Boston's com prehensive document. It will not provide sufficient solace to victims of future events or the loved ones of those lost in past events to muddle through or offer the fact that we're making progress as an excuse for the state of our emergency plans. As Toft and Reynolds [190, p. xii] have pointed out: "[w]e owe it to those who have lost their lives, been injured, or suffered loss to draw out the maximum amount of information from those lessons and to apply it to reduce future suffering."

References

1. Von Clausewitz C, Graham J (1909) On war. Project Gutenberg, London. Found at: http://www.gutenberg.org/files/1946/1946-h/1946-h.htm
2. Fessenden-Raden J, Fitchen JM, Heath JS (1987) "Providing risk information in communities: factors influencing what is heard and accepted", science, technology & human values, Summer/Fall, pp 94–101

3. Drucker P (1998) The coming of the new organization (Harvard Business Review). In: Harvard business review on knowledge management. Harvard Business School Press, Boston
4. Carey J (1991) Getting business to think about the unthinkable. Bus Week 24:104–107
5. Abe K (1978) Levels of trust and reactions to various sources of information in catastrophic situations. Disasters: Theory and Research, 147–158. Retrieved 4 July 2016, from: http://psycnet.apa.org/psycinfo/1979-33466-001
6. Kasperson RE et al (1988) The social amplification of risk: a conceptual framework. Soc Risk Anal 8(2):177–187
7. Sandman P (1987) Risk communication: facing public outrage. EPA J, 21–22
8. Sandman PM, Sachman (1985) Getting to maybe: some communications aspects of siting hazardous waste facilities. Seton Hall Legislative Journal 9:437
9. Slovic P (1987) Perception of risk. Science 236, 280–285
10. Slovic P, Kraus N, Covello VT (1990) What should we know about making risk comparisons? J Soc Risk Anal 10(3):389–392
11. Roth E (1990) What do we know about making risk comparisons? J Soc Risk Anal 10 (3):375–387
12. Pidgeon NF (1992) The psychology of risk. In: module 1, unit 3 ('Psychological Approaches to Risk Management') of M.Sc. in risk, crisis and disaster management. Leicester: SCSPO, pp 89–107
13. Ruckelshaus WD (1983) Science, risk, and public policy. Science 221:1026–1028
14. Ruckelshaus WD (1984) Risk in a free society. J Soc Risk Anal 4(3):157–162
15. Douglas M, Wildavsky A (1982) How can we know the risks we face? Why risk selection is a social process. J Soc Risk Anal 2(2):49–51
16. Clarke L (1993) The disqualification heuristic: when do organizations misperceive risk? Res Soc Probl Publ Policy 5:289–312
17. Morgan MG, Slovic P, Nair I, Geisler D, MacGregor D, Fischhoff B, Lincoln D, Florig K (1985) Powerline frequency electric and magnetic fields: a pilot study of risk perception. J Soc Risk Anal 5(2):139–149
18. Kahneman D, Slovic P, Tversky A (1982) Judgment under uncertainty: heuristics and biases. Cambridge: Cambridge University Press
19. Fischhoff B (1985) Managing risk perception. Issues Sci Technol, 84–96 (Fall)
20. Wilson R (1979) Analyzing the daily risks of life. Technology Review, February.
21. Bonner A (n.d.) Gambling with food. Calgary Herald. Retrieved 15 Dec 2015, from: http://allanbonner.com/articles/Gamblingwithfood.pdf
22. Bonner A (2003) Scared by SARS. The Ottawa Citizen, 16 April [Online]. Available from: http://allanbonner.com/articles/ottawacitizenchanged.pdf. Accessed on 4 Jan 2015
23. Covello V (1983) The perception of technological risks: a literature review. Technol Forecast Soc Chang 23:285–297
24. Slovic P (1986) Informing and educating the public about risk. Risk Anal 6(4)
25. CTV Barrie (2015) Timeline of the devastating tornadoes that hit 30 years ago. Retrieved 23 Dec 2015, from; http://barrie.ctvnews.ca/timeline-of-the-devestating-tornadoes-that-hit-30-years-ago-1.2390681
26. Raymond CA (1985) The language of risk, conflicting perspectives on occupational health. In: Nelkin D (ed) Beverly hills. Sage Publications, Inc., California
27. Fischhoff B (1987) Treating the public with risk communications: a public health perspective. Sci Technol Human Values 12(3&4):13–19
28. Cumming RB (1981) "Is risk assessment a science?" J Soc for Risk Anal 1(1):1–3
29. Kasperson RE (1986) Six propositions on public participation and their relevance for risk communication. Risk Anal 6(3):275–281
30. Mazur A (1984) The journalists and technology: reporting about love canal and Three Mile Island. Minerva 22:45–66
31. Royal Society (1992) Risk: analysis, perception and management. Royal Society Study Group, The Royal Society, London, pp 135–192

32. Nelkin D (1985) "Ethical conflicts in occupational medicine" in the language of risk, conflicting perspectives on occupational health, Nelkin, Dorothy, Ed., Beverly Hills, California, Sage Publications, Inc.
33. Wynne B (1996) May the sheep safely graze? A reflexive view of the expert-lay knowledge divide. In: Lash et al (ed) Risk, environment and modernity: towards a new ecology. Sage Publications, Inc., London, pp 44–83
34. Covello V (2002) Risk communication slides. Powerpoint presentation for the Center for Risk Communication, New York [Online]. Available from: http://nwcphp1.cac.washington.edu/docs/pdf/april02color.pdf. Accessed 3 Jan 2016
35. Nixon R (1974) I am not a crook. Retrieved 16 Dec 2015, from: http://www.history.com/speeches/nixon-i-am-not-a-crook
36. Bush G (1988) Read my lips: no new taxes video. Retrieved 15 Dec 2015, from: http://www.history.com/topics/us-presidents/george-bush/videos/read-my-lips-no-new-taxes
37. Sandman P (1993) Responding to community outrage: strategies for effective risk communication. American Industrial Hygiene Association, Fairfax, VA
38. Stanton A (1989) Management can keep a crisis from turning into a calamity. Oil & Gas Journal 8:15–16
39. Bonner A (2009) Political columns. Sextant Publishing, Edmonton
40. Barton L (1983) Crisis in organizations: managing and communicating in the heat of chaos. South-West Publishing, Cincinnati 1993
41. Conner DR (1995) Managing at the speed of change. Villard Books, New York
42. Tlingit First Nation v. British Columbia (Project Assessment Director), Supreme Court of Canada (2004). Retrieved 15 Dec 2015, from the Judgments of the Supreme Court of Canada website: http://scc-csc.lexum.com/scc-csc/scc-csc/en/item/2190/index.do?r=AAAAAQAKdGFrdSByaXZlcgE
43. Ontario Ministry of Municipal Affairs and Housing (2015) Municipal Act. Retrieved 15 Dec 2015, from: http://www.mah.gov.on.ca/Page184.aspx
44. Ontario (1990) Planning Act, R.S.O. 1990. Retrieved 30 Dec 2015, from: http://www.ontario.ca/laws/statute/90p13
45. Law Society of Upper Canada (2000, Amended 2014) Rules of professional conduct. Toronto, Ontario [Online]. Available from: http://www.lsuc.on.ca/WorkArea/DownloadAsset.aspx?id=2147501952. Accessed on 4 Jan 2016
46. Quebec (2015) Code of ethics of architects. Retrieved 18 Nov 2015, from http://www2.publicationsduquebec.gouv.qc.ca/dynamicSearch/telecharge.php?type=3&file=/A_21/A21R5_1_A.HTM
47. Ontario Professional Planners Institute (n.d.) Professional code of practice [Online]. Available from: http://ontarioplanners.ca/Knowledge-Centre/Professional-Code-of-Practice. Accessed 4 Jan 2015
48. Ontario (1990) Professional Engineers Act, R.S.O. 1990. Retrieved 18 Nov 2015, from: http://www.ontario.ca/laws/statute/90p28
49. Pue K (2013) A "Big City Charter" for Edmonton and Calgary: explaining the role of municipalities in Canada's federal framework. Retrieved 15 Dec 2015, from: http://ualawccsprod.srv.ualberta.ca/ccs/index.php/constitutional-issues/federalism/729-a-big-city-charter-for-edmonton-and-calgary-explaining-the-role-of-municipalities-in-canada-s-federal-framework
50. Makuch S et al (2004) Canadian municipal and planning law, 2nd edn. Carswell, p 273
51. O'Connor R, Swift N (2015) US local government and mayors of largest cities. Retrieved 15 Dec 2015, from: http://www.citymayors.com/mayors/us-mayors.html
52. Sellnow TL (1993) Scientific argument in organizational crisis communication: the case of Exxon (Exxon Corp.). Argumentation Advocacy 30(1):28 (Summer(15))
53. Bonner A (2005) What constitutes professional, legal, ethnographic research? Submitted for GS Law 6301 P, December 12 (Unpublished)
54. Blalock HM (1970) An introduction to social research. Prentice-Hall, Englewood Cliffs, New Jersey

55. Phoenix Strategic Perspectives (2012) Secondary research into cell phones and telephone surveys. Retrieved 23 Nov 2015, from: http://www.tpsgc-pwgsc.gc.ca/rop-por/rstcst-srcpts-eng.html#fig6
56. Relish Communication (2014) The end of the line for the landline. Retrieved 18 Nov 2015, from: https://www1relish.net/mediacentre#An-end-to-the-landline
57. Sharlin H (1987) Macro-risks, micro-risks, and the media: the EDB case. In: Johnson B, Covello V (eds) The social and cultural construction of risk: essays on risk selection and perception. Dordrech, D. Reidel Pub
58. Tolman C (1996) Problems of theoretical psychology. Captus Press, North York
59. Sandman PM, Sachsman DB, Greenberg MR, Gochfeld M (1987) Environmental risk and the press: an exploratory assessment. Transaction, Inc., New Brunswick, New Jersey
60. Nelkin D (1985) The language of risk: conflicting perspectives on occupational health. Sage Publications, Beverly Hills
61. Nelkin D (1985) Introduction: analyzing risk. In: Nelkin D (ed) The language of risk, conflicting perspectives on occupational health. Sage Publications, Inc., Beverly Hills
62. Williams M (n.d.) The museum of broadcast communications-encyclopedia of television—Milton Berle show. Retrieved 17 Dec 2015, from: http://www.museum.tv/eotv/miltonberle.htm
63. Biography.com (ed) (n.d.) Bing crosby biography. Retrieved 17 Dec 2015, from: http://www.biography.com/people/bing-crosby-9262159#early-life
64. Goodard T (ed) (2010) Checkers speech. Retrieved 18 Dec 2015, from: http://politicaldictionary.com/words/checkers-speech/
65. Bonner A (1999) Given its size and impact why is the Exxon Valdez oil spill so prominent in disaster literature? Unpublished M.Sc. thesis, University of Leicester [Online]. Available from: http://allanbonner.com/products-papers/. Accessed on 30 Dec 2015
66. Bonner A (2004) Exxon Valdez: was the impact of the notorious oil spill overblown? The Calgary Herald, 19 March, p A13
67. Keeble J (1991) Out of the channel: the Exxon Valdez oil spill in Prince William Sound. Harper Collins, New York
68. Peaks MH (ed) (1990) The Alaskan oil spill: lessons in crisis management. Manage Rev (April)
69. McAlpine S (1990) The Alaskan oil spill: lessons in crisis management. Manag Rev
70. Busch L (1991) Science under wraps in Prince William Sound. Science 252
71. Daley P, O'Neill D (1991) 'Sad is too mild a word': press coverage of the Exxon Valdez Oil Spill. J Commun 41(4) (Autumn)
72. Davidson A (1990) In the wake of the Exxon Valdez. Sierra Club Books, San Francisco
73. The Economist (1990) Of oil and sea-otters. Davidson, Art, May 19
74. Seibert FS, Peterson T, Schramm W (1974) Four theories of the press. University of Illinois Press, Urbana
75. Palen L (2008) Online social media in crisis events. Edu Quart (3)
76. Lee D (2013) Boston bombing: how internet detectives got it very wrong. BBC News, 19 April [Online]. Available from: http://www.bbc.com/news/technology-22214511. Accessed on 4 Jan 2016
77. Langer J (1998) "Tabloid television. Popular Journalism and the 'Other News'," Routledge, New York
78. Mislove A, Gummadi K, Druschel P (2007) Exploring social networks for internet search. Retrieved from: http://www.mpi-sws.org/~amislove/publications/PeerSpective HotNets.pdf
79. Sonderman J (2012) One-third of adults under 30 get news on social networks now. Poynter. Retrieved 16 Dec 2015, from: http://www.poynter.org/news/mediawire/189776/one-third-of-adults-under-30-get-news-on-social-networks-now/
80. Blodget H (2012) 90 % of 18–29 year-olds sleep with their smartphones. Business insider. Retrieved on 16 Dec 2015, from: http://www.businessinsider.com/90-of-18-29-year-olds-sleep-with-their-smartphones-2012-11

81. Google (2013) Our mobile planet: Canada. Retrieved 16 Dec 2015, from: http://services.google.com/fh/files/misc/omp-2013-ca-en.pdf
82. Yablonski C (2013) The impact of social media on the calgary flood. [Blog] Available from: http://www.inboundinteractive.ca/the-impact-of-social-media-on-the-calgary-flood/. Accessed on 30 Dec 2015
83. Canadian Red Cross (2012) Social media during emergencies [Online]. Available from: http://www.redcross.ca/cmslib/general/pub_social_media_in_emergencies_survey_oct2012_en.pdf. Accessed 30 Dec 2015
84. Kiesling E (2001) On war without the fog. Mil Rev, 85–87
85. United States Marine Corps (1997) Warfighting: the U.S. Marine Corps book of strategy. Doubleday, New York
86. Apker J (2013) Systems theory as scaffold for storytelling…and other reflections on healthcare organizing. Health Commun 28(7):208
87. Bertalanffy L von (1968) General systems theory: foundations, development, applications. New York: George Braziller
88. Bonner A (2014) Wounded leaders. Sextant Publishing, Edmonton
89. Parker R (1975) The flixborough disaster: report of the court of inquiry. H.M.S.O, London
90. Griffiths RF (ed) (1981) Introduction: the nature of risk assessment. In: Dealing with risk: the planning management and acceptability of technological risk. Wiley, New York, pp 1–19
91. Fennell D (1988) Investigation into the King's cross underground fire. [For] Department of Transportation (by) H.M.S.O, London
92. University of Leicester (2006) Case studies of crises and disasters, module 4, pp 3.1–3.34. University of Leicester
93. Aptekar L (1990) A comparison of the bicoastal disasters of 1989. Behav Sci Res 24:73–101 (HRAF)
94. City of Guelph (2014) City of Guelph emergency response plan: 2015. Guelph, Ontario
95. Statistics Canada (2013) Canadian Internet Use Survey, 2012. Retrieved 22 Dec 2015, from: http://www.statcan.gc.ca/daily-quotidien/131126/dq131126d-eng.htm
96. Berke PR, Beatley T (1992) Planning for earthquakes, risk, politics and policy. The Johns Hopkins University Press, Baltimore and London
97. City of Baltimore (2004) All-hazards plan for Baltimore city. Baltimore, Maryland
98. City of Brisbane (2013) Disaster management plan. Brisbane, Australia
99. City of Philadelphia (2012) Natural hazard mitigation plan. Philadelphia, Pennsylvania
100. City of Pittsburgh (2007) Pittsburgh emergency operations plan. Pittsburgh, Pennsylvania
101. City of Kitchener (2012) Corporation of the city of Kitchener emergency plan. Kitchener, Ontario
102. City of Sherbrooke (Undated pamphlet) City of Sherbrooke: what to do in case of an emergency? Heavy rain and flooding. Sherbrooke, Quebec
103. James D (1985) 1985 Barrie Tornado. Retrieved 23 Dec 2015, from: http://www.cbc.ca/archives/entry/1985-barrie-tornado
104. City of Barrie (2007) City of Barrie emergency plan. Barrie, Ontario
105. McLuhan M (1964) Understanding media. Mentor, New York
106. City of Detroit, Department of Homeland Security (2012) City of Detroit evacuation plan. Detroit, Michigan
107. Bowling D, Hoffman D (eds) (2003) Bringing peace into the room: how the personal qualities of the mediator impact the process of conflict resolution. Wiley, San Francisco
108. The Economist (1995) A survey of earthquake engineering, fear of trembling April 22, pp 3–12
109. Giulliani RW (2002) Leadership. New York: Miramax Books, p 171
110. Richardson B (1994) Socio-technical disasters: profile and prevalence. Sheffield Business School, Sheffield, UK
111. Sood R, Stockdale G, Rogers EM (1987) How the news media operate in natural disasters. J Commun 37(3):27–41
112. The Economist (1995) Aftershocks in Japan (January 21) 34:1
113. City of Leeds (2012) Leeds City council: planning for emergencies. Leeds, UK

114. City of Louisville (2007) Louisville/Jefferson county emergency operations plan. Louisville, Kentucky
115. City of Birmingham (2013) Birmingham city council corporate emergency plan. Birmingham, UK
116. City of Halifax (2009) Halifax regional municipality master emergency plan annex C: emergency evacuation plan. Halifax, Nova Scotia
117. City of Leeds (2012) Leeds city council advice to businesses: mailroom procedures for suspicious packages. Leeds, UK
118. Bernalillo County/City of Albuquerque/Albuquerque Metropolitan Arroyo Flood Control Authority (AMAFCA)/Village of Los Ranchos de Albuquerque/Village of Tijeras Hazard Mitigation Plan (2014) New Mexico
119. Fort Worth-Tarant County Office of Emergency Management (2008) Fort Worth hazard mitigation action plan. Fort Worth, Texas
120. City of Milwaukee (n.d.) Office of emergency management and homeland security [Online]. Available from: http://city.milwaukee.gov/OfficeofHomelandSecurity. Accessed on 19 Sept 2013
121. City of Oklahoma (August 2004, Revised May 2006) Hazard mitigation plan. Oklahoma City, Oklahoma
122. New Zealand Ministry of Civil Defence and Emergency Management and Auckland Council (2011) Emergency management group plan 2010–2015. Auckland, New Zealand
123. City of Bradford (n.d.) Don't panic, prepare [Online]. Available from: http://www.bradford.gov.uk/NR/rdonlyres/370ED151-AFED-458A-8632-D61AF7FD6B65/0/DontPanicBooklet.pdf. Accessed 29 Dec 2015
124. Jacksonville/Duval County, (Consolidated City of) (2010) Comprehensive emergency management plan. Jacksonville, Florida
125. City of Boston (2014 update) Natural hazard mitigation plan. Boston, Massachusetts
126. City of Oshawa (n.d.) Emergency preparedness [Online]. Available from: www.oshawa.ca/fire/eplanning.asp. Accessed 7 Sept 2013
127. New Orleans Office of Homeland Security and Emergency Preparedness (2013) Invitation to evacuspots unveiling [Online]. Available from: http://www.nola.gov/homeland-security/. Accessed 9 Sept 2013
128. City of Houston, Office of Emergency Management (2014) Annexes to city of Houston emergency management plan: public information. Houston, Texas
129. City of Burnaby (n.d.) 26 Steps to emergency preparedness [Online]. Available from: https://www.burnaby.ca/City-Services/Public-Safety/Emergency-Program/Emergency-Preparedness/26-Steps-To-Emergency-Preparedness.html. Accessed 29 Dec 2015
130. City of Surrey (2009) Surrey emergency management plan: 26 steps to emergency preparedness. Surrey, British Columbia
131. The city of long beach 2014 hazard mitigation plan (2014, August 10) Retrieved June/July 2016, from: http://www.longbeach.gov/disasterpreparedness/media-library/documents/home/long-beach-hazmit-plan-08_10_2014-v4/
132. City of Kansas, Missouri Evacuation and Transportation Services Plan (2008) Kansas City, Missouri. Retrieved 23 Dec 2015, from: http://kcmo.gov/citymanagersoffice/wp-content/uploads/sites/11/2013/09/KCMO-Evacuation-Plan.pdf
133. Moynihan D (2009) The response to Hurricane Katrina. International Risk Governance Council
134. Litman T (2006) Lessons from Katrina and Rita: what major disasters can teach transportation planners. Victoria Transport Policy Institute
135. Hess D, Gotham J (2007) Multi-modal mass evacuation in upstate New York: a review of disaster plans. J Homel Secur Emerg Manag 4(3) (Article 11)
136. Kendra J, Rozdilsky J, McEntire DA (2008) Evacuating large urban areas: challenges for emergency management policies and concepts. J Homel Secur Emerg Manag 5(1) (Article 32)

137. Wolshon B (2001) Contra flow freeway operation for hurricane evacuation in the United States. In: Proceedings of the Eastern Asia Society for transportation studies, vol 3
138. City of Sacramento, Office of Emergency Services (2008) City of Sacramento plan: floods and other emergencies. Sacramento, California
139. Klasson J (1976) Civilizing the machine. In: Ethical conflicts in occupational medicine. The language of risk: conflicting perspectives on occupational health. Grossman, New York, p 48
140. Nichols R, Stevens L (1957) Listening to People (Harvard Business Review. September-October). In: Harvard business review on effective communication. Harvard Business School Press, Boston (1999)
141. Roush W (1993) Learning from technology. Technol Rev (MIT, Cambridge, MA, August/September)
142. Interactive Emergency Evacuation Guidebook (2003, September 07) Charlotte, NC's Emergency Plan, 1–35. Retrieved May/June 2016
143. Kaplan R, Norton D (2000) Having trouble with your strategy? Then map it (Harvard Business Review, September October). In: Harvard business review on advances in strategy. Harvard Business School Press, Boston (2002)
144. Dooley D, Catalano R, Mishra S, Serxner S (1992) Earthquake preparedness: predictors in a community survey. J Appl Soc Psychol V.H. Winston & Son, Inc. pp 451–470
145. City of Nashville, Mayor's Office of Emergency Management (2012) Metropolitan Nashville-Davidson county multi-hazard mitigation plan. Nashville, Tennessee
146. City of Long Beach (2015) 2015 hazard mitigation plan. Long Beach, California
147. City of Kansas (2014) Missouri emergency operations plan, Kansas city. Missouri base plan, Kansas City, Missouri
148. City of London (2011) City of London emergency response plan. London, Ontario
149. City of Houston, Office of Emergency Management (2010) City of Houston emergency management plan: volume I, basic plan. Houston, Texas
150. Columbus Public Health (n.d.) Emergency preparedness: program information [Online]. Available from: publichealth.columbus.gov/emergency-preparedness.aspx. Accessed 28 Sept 2013
151. Baltimore County Government (2011) Emergency operations plan. Baltimore, Maryland
152. City of Toronto, Toronto Office of Emergency Management (2013) City of Toronto emergency plan. Toronto, Ontario
153. Milwaukee County (2013) Comprehensive emergency plan. Milwaukee, Wisconsin
154. Oakville Town (2013) Oakville emergency plan: public copy. Oakville, Ontario
155. Leivesley S (1995) Security into the 21st century. Paper given at the Risk and Security Forum, Police Training College, Bramshill
156. Flood Recovery (2015) Retrieved 23 Dec 2015, from: http://www.calgary.ca/General/flood-recovery/Pages/FloodRecoveryHome.aspx?redirect=/floodrecovery
157. Aiello R, Watkins MD (2000) The fine art of friendly acquisition. Harvard business review (November–December 2000). In: Harvard business review on mergers and acquisitions. Harvard Business School Press, Boston
158. Brenneman G (1998) Right away and all at once: how we saved continental. Harvard Business Review (September–October 1998)
159. Cassidy J (2001) Gut punch: how great was Jack Welch? The New Yorker (October 1)
160. Collins J (2001) Level five leadership: the triumph of humility and fierce resolve. Harvard Business Review
161. Mackenzie JK (1989) It's show time. Dow Jones-Irwin, Homewood, Illinois
162. Williams P (Various formal and information conversations during approximately 18 years of study, ending in about 2013) Gojy Ryu Karate personal studies. Studies taken through CMAC Dojo, Toronto
163. Faraday A, Plummer K (1979) Doing life histories. Sociol Rev 27(4):773–792 (Module 3, Unit 5 ('Life History') of M.Sc. in the Study of risk, crisis and disaster management. SCSPO, Leicester, pp 183–187)
164. McLaglen A (1968) Hellfighters. Motion picture on DVD, Universal

165. Kim R (1982) The classical man. Master's Publication, Hong Kong
166. Dupuy TN (1989) Combat data and the 3:1 Rule. Int Secur 14(1):195–201. Retrieved 23 Dec 2015, from: http://www.jstor.org/stable/2538771?seq=1#page_scan_tab_contents
167. Swainson G (2010) Province buries Site 41 dump site in Simcoe. The Toronto Star. Retrieved 18 Nov 2015, from: http://www.thestar.com/news/ontario/2010/05/26/province_buries_site_41_dump_site_in_simcoe.html
168. City of St. Catharines (2009) City of St. Catharines emergency management plan. St. Catharines, Ontario
169. City of Waterloo (2012) City of Waterloo—emergency management plan. Waterloo, Ontario
170. City of Mississauga (2011) City of Mississauga emergency plan. Mississauga, Ontario
171. James M (2007) The permanent-emergency compensation state: a "Postsocialist" tale of political dystopia. In: Smith M (ed) Critical policy studies. UBC Press, Vancouver, pp 321–346
172. Pauly J, Hutchinson L (2005) Moral fables of public relations practice: the Tylenol and Exxon Valdez cases. J Mass Media Ethics 20(4):231–249
173. Associated Press (2009) India Court Issues Arrest Warrant for Former CEO of Union Carbide in Gas Leak Case. The guardian. [Online] 31st July. Available from: http://www.theguardian.com/world/2009/jul/31/warren-anderson-arrest-warrant. Accessed 30 Dec 2015
174. Anderson Brower K, Dodge C (2010) Bush calls New Orleans Flyover a 'Huge Mistake' Bloomberg Business. Retrieved on 18 Nov 2015, from: /news/articles/2010-11-05/bush-calls-new-orleans-flyover-in-wake-of-hurricane-katrina-huge-mistake
175. World Health Organization (2015) Road traffic deaths data by country. Retrieved 18 Nov 2015, from: http://apps.who.int/gho/data/node.main.A997?lang=en
176. Canadian Press (2008) Small N.B. city in disbelief after 8 killed in crash. Retrieved 18 Nov 2015, from: http://www.cbc.ca/news/canada/small-n-b-city-in-disbelief-after-8-killed-in-crash-1.695137
177. Gaissmaier W, Gigerenzer G (2012) 9/11, Act II: a fine-grained analysis of regional variations in traffic fatalities in the aftermath of terrorist attacks. Psychol Sci 23(12):1449–1454. doi:10.177/0956797612447804
178. Patriquin, M (2008) Of meat, disease and death. Maclean's Magazine. Retrieved 18 Nov 2015, from: http://www.macleans.ca/general/of-meat-beer-disease-and-death/
179. Hannaford P (1986) Talking back to the media. Facts on File Publications, New York
180. Lyons R (1993) Hoaxes are found in Pepsi case. The New York Times, 18 June [Online]. Available from: http://www.nytimes.com/1993/06/18/us/hoaxes-are-found-in-the-pepsi-case.html. Accessed 31 Dec 2015
181. Harris V et al (n.d.) Case study: the Johnson & Johnson Tylenol crisis. Retrieved 23 Nov 2015, from: http://www.ou.edu/deptcomm/dodjcc/groups/02C2/Johnson%20&%20Johnson.htm
182. Earle TC, Cvetkovich G, Slovic P (1990) The effects of involvement, relevance and ability on risk communication effectiveness. In: Borcherding K, Larichev OI, Messick DM (eds) Contemporary issues in decision making. Elsevier, Amsterdam, pp 271–289
183. Egan M (2007) Barry commoner and the science of survival: the remaking of American environmentalism. The MIT Press, Cambridge, MA
184. Klein G (1993) p.viii "A recognition-primed decision (RPD) model of rapid decision making" was taken from Flin. Citations correspond
185. Kierans E, Stewart W (1989) Wrong end of the rainbow: the collapse of free enterprise in Canada. Harper & Collins, Toronto
186. Schon DA (1983) The reflective practitioner: how professionals think in action. Basic Books, New York
187. Bonner A (2015) Municipalities often least responsive to meeting citizen needs. Troy Media, July 16 [Online]. Available from: http://www.troymedia.com/2015/07/16/municipalities-often-least-responsive-to-citizen-needs/. Accessed on 23 Dec 2015
188. Andersen T (2012) The station fire takes big toll on survivors. The Boston Globe [Online]. 17th October. Available from https://www.bostonglobe.com/metro/2012/10/16/the-station-fire-took-emotional-toll-survivors-study-finds/bjY7OK5EHUujmP58kwgQFK/story.html. Accessed 30 Dec 2015

189. Nash JR (1976) Darkest hours. Nelson-Hall, Chicago
190. Toft B, Reynolds S (1994) Learning from disasters: a management approach. Butterworth Heinemann, Oxford, UK
191. Anderson K, Gardenio M (1994) Gender differences in reported stress response to the Loma Prieta Earthquake. Sex Roles 30(9/10) (Plenum Publishing Corporation)
192. Becker HS (1971) Sociological work: method and substance. SCSPO, Leicester
193. Bonner A (2011) An ounce of prevention. Sextant Publishing, Edmonton
194. Bonner A (2014) Predicting technological advances a mug's game. Winnipeg Free Press. Retrieved 15 Dec 2015, from: http://www.winnipegfreepress.com/opinion/analysis/Predicting-technological-advances-a-mugs-game-244378701.html
195. Bonner A (2014) Interview on "The Current," CBC Radio, October 21 [Online]. Available from: http://www.cbc.ca/radio/thecurrent/ebola-scare-sparks-pr-blitz-from-hospitals-businesses-and-government-agencies-oct-21-2014-1.2807232?autoplay=true. Accessed on 4 Jan 2016
196. Bonner A (2015) Safer cities of the future. Sextant Publishing, Edmonton
197. Browning J (1993) Union carbide: disaster at Bhopal. In: Jottschalk J (ed) Crisis response: inside stories on managing under seige. Visible Ink Press, Detroit
198. Covello V (2009) Risk and crisis communication: communicating effectively in high concern, high stress, or low trust situations. Lecture presented in Center for Risk Communication, New York
199. Covello V, Menkes J, Nehnevajsa J (1982) Risk analysis, philosophy, and the social and behavioral sciences: reflections on the scope of risk analysis research. J Soc Risk Anal 2 (2):53–57
200. Davis T et al (2006) A failure of initiative: final report of the select bipartisan Committee to investigate the preparation for and response to Hurricane Katrina. US. G.P.O, Washington
201. Ethylene Dibromide (EDB) (1993) Retrieved 15 Dec 2015, from: http://pmep.cce.cornell.edu/profiles/extoxnet/dienochlor-glyphosate/ethylene-dibromide-ext.html
202. Federak News Service (1998) Clinton: 'There Is In Improper Relationship'. Retrieved 15 Dec 2015, from: http://www.washingtonpost.com/wp-srv/politics/special/clinton/stories/excerpts012298.htm
203. Fessenden-Raden J, Fitchen JM, Heath JS (1987) Providing risk information in communities: factors influencing what is heard and accepted. Sci Technol Hum Values, 94–101 (Summer/Fall)
204. Funakoshi G (1975) Karate-Do—my way of life. Kodansha, New York
205. City of Guelph (2015) Your personal preparedness guide [Online]. Available from: http://guelph.ca/wp-content/uploads/YourPersonalPreparednessGuide.pdf. Accessed 29 Dec 2015
206. Guskin E, Rosensteil T, Moore P (2011) Network news: durability and decline. Retrieved 17 Dec 2015, from: http://www.stateofthemedia.org/2011/network-essay/
207. Haida Nation v. British Columbia (Minister of Forests), Supreme Court of Canada (2004). Retrieved 15 Dec 2015, from the Judgments of the Supreme Court of Canada website: http://scc-csc.lexum.com/scc-csc/scc-csc/en/item/2189/index.do
208. Hay T (2003) Public be damned. In: Encyclopedia.com. Retrieved 15 Dec 2015, from: http://www.encyclopedia.com/doc/1G2-3401803443.html
209. Kahneman D, Tversky A (1979) Prospect theory. An analysis of decision making under risk. Econometrica 47:2
210. Low A (1976) Zen and creative management. Anchor Books, Anchor Press/Doubleday, New York
211. Plans-Emergency Services/9-1-1 (2015) Retrieved 23 Dec 2015, from: http://marc.org/Emergency-Services-9-1-1/Plans.aspx
212. Richard E (2008) Cause and effect. Adweek, pp 53–62
213. Robertson J (2003) Where is the knowledge in a CMS? [Blog] column two. Available from: http://www.steptwo.com/au/papers/kmc_wherek/index.html. Accessed 10 June 2014
214. Takuma T (undated) Human behavior in the event of earthquakes. In: Disasters: theory and research. Tokyo Metropolitan University, Japan

215. Transport Canada (2015) Canadian motor vehicle traffic collision statistics. Retrieved 18 Nov 2015, from https://www.tc.gc.ca/eng/motorvehiclesafety/resources-researchstats-menu-847.htm
216. USA Work (n.d.) Fatality analysis reporting system. Retrieved 18 Nov 2015, from: http://www-fars.nhtsa.dot.gov/Main/index.aspx
217. Victims of deadly Rhode Island nightclub fire lament deadly Romanian blaze: 'eerily similar' (2015) Chicago Tribune. Retrieved 23 Dec 2015, from: http://www.chicagotribune.com/news/nationworld/ct-bucharest-club-fire-rhode-island-great-white-20151031-story.html
218. Wilson WE (n.d.) Generations of GM history. Retrieved May/June 2016, from: https://history.gmheritagecenter.com/wiki/index.php/Wilson,_Charles_E

Disaster Forensics: Governance, Adaptivity and Social Innovation

Dorte Jessen

Abstract This chapter seeks to contribute to the discourse on disaster forensics, by arguing that the root cause and complex causality is ultimately governance, ideally cultivating the collective ability to navigate disasters rather than to command control. The focus will be on the social dimension and its impact on disasters. Governance theory, combined with complex adaptive systems theory (Duit and Galaz in Gov Int J Policy Adm Inst 21(3):311–335, 2008 [12]), will provide the analytical foundation for the examination of Hurricane Katrina and the Fukushima Daiichi nuclear disaster. The theoretical deconstruction will reveal that the traditional virtues embedded in the social amplification of risk (Kasperson and Kasperson in The social contours of risk, volume I: publics, risk communication and the social amplification of risk. Earthscan, London, 2005 [21]), remain at the heart of complex causality. With this insight, it is observed that social innovation, with its inherent positive connotation (Matei and Antonie in Soc Behav Sci 185:61–66, 2015 [28]), is expanding the horizon for how social divisions, vulnerabilities and resilience are measured. Optimistically, it is suggested that social innovation, driven by civil society, may prove a vital component in the creation of a new social narrative.

Keywords Governance · Social amplification of risk · Reflexivity · Social innovation

1 Introduction

Undeniably, science has equipped us with a growing understanding of the discernible patterns of natural destructive phenomena [42: 53], while complexity theory has taught us that understanding is not the same as being able to predict [25: 409]. Thus the obsession in the pursuit of disaster aetiology forensics is driven by

D. Jessen (✉)
Risk, Crisis and Disaster Management, University of Leicester, Leicester, UK
e-mail: dj103@leicester.ac.uk

A.J. Masys (ed.), *Disaster Forensics*, Advanced Sciences and Technologies for Security Applications, DOI 10.1007/978-3-319-41849-0_6

the innate need and human desire to apply order and structure, effectively increasing the collective *illusion* of control. Efforts which to some are rendered futile as accidents will happen [33]—it is only a matter of time. And once they do, it is only a question of how big a role human error played in its demise and the extent of its destruction [34]. The purpose of this chapter is to contribute to the discourse on disaster forensics, arguing that the root cause and complex causality is ultimately governance, ideally cultivating the collective ability to *navigate* disasters rather than to command control. The focus will, in particular, be on the social dimension and its impact on disasters.

The opportunities and latitude for collective organizational growth will be demonstrated through Governance Theory with a Complex Adaptive Systems lens [12]. The discourse will establish that at the heart of complex causality, traditional virtues such as the social aetiology of disasters [42] and the social amplification of risk [21] remain. This is coupled with High Reliability Theory [23]; the art of challenging weaknesses within the system to expose 'black swans' [27]. This will ideally foster behavioural change; enable early warnings, and minimise the impact by turning these majestic black creatures into grey swans (less black more white).

Governance as the root cause of complex accident aetiology, will be argued through Hurricane Katrina, the embodiment of complex causality. Katrina taught us that in the wake of catastrophe, the root cause can be altered in the name of re-examination [37]. This means, that disaster aetiology cannot be separated from the public interest and politics, and existing power structures. Looking at other cultures with different governance structures in place, but where the patterns of disaster aetiology in principle remain the same, will add texture to the examination. To this end, the Japanese Fukushima Daiichi nuclear power disaster, where avoidance, selective exposure and information rejection [38] and a deeply culturally anchored respect for authority and protocol outranked that of safety, will serve as an example.

Increasingly, governance involves non-governmental actors hailing from the private sector and civil society. Society, in its pursuit of social protection is converging to strengthen civil society, social capital and relational dynamics, through social innovation dynamics.

The argument will draw to a close by optimistically suggesting that social innovation fuelled by civil society, may prove a much needed agent of change in the creation of a new social narrative.

2 Complex Adaptive System Traits in Governance Theory

There is a discernible world pattern of disaster proneness and chain of events precipitating a disaster, ideally enabling us to treat the underlying root causes precipitating disasters, rather than the symptoms [42: 55]. Scholars have long come to realise, that although detectable, these patterns are not predictable, let alone linear [12: 311], and that these chains of events are compounded or diffused by the type and style of governance resident in a given context. This realization has spurred connections to complexity theory, which offers a far broader scope for analysing non-linear cause and effect. In particular, Complex Adaptive Systems (CAS) Theory has proven beneficial for increasing the understanding of governance in relation to disaster management and complex causality, as it includes a much greater range of variables [12]. Its multiple strands of complexity paradigms, self-organised criticality, social equilibrium and stability serve to advance theoretical analysis of social dynamics and policy [35].

While governance theory explains how various governance systems and the capacity of their buffering and amplifying abilities are critical in determining the impact of disturbances (i.e. disasters), and how multi-level governance coexist and interact across societal levels [12: 312], complexity theory aims to deepen our understanding of unpredictable systems with multiple temporary equilibria, self-organisation through integration and disintegration [9]. The latter does so without insisting that all aggregate outcomes should be fully understood, with futile efforts of delineating variables and causal effects through various paths of opportunity and abstracting away their interdependencies and non-linear interactions. Instead, CAS operates at an inherently multi-level of abstraction, because order is dependent on lower-level behaviours as part of their constantly shifting integrative cross-level foundation, making CAS models and ordinary causal models complementary, not rivals.

As such, there is no single component that dictates the collective behaviour of a complex adaptive system. Instead it is possible to focus on an agent in its local environment, as part of a system that self-sustains by importing energy, enabling agents within the system to self-organise, co-evolve and adapt to the environment over time, and through these transformations constantly create temporary equilibria, conceived by shifts in the patterns of interconnectedness [3: 219–220]. This is, as it turns out, very useful in the enquiry into the hidden sources of order [35: 116] embedded in governance and complexity theory, which for the purposes of this discourse will focus on the institutional and collective social behaviour and its impact of disasters.

The quintessential role of government is to provide broad social protection to its citizens, an obligation which includes catastrophic events [10: 336]. Yet, the classic symbiosis of governance, social protection and constituency is mutually dependent on 'the conflict between the stability-inducing role of institutions and their capacity to experiment, innovate, and learn from changing circumstances' [12: 319]. This institutional flexibility and robustness can be represented by the constant tension between *exploration* and *exploitation*. Exploration denominates the 'search, variation, risk taking, experimentation, play, flexibility, discovery, innovation', while exploitation is defined by 'refinement, choice, production, efficiency, selection, implementation and execution' [26: 71]. The strength of a governance system's capacity for exploitation and financial viability is contingent on cooperation, and depends on social acceptance of institutional rules, including norms of force, hierarchy, trust, network structures, reciprocity and belief-systems [12: 319]. Especially this *voluntary* acceptance of social control [13: 790] is a critical dimension in building trust and stability in relation to social protection in vulnerable communities.

Exploration is determined by a community's capacity to learn, experiment, trial-and-error new policies and institutional configurations, evaluate and to gather and analyse information. These processes are all known to be costly, both in terms of physical, monetary, human and social capital, rendering the capacity for exploration contingent on available resources and 'is reflected by the quality of its educational system and informational infrastructures such as the existence of independent universities, research institutes as well as in arenas for public debate and science-policy dialogues and unbiased mass media' [12: 320]. Intuitively, in the pursuit of predictability and stability, people seek to establish institutions and norms of reciprocity. But stability is accompanied by rigidity, and while it is necessary for improving exploitive activities and to raise overall welfare, it is counter-productive in terms of maintaining a flexible and dynamic society. Therefore, building on March's delineation of exploration versus exploitation, Duit and Galaz argue that *the adaptive capacity of a governance system* is a function of the trade-off between exploration and exploitation, rooted in the fundamental tension between the *mutually opposing needs for institutional stability and change* [12: 320], emphasis added). Accepting this realisation, assisted by a diagram dominated by two parameters: exploration and exploitation, four governance types can be distilled: rigid, robust, fragile and flexible, each representing the adaptive capacity within each governance system [12: 321–322].

Governance in societies with a high level of exploitation and a low level of exploration are considered rigid as it maximises stability, but lacks flexibility. As long as no surprises occur, this state-dominated, centralised governance model is the most efficient form, as it capitalises on the stability and predictability necessary for keeping transaction costs low. Japan is considered an example of a state-centric governance system [12: 320].

Societies with high levels of both exploitation and exploration are considered robust as it is equally apt to provide firm state governance, long-term transformation processes, and navigate sudden changes. This is 'an ideal state in which the rigidity-inducing effects of institutions are kept from obstructing necessary processes of exploration', but does not exist in its purest form. 'The robust governance type is the only governance type that has a sufficiently high level of adaptive capacity to be able to respond to all sorts of complex processes' [12: 321]. The closest proxy would be High Reliability Organisations (HROs) [12: 321] characterised by 'early detection of change, flexibility in decision making in combination with dense patterns of cooperative action, and the ability to reorganize' [22].

One may wonder why this pure form does not exist, and what it would take for it to come to life. The answer may be found in the mutually opposing tension—perhaps even mutually exclusive tension—between exploration and exploitation. The value of imminent danger and potential annihilation (nuclear power-plants and space programs) should also not be underestimated as a catalyst for introducing direction at all levels of governance. Clarity of the *purpose of compliance* in a population, may be one of the reasons why the robust governance style not only does not exist in its purest form among sovereign nations or societies, but why the examples or proxies are borrowed from organisational structures, constructed by the single purpose and motivation, with a well-defined goal such as safely running a nuclear plant, or launching a space mission—without the ambiguity of running a country.

In other words, the ability to reorganise at different levels due to the high element of network-based connections with apt decision-making capacity, risk of sub-optimisation or to compromise the overall goal is minimised as decisions taken at all levels of the organisational structure are guided by the pursuit of a clearly articulated goal. It is not a dictatorship—it is not a democracy either. It is a collective consciousness, guided by a clear goal and purpose of compliance. It may be naive to imagine that something as complicated as running a country could ever fall into this category, but it does not mean that one cannot speculate what it would take to at least find inspiration from the application of the purity of its principles.

Returning to the review of governance types; *fragile* denotes low levels of both exploitation and exploration. It is, by and large, traditionally observed in developing nations, as they face difficulties building institutional knowledge and capital due to high transaction costs. This compromises their capacity to reorganise and adapt to changes in the environment, with little resilience to buffer effects of shocks, inadvertently fuelling a vicious cycle, rendering them unable to 'achieve even moderate levels of economic development and human well-being' [12: 322].

Finally, the flexible governance system with high exploration and low exploitation, enjoys well-developed capacities for exploration (e.g., learning processes, feedback loops, monitoring schemes, resources, and capital) but is lacking in the capacity to transform these gains into economic growth and long-term opportunity. Flexible governance models are suggested to thrive in the welfare regimes in France, Germany, and the United Kingdom and have a certain level of capacity (incl. finances) and ability to adapt, yet hampered by their deep rooted individual and democratic heritage as the exploration factor lacks direction and is uncoordinated, and in the spirit of individualism, creating niche upon niche, thereby simultaneously presenting innovative dynamics and systematic organizational failure [12: 322].

Germany, however, might have distanced itself from this classification with a prominent production component in their sound national economy as well their adamant and broad support to the escalating refugee crisis in Europe during the course of 2015. It bears witness to the ability to adapt to a new situation, promptly. Whether their leadership by way of managing the refugee flows will have a long-term effect not only on Germany but also the rest of Europe, is yet to be seen. No doubt, the solution will be Trans-European, and perhaps this will be an opportunity (or risk, depending on the lens) to further merge the governance models in Europe, this time, with a clearly articulated common goal. Suffice to say, the unprecedented refugee streams into Europe in 2015 are likely to become a defining moment and a game-changer in European governance, the historic significance of which is yet to be fully realised.

The adaptive capacity within each of the four governance types, depends on the rate of change and the degree of predictability rendered in a conceptual space (refer to Fig. 1). Not surprisingly, the ideal (and possibly non-existing) robust governance

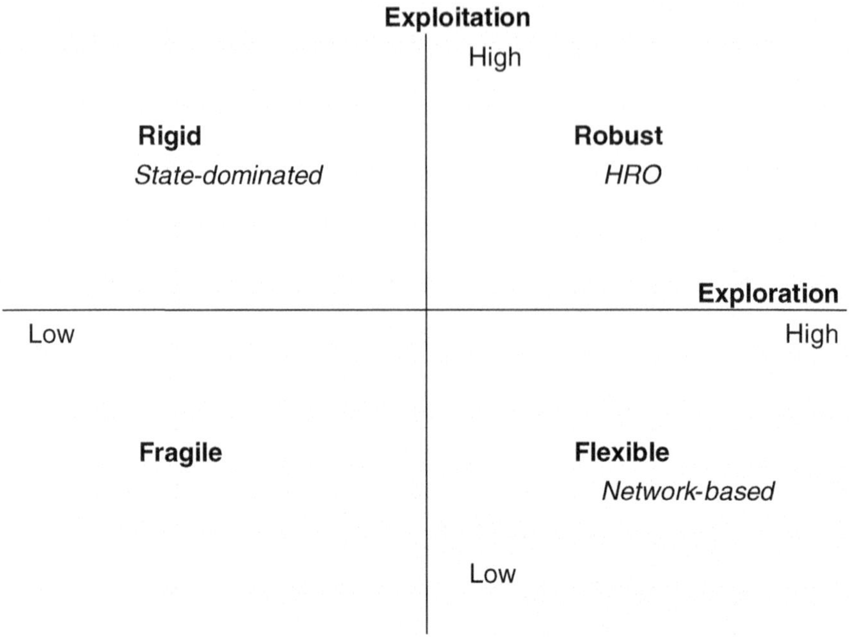

Fig. 1 Adaptive capacity of four governance types (*Source* adapted from Duit and Galaz [12] 'Governance and Complexity: Emerging Issues for Governance Theory', p 323. *Note* governance types added)

model can handle rapid/high rate of change with high unpredictability, steadily and without losing balance, as it is equally equipped to handle surprises, high impact and status quo. Meanwhile, the fragile governance system is equally unequipped to handle either, and will easily lose its equilibrium with even a low impact or small change [12: 323].

More interesting perhaps, are the conditions surrounding flexible—or Network-Based Governance (NBG)—as it, with its multiple governance levels, is able to harness changes and apply learning capacities and instant decision-making capacity, based on both formal and informal linkages with institutional diversity. While this type may be very efficient in responding to rapid changes (i.e. disasters), as long as the disaster is concentrated in a limited space where the informal and repeated social interactions can sustain themselves, it is less suited to convert this into long-term economic opportunity due to its low ranking on the exploitation dimension. It is worth noting that when dealing with a large-scale rapid onset disaster, the NBG model may lose its effectiveness as a disaster may demand quick, unilateral responses outside of the immediate vicinity of the NBG decision-making arena [12: 324].

This, however, is the State-Dominated Governance model's forte, given the legitimacy of the *democratic* state amongst its citizens and its ability to distribute powers both within different levels of government and civil society with accepted authority, it can be argued that it may provide for the stability of governance infrastructure to soundly navigate disasters [18]. Although it may be somewhat hampered by its limited capacity to deal with information deficits, time of travel of information to decision makers and biased information management once it gets there. Given the path dependency with its strong institutions and norms, this governance model is not well equipped to deal with novel and fast changes [12: 324], but theoretically well equipped to deal with disaster, with the understanding that decisions need to be taken abruptly perhaps even abrasively within the legitimacy of the state, to limit the impacts of disaster. The US federal emergency management system can be considered an example of a rigid governance type [14: 235].

With this combination of complex adaptive systems traits in governance theory, a framework for analysing disaster forensics emerges. With an infinite variety of possibilities, it is virtually impossible to predict an outcome of an event, as they are inter-twined and linked with an endless number of feedback loops and paths of destruction. As are the social adaptations—equally unpredictable—which further enable these capricious patterns. Through the case examples of Hurricane Katrina in 2005 and Fukushima Daiichi Nuclear Power Disaster 2011, it is suggested that governance, regardless of culture, is at the core of disaster forensics, and with better understanding to enable a more suitable governance, disasters can be avoided, or at the very least, the social impact reduced.

3 Hurricane Katrina, 10 Years on

Hurricane Katrina made landfall on the Gulf Coast August 29, 2005 and is regarded 'the most destructive natural disaster in American history' [41: 1]. Havoc and destruction followed in her path, with entire coastal communities obliterated by the storm surge, killing more than 1300 people. New Orleans was one of the worst hit areas. Its 350 mile levee system was stressed past breaking point with overflowing and breaching the levees, flooding the city, much of which is below sea level [41: 34–35]. The vast majority of the fatalities (80 %) hailed from the New Orleans Metropolitan area, many of whom were elderly or infirm [41: 8]. At the height of the disaster, approximately 80 % of New Orleans flooded, transforming Hurricane Katrina into a 'catastrophe within a catastrophe'; devastating the lives of countless residents and presented state and local officials with challenges vastly exceeding their capabilities [41: 36].

A mass evacuation was called during August 30, mainly due to massive flooding. Despite concerns, the Superdome stadium was opened for the general population as a shelter of last resort, and by midnight, 12,000 people had arrived [41: 29]. Footage of an overcrowded and uninhabitable Superdome, and startling images of desperate residents marooned on rooftops were broadcast [37: 3] and etched into the public memory. Emergency preparedness and plans were all put to the ultimate test, and fell short.

The wider American public responded by directing disappointment and frustration at the local, state and federal government and their apparent inability to respond effectively to the crisis. In the wake of 9/11 five years earlier, and the associated structural changes with the Federal Emergency Management Agency's (FEMA) move into the Department of Homeland Security, which was widely perceived to hamper FEMAs capabilities in dealing with natural disasters [14: 227], 'millions of Americans were reminded of the need to protect themselves and their families' [41: 1].

Ten years on, Hurricane Katrina, remains part of the public domain and consciousness. She has been extensively researched, and has virtually become the embodiment of complex causality. In the study of causal evolution, Katrina taught us that in the wake of catastrophe, the root cause can be altered before one's eyes in the name of re-examination [37: 6]. The complex causality of Katrina was the centre of extensive post disaster controversy, much of which was motivated by establishing culpability and with that, liability [37: 10], leveraging existing power structures and capitalist interests, proving once again, that disaster aetiology cannot be separated from public interest and politics. While it is not unusual for the causal chain to be refined and adjusted as the dust settles after impact, and the disaster aetiology forensics process and legal apparatus is engaged, it is unusual to see a dramatic causal shift.

In the immediate aftermath of Katrina, it was widely accepted that a massive surge of water caused by a hurricane engulfed the city [37: 7], attributing it as a *natural disaster*. Engineering examinations would later identify defective, breached levees 'catastrophic structural failure', as the cause of flooding (as opposed to the storm), essentially a geo-technical failure. Now attributed a *man-made disaster*, it was far easier to assign blame, which was directed at the Army Corps of Engineers [37: 12]. Finally, the controversy moved on to attribute the flooding as an *environmental disaster*, by virtue of the *Mississippi River Gulf Outlet Shipping Channel*, which, after 40 years of increased salinity was slowly eroding the protective environment. The wetlands and barrier islands were critical in protecting the coastal areas from a storm surge. Yet, long-standing political factors inhibited wetland and barrier restoration due to conflicting political and economic interests in the oil and gas extraction in the area.

This dissonance between ideology and policy resulted in erosion and stark decrease of the Louisiana wetlands and marsh, making the region more vulnerable to hurricanes [43: 24–25]. Less swamp and fewer trees were noticeable to the naked eye, while a more virtual depiction of its importance was that the shipping channel alterations in topography had provided for a 'hurricane highway' said to funnel water directly into the heart of the city. As with the levees, the Army Corps was kept accountable for the extended lack of maintenance of the shipping channel, resulting in erosion [37: 19, 43: 30].

Information management and public perception played a central role in the after-math of Katrina, as did governance, especially in the interaction between the different levels of government. As previously established, the adaptive capacity of a governance system is amplified or attenuated by the interplay between various levels of government. In theory, Katrina would have enjoyed the benefits of all of the above, except she didn't. Instead, the lack of structured governing arrangements exacerbated the lack of preparedness, impeding New Orleans' ability to respond, effectively exposing 'the failure of a nonregime' [6: 517]. But first, before examining these dichotomies, a review of Fukushima nuclear disaster, which like Katrina, had an incubation time of 40 years.

4 'Myth of Nuclear Safety' and Black Swans

A culture of complete nuclear safety had developed in the Japanese nuclear industry. Natural disasters were considered low risk, thus only limited resources were allocated to mitigating measures and disaster preparedness. And in accordance with classic group behaviour of minimising cognitive dissonance, information not conforming to pre-existing attitudes pro-nuclear power was avoided, ignored or distorted, ultimately contributing to a 'myth of nuclear safety' [38: 60].

This was a contributing factor to the lack of preparedness, when a powerful earthquake struck the east coast of Japan in March 2011. The earthquake generated a major tsunami, killing almost 20,000 people and causing multiple meltdowns at the Fukushima Daiichi nuclear power plant [2: 2], causing 'the worst nuclear disaster in history' [17]. Explosions at the nuclear power plant caused the release of radiation into the environment, affecting thousands of people living in the vicinity [2: 11], 150,000 of whom were evacuated. It was regarded by the Chairman of the subsequently established Nuclear Accident Independent Investigation Commission, as a 'profoundly manmade disaster—that could and should have been foreseen and prevented' [32: 9].

The Japanese nuclear industry was promoted with unwavering clarity throughout the past four decades, and was, for all intents and purposes, relatively accident-free. Substantial resources were invested in nuclear power compared to other sources of energy, more importantly, personal guarantees and social capital had been put forward in the promotion of nuclear energy. In Japan, personal endorsements are not offered willy-nilly, and certainly not at that level of government. The nuclear industry structure was institutionally flawed as the regulatory bodies did not have legislative power to implement safety measures, and the main regulatory body Nuclear and Industry Safety Agency (NISA), was part of the Ministry of Economy, Trade and Industry, which was responsible for promoting nuclear power [38: 60, 32: 9]. While these structural observations in themselves were concerning, the intimate, 'totally inappropriate' relationships between corporate and political sectors [32: 43], including authorities who were responsible for nuclear safety, further burdened the structural impartiality. In fact, it was so common for a retiring government official to accept highly paid jobs in the industry, that a term was coined to express this phenomenon of *amakudari*—or 'descent from heaven' [2: 27]. 'With such a powerful mandate, nuclear power became an unstoppable force, immune to scrutiny by civil society' [32: 9]. Needless to say, if it is unthinkable for a mere mortal to lose face in Japan, then let alone an affluent member of the business and society with an air of deity.

Collectively, the nuclear industry could not afford for nuclear power to fail. Not financially, not personally. Failure in either avenue would be detrimental to the identity of the industry, and it would be seen as 'losing face' which is unthinkable in Japanese culture, especially in the elite. This devotion to obedience and reluctance to question authority [32: 9] resonated with the state-centric governance system, which fuelled predictability and stability, at the cost of flexibility, which became apparent in the subsequent disaster forensics and investigatory process.

A common misperception is that an accident-free track record is an indicator of safety. One problem with that is, that systems are inadvertently inhibited with latent failures or 'resident pathogens' [34: 74] thus it does not take into account the unrecorded adaptations, adjustments and tweaks constantly applied to keep an often imperfect system operating safely [34: 84], or inadequate tolerance built into the system of an external event. This is true, regardless of culture and governance structures, where the patterns of disaster in principle remain the same, and where predictable chains of error are left unattended, rendering the organisation unable to detect that an incremental mistake is compounding [29]. It can be the result of *taught oblivion* (much like group think, with a cultural dimension), or simply a flawed safety-culture.

Therefore, the operator of the Fukushima Daiichi Nuclear Power Plant, Tokyo Electric Power Company's (TEPCO) argument that 'severe accidents only occur every 100 years, and the lifespan of a reactor is shorter than that' [38: 60] as TEPCO's justification for refraining from implementing recommended safety procedures [32: 28], seems not only flawed, but alarmingly naive. And that is coming from a high risk industry, which is trusted nationally and internationally to honour their responsibility of constantly testing and adapting their high risk systems to changes in the environment, and most importantly: relentlessly and systemically challenging the system for weaknesses and latent errors to expose blind spots or 'black swans', effectively expanding the risk horizon and realm of possibility [27: 330], thereby stimulating the chances of *active foresight* [40: 65].

Perrow maintains that no matter how hard we try, accidents cannot be prevented. At best the frequency can be reduced [33]. High Reliability Theory complements in its continuous quest in search of systemic weaknesses, stemming latent errors before causing disruptions and seeking to de-couple elements and allow for more flexibility [23]. This view is supported by Reason, who discerns that it is not the pursuit of excellence which will bring the best outcomes, but instead an ability to detect and correct mistakes en route, within the given level of flexibility [34: 97], essentially navigate rather than control.

This is exactly what Complex Adaptive Systems theory promotes, although with many more variables. But a prerequisite for this to happen is that the governance allows for this flexibility. During Fukushima it did not. Let alone Katrina.

5 Navigating the Social Dimension of Disaster Forensics

Hurricane Katrina, without question, impacted the lower levels of society at a far greater rate than the more affluent members. In the case of Fukushima, vulnerability was impacted by filtered information flows, biased signals and a collective disregard or acceptance of risk. How is this reflected in governance? And how can governance address these issues? Typically too great an emphasis is placed on the scrutiny of the natural hazard itself, with only fleeting attention to the underlying sociological root causes that create vulnerability to natural disasters, and how social initiatives engage community and civil society [44: 18–20]. While it pre-dates both Katrina and Fukushima, it is a fair and very applicable observation, especially when contrasted with the controversies surrounding Fukushima, which was afforded a far more nuanced social lens in the theoretical deconstruction of its disaster aetiology.

In the late-modern, reflexive risk society, socio-technical and industrial disasters have become the norm [4]. While the environment is steadily turning more volatile and unpredictable, the public expectations for social protection has increased, and the tolerance for disasters and impact diminished, approaching zero. In other words, the two poles are headed in extreme opposite directions, leaving little room for reconciliation. Reflexivity has so far been the coping mechanism, 'keeping society honest' through an equilibrium held in check by multiple strands of constant opposing tension. While this has been the cautiously optimistic accepted locus, the question is now, whether a new state can be born from this condition that may inspire social innovation.

The increasing frequency of disasters is accompanied by an increase in the vulnerability of the population at risk, thus the social causes of disaster commands attention, especially underlying vulnerabilities and predispositions to disaster associated with rapid population growth and population density [42: 53–54]. Disaster vulnerability is a social construct. It is the economic and political power disparities between groups, disparities in the distribution of assets (i.e. knowledge and information) and disparities in social protection (i.e. disaster relief and recovery resources). Inequities in societal arrangements are likely to replicate themselves during and after disaster events and makes addressing those inequities a difficult political proposition and effective preparedness often is hampered by political and behavioural constraints [14: 237], a basic vulnerability premise easily applied to Katrina. According to Kasperson, the way risk is *collectively* experienced is referred to as the social amplification of risk. It denotes the social structures and group behaviour that shape the *perception* of risk, how they result in individual and collective responses and their effect on community, society and economy [21: 101]. This means that vulnerability is spawned by social, economic and political processes; all of which influence how hazards affect people in varying ways and with differing intensities [44: 7]. Governance in its purest form is able to address these issues, and inherent in all governance types, is an element of social amplification—or attenuation—of risk, depending on the gearing. Risk communication and information flows are integral parts of forming an individual or group's perception, as they can intensify or weaken the information available, and filter what is attributed to a risk, essentially distorting the signal [21: 102–105]. Thus, the social amplification of risk along with social protection provides a framework for analysing the strength and type of governance, in relation to disaster aetiology.

The impact of Hurricane Katrina devastated the poor neighbourhoods of New Orleans—especially in the Lower Ninth Ward. Evidently, in terms of lives at peril, the residents unable or unwilling to leave were at greatest risk. While this seems rudimentary, a large element of risk perception, which is framed by the social amplification of risk lens, determined the actions of the population. Although free busses and basic accommodation were made available, residents stayed for a number of reasons, including being unable to leave due to unavailability of money, transportation and simply no place to go, as well as lack of clear guidance [36: 516]. There was a fear that by leaving, they would be left with nothing and due to poverty or old age would not be able to start afresh, either in another place, or in New Orleans. There was also a sentiment that by staying, they would be able to better protect their homes and belongings. These fears may have been powerful enough to drown out—or attenuate—the perception of the real risk. Naturally there was a portion of the population that simply stayed as they had an emotional attachment to their homes, and decided to stick to their homes, no matter what. Some people changed their mind last minute, for some, too late.

Many factors influence a population's propensity to accept risk. Where populations are poor, uneducated and uninformed, their daily struggle for subsistence would greatly influence their propensity to accept risk—either knowingly or unknowingly. Adams' risk thermostat outlines that everyone has a propensity to take risk, which varies from person to person and individual risk-taking decisions represent a balancing act against *perceived* danger, outcome expectancy along with the consequential rewards or accidents, and the severity of each. Add to this a cultural filter which denotes the characteristics or rationalities from cultural theory: fatalist, hierarchist, individualist and egalitarian [1: 42–45].

These elements all in turn represent the balancing act that influences each individual choice and response or in some cases lack of response as it may be. The perception of risk, taking into account an element of controllability and personal influence on the outcome, will inherently influence the decision making process and thereby the tolerance and in turn: the risk thermometer. A risk adverse culture increase risk seeking behaviour once commitment to a course of action has begun, especially in group decisions as those in favour of risky decisions tend to be more committed to create a consensus which is in favour of their position, which increases the social pressure of conforming to the dominant position [38: 60], and thus reinforcing a bias.

The motivation behind information avoidance, selective exposure to information, ignoring or rejecting certain kinds of information may be to minimise 'cognitive dissonance' [7], caused by information not aligning to the existing worldview. Information avoidance and attenuation of risk may even serve as coping mechanisms towards an incomprehensible level of risk and can foster serious adverse consequences from underestimation of risk in the shape of lack of safety precautions or mitigating response [21: 102–103], as was the case during Hurricane Katrina. In other cases, such as Fukushima Daiichi, the disregard of risk may simply be born out of ignorance, enabled by a culture of reflexive obedience and insularity [32: 9], fostering a *taught oblivion.*

In line with Kasperson's observations of collective social behaviour in relation to risk communication 'Information behaviour plays an important role in information failures', particularly of interest are the barriers to either releasing, seeking or acknowledging information. Information behaviour will be affected by coping strategies in relation to stressful situations which will impact the perceived risk and usefulness of the information available [38: 57]. Affective Load Theory digs deeper into these risk communication behaviours and how they manifest collectively and culturally. It posits that by virtue of belonging to a cultural group, people are bound to develop 'learned affective norms' which influence how information is perceived, i.e. which cognitive and emotional (affective) strategies are employed [31: 191].

Information behaviour is habitually organised into patterns to cope with situational requirements. These are socially created and shared, and taking into account social and political values, paving the road for the establishment of 'learned affective and cognitive norms'. This helps create a reference point or inventory of jointly held attitudes, against which information is validated, used, avoided or ignored, which in turn reinforces the information behaviour [38: 58]. When these norms are optimistically geared, it is likely that a variety of different search strategies will be engaged to openly source information; however, when pessimistically geared, only limited information searching strategies will be employed and information behaviour will be rigid [30: 194].

6 Social Innovation as an Agent of Change?

The welfare state is undisputedly undergoing a severe crisis. While the underlying causes of lower productivity, erosion of 'normal' employment regimes and demographic changes including an ageing population have been widely accepted as underlying reasons, a less acknowledged factor is the erosion of the *moral* foundation. In step with the ongoing process of social individualization and financial independence, individualist ethics have followed in its wake [15: 2008], in turn manifesting itself in governance through political, intellectual and moral leadership [20: 455]. And it would have to, as it has become evident that the state-centric constellation with the custodianship for post-event relief insurance creates a fundamental dilemma with its disincentive to voluntary efforts for pre-disaster risk reduction [11: 3]—a paradigm that has been challenged past the point of return by civil engagement and social innovation dynamics.

Inherently, in any governance system, vulnerabilities and pre-dispositions to disaster are related to administrative and institutional arrangements [42: 54]. Add to that a discernible pattern of disaster proneness [42: 55], which is compounded by the type and style of governance in a given context. In terms of social amplification of risk, it is clear that marginalised groups tend to concentrate in poor quality housing, separated not only spatially, but also socially. Given housing relies on free market economy processes, houses in low-lying flood-zones are more affordable, rendering the state unable to control an equitable socio-spatial distribution of households [15: 2005–2012]. Vulnerabilities are reinforced, as the population in these areas experience a higher affinity of families with limited coping strategies and absorption capacity for shocks.

Hurricane Katrina primarily suffered from the ripple-effects of these social inequities, secondarily by the risk communication in the economic and political agenda which demonstrated that the cause and effect of a disaster cannot be separated from perception—and by managing information and feedback, it can be willed into existence. Meanwhile, in the case of Fukushima, this was the exact opposite. Here, while vulnerable families were affected, the distortion of risk communication filtered through a deep-rooted cultural notion of infallibility was considered to have caused more damage than that of social inequities, city planning and zoning.

Both Katrina and Fukushima were governed at several levels (state, local and civil society) with a combination of robust state-legitimacy and a flexible network-based governance with local application of civil society and law. Theoretically, this combination of governance types would provide for buffering and amplifying characteristics, ideally assisting in navigating these disasters. Nonetheless, due to longstanding precipitating root causes, in the case of Katrina, environmental erosion across decades, and a nesting 'myth of nuclear safety' in the case of Fukushima, evidently, insufficient buffering capacity was available within the governance apparatus, stifling the responses. The general public was exposed, needless to say, the vulnerable part of the public, even more so.

It has been deduced that the society ideally equipped to withstand and navigate disasters would be administered by network-based-governance with high-reliability traits. Regardless of culture. Problem is, it does not exist. At least not at a sovereign state level. Then what is the solution? Is it possible, in a risk society where disasters are becoming more volatile and more frequent [4], rendering the ability to stay accident free is diminishing by the hour, and where the public expectation for social protection is challenging the status quo, that the answer is to be found in the application of complex adaptive systems? They do not rely on a central controller, but instead energy imported by independent agents, and self-organisation crystallising into patterns of regularity [3: 221]. Social (voluntary) entities thrive and self-organise as long as their members continue to contribute to work. In this milieu, informal—often persistent—structures emerge. The more turbulent the environment, the more energy is required to uphold a critical level of system or organisational sustenance [3: 222]. How can sufficient energy consistently be introduced into a massive civil protection system, and still remain nimble?

Perhaps the solution will emerge from society itself. Not as a revolution, but as an order spontaneously forming within larger scale systems, spurred by energy, born of social innovation. Bearing in mind that a defining feature of complexity is that self-organisation is a natural consequence of interaction between at least two simple agents [3: 222], is it really that unthinkable that a public, reminded of the need to protect themselves and their families—without alluding to anarchy—in free and voluntary association has decided to engage? To spontaneously converge and contribute with their level of energy, investing in civil society, creating yet another equilibrium.

Social innovation has an explicit positive connotation and is fuelled by the desire to improve the quality of life of individuals and communities through new ideas developed to fulfil unmet social needs, either underserved or uncovered by services traditionally provided by the state [28: 32]. It builds on the principle of strengthening civil society with a moral foundation fuelled by the purity innovation, entrepreneurship and social capital. Without bordering on the naive, social innovation follows the principle of continuity, gradually optimizing the condition of society powered by optimism [24], in this context with the dimensions of individualism and civil society bridging the two. Building on governance theory, social innovation is, without a doubt, faring well in the exploration dimension [26: 71], and time will tell if it will find itself restricted from the lack of exploitation—or if a new equilibrium can be created, one that overcomes this dilemma.

Increasingly, governance involves non-governmental actors hailing from the private sector and civil society. Simultaneously, social innovation in governance relies on network-based relations and ties across fragmentations which can all be found in the labour market, the political, cultural and civil society, creating a hybrid institutional arena, by which actions and identities are formed and structures shaped [15: 2015]. Although civil society is far from a homogeneous entity, it cannot be considered independently from its historical and cultural context and the prevailing values, as the spirit and ethics of society is informed by a common reference to a collective experience [16].

Add to that, resilience: an indicator of the capacity to endure the impact of disaster, to cope and rapidly recover. Resilience has an *inherent* component, which is the classic delineation of the ability to with-stand disasters without major disruption; and an *adaptive* component, which refers to the ability to adapt, improvise, and access resources after disasters in order to cope and eliminate uncertainty.

It is relevant to recognise the drivers behind resilience, as it cannot be separated from wealth, social and cultural capital and political influence [39: 121–122]. Even then, it witnesses of a time where sociological and socioeconomic dimensions of community vulnerability and resilience has been given a far greater scope and promise of societal creativity in the inception of coping mechanisms. It would appear that the individualism has finally peaked, leaving space for a more collectivistic approach. For instance, vulnerability is increasingly becoming about 'dealing with the awkward issue of poverty in society' [44: 56].

Social divisions in society are changing their character, and vulnerability is no longer equated to a one-dimensional distributional notion of poverty and disposable resources [15: 2005–2010], although liquidity and social capital are fundamentally different, poverty in its traditional sense, has been known to dampen the ability to contribute and participate alongside more resilient members of society. Nonetheless, a multi-facetted form is emerging, where social exclusion is considered in terms of *relational* dynamics such as 'inadequate social participation, limited social integration and lack of power' [15: 2010], ultimately resulting in *disaffiliation* [8: 2010], non-integration into social and institutional relations and the *absence of interdependence*, and with that, non-participation and non-affiliation with various dimensions of social life [15: 2005–2010].

It is interesting to note, how this constellation of social exclusion is easily reconcilable and traceable within complex adaptive systems theory, as the concerns raised in terms of *lack of participation* and *lack of interdependence*, seems the exact antithesis to a complex adaptive system, thriving on introduction of free agents and energy. The motivation for this reorientation of terminology, the optimist may wonder, is whether this is due to a desire to soften the blow or to take the stigma out of 'poverty', essentially removing the shame, with an understanding that poverty is not the same as unable to contribute. Certainly, a non-denotational treatment of poverty as an equaliser for social *inclusion* will provide for an important step, from a social innovation perspective.

It may even be, that in the wake of this intentional societal change, with its adaptive, dynamic and non-linear nature [5: 610] a measure of societal self-esteem [19: 51] is emerging. A value system with a consciousness that thrives on reflexivity, and perhaps even outgrows the transient anxiety of the risk society [4]. Either way, social innovation, with its distinct element of exploration, is likely to be part of the solution.

7 Conclusion

Clearly, understanding root cause and complex causality of disasters is not without difficulty. The complex adaptive systems perspective enhances the analytical leverage of governance theory by acknowledging a much greater variety of systems behaviour, with its application of multiple, simultaneously moving systems equilibria [12], energised by their thriving inter-dependencies.

On this basis, it has been demonstrated that a centralised governance system enjoying the legitimacy of the state, combined with a flexible network-based local administration, with informal network-based qualities, in principle, would provide for a robust emergency response. Through the examples of Hurricane Katrina and Fukushima Daiichi nuclear disaster, it has been deduced that in reality, this is not always the case. Instead, supported by complexity theory, a society ideally equipped to withstand and navigate disasters would be governed by network-based-governance with high-reliability traits [12]. Although this form does not exist at a sovereign state level, in its quest for reliable social protection, the risk society may, inadvertently, already be in wild pursuit of a solution through social innovation. It is only a natural progression, given the realisation that the state-centric system no longer is able to offer social protection on par with the expectations of a late-modern constituency. Instead, a reorientation towards a collective civil society, social inclusion emerges [15], as for most humans, safety comes in numbers. Or at the very least; the illusion of safety.

The contemporary risk society [4] will need to learn to navigate a rampant risk-infused reality, with increasing aptitude. Perhaps this is the reflexivity the late Ulrich Beck was *really* referring to. It will be the gauge of the success of our collective global societal resilience. We will need to constantly bounce, search for weaknesses, always with a fresh perspective—and accept that accidents *will* happen [33]—the key is to minimize the *impact*. Turning black into grey. That comes down to (safety-) culture, politics and essentially: governance.

Social innovation, with its vigour and energy with an inherent positive connotation [28], extends the scope for new ideas in this societal state of being. It is key to creating a new condition in which social divisions, vulnerabilities and resilience are considered from a far more nuanced vantage point—one that builds on the principles of both reflexivity *and* self-organisation. This time, in the creation of the social narrative, by virtue of evolution by inclusion, it might even replace the methodical scepticism with optimism.

References

1. Adams J (1995) Risk. UCL Press, London
2. ANS—American Nuclear Society (2012) Fukushima Daiichi ANS Special Committee Report. American Nuclear Society, Illinois
3. Anderson P (1999) Complexity theory and organization science. Organ Sci 10(3):216–232
4. Beck U (1992) Risk society: towards a new modernity. Sage Publications, London
5. Boyatzis R (2006) An overview of intentional change from a complexity perspective. J Manag Dev 25(7):607–623
6. Burns P, Thomas M (2006) The failure of the nonregime: how Katrina exposed New Orleans as a regimeless city. Urban Aff Rev 41(4):517–527
7. Case D, Andrews J, Johnson D, Allard S (2005) Avoiding versus seeking: the relationship of information to avoidance, blunting, coping, dissonance and related concepts. J Med Libr Assoc 93(3):353–362
8. Castel R (1995) La me´tamorphose de la Question Sociale. Fayard, Paris
9. Cham K, Johnson J (2007) Complexity theory: a science of cultural systems? M/C J 10(3)
10. Comfort L (2005) Risk, security, and disaster management. Annu Rev Polit Sci 8:335–356
11. Daniels R, Kettl D, Kunreuther H (eds) (2006) On risk and disaster: lessons from Hurricane Katrina. University of Pennsylvania Press, Philadelphia
12. Duit A, Galaz V (2008) Governance and complexity: emerging issues for governance theory. Gov Int J Policy Adm Inst 21(3):311–335
13. Foucault M (1982) The subject and power. Criti Inq 8(4):777–795
14. Gerber B (2007) Disaster management in the United States: examining key political and policy challenges. Policy Stud J 35(2):227–238
15. Gerometta J, Haussermann H, Longo G (2005) Social innovation and civil society in urban governance: strategies for an inclusive city. Urban Stud 42(11):2007–2021
16. Gosewinkel D, Rucht D (2004) 'History meets sociology': 'Zivilgesellschaft als Prozess'. In: Gosewinkel D, Rucht D, Van Den Daele W, Kocka J (eds) *Zivilgesellschaft - National und Transnational*. Edition Sigma, Berlin, pp 29–60

17. Gutierrez D (2015) Fukushima disaster caused at least 1,232 fatalities last year as radiation death rate accelerates. Natural news. Available online at: http://www.naturalnews.com/049277_Fukushima_disaster_radiation_deaths_thyroid_cancer.html; accessed 7 December, 2015
18. Hirst P (2000) Democracy and governance. In: Pierre J (ed) Debating governance: authority, steering, and democracy. Oxford University Press, Oxford, pp 13–35
19. Jessen D (2015) A case study of governance in relation to private space exploration, exemplified by Mars one: private one-way colonisation of Mars. London Institute for Space Policy and Law. Available online at: http://www.space-institute.org/app/uploads/1448556892_Post_this_Dissertation_Dorte_Jessen__Mars_One_February_2015.pdf. Accessed 27 Nov 2015
20. Jessop B (2002) Liberalism, neoliberalism and urban governance: a state-theoretical perspective. Antipode 34(2):452–472
21. Kasperson J, Kasperson R (2005) The social contours of risk, volume I: publics, risk communication and the social amplification of risk. Earthscan, London
22. La Porte T (1996) High reliability organizations: unlikely, demanding, and at risk. J Contingencies Crisis Manag 4(2):60–71
23. La Porte T, Consolini P (1991) Working in practice but not in theory: theoretical challenges of 'High-Reliability Organizations'. J Public Adm Res Theor 1(1):19–48
24. Leibniz G (1991) [1686] Discourse on metaphysics and other essays: on the ultimate origination of things. Hackett Publishing Company Inc, Indiana
25. Manson S (2001) Simplifying complexity: a review of complexity theory. Geoforum 32 (3):405–414
26. March J (1991) Exploration and exploitation in organizational learning. Organ Sci 2(1):71–87
27. Masys AJ (2012) Black swans to grey swans: revealing the uncertainty. Disaster Prev Manag Int J 21(3):320–335
28. Matei A, Antonie C (2015) Complexity theory and the development of the social innovation. Soc Behav Sci 185:61–66
29. Mittelstaedt R (2005) Will your next mistake be fatal? Avoiding the chain of mistakes that can destroy your organization. Pearson Education Inc, Upper Saddle River
30. Nahl D (2004) Measuring the affective information environment of web searchers. In: Proceedings of the 67th ASS&T annual meeting, vol 41(1), pp 191–197
31. Nahl D (2009) Affective load. In: Fisher K, Erdelez S, Mechnie L (eds) Theories of information behaviour. ASIST, New Jersey, USA, pp 39–43
32. NAIIC—Nuclear Accident Independent Investigation Commission (2012) The official report of The Fukushima Nuclear Accident Independent Investigation Commission. Available online at: http://reliefweb.int/sites/reliefweb.int/files/resources/NAIIC_report_lo_res2.pdf. Accessed 15 Dec 2015
33. Perrow C (1984) Normal accidents: living with high-risk technologies. Princeton University Press, Chichester
34. Reason J (2013) A life in error: from little slips to big disasters. Ashgate Publishing Limited, Surrey
35. Room G (2011) Social mobility and complexity theory: towards a critique of the sociological mainstream. Policy Studies 32(2):109–126
36. Schneider S (2005) Administrative breakdowns in the Governmental response to Hurricane Katrina. Public Adm Rev 65(5):515–516
37. Shrum W (2014) What caused the flood? Controversy and closure in the Hurricane Katrina disaster. Soc Stud Sci 44(1):3–33
38. Thatcher A, Vasconcelosa A, Ellis D (2015) An investigation into the impact of information behaviour on information failure: the Fukushima Daiichi nuclear power disaster. Int J Inf Manage 35(1):57–63
39. Tierney K (2006) Social inequality, hazards, and disasters. In: Daniels RJ, Kettl DF, Kunreuther H (eds) On risk and disaster: lessons from Hurricane Katrina. University of Pennsylvania Press, Philadelphia, pp 109–128

40. Toft B, Reynolds S (1994) Learning from disasters: a management approach, 3rd edn. Perpetuity Press Ltd., New York
41. Townsend F (2006) The federal response to Hurricane Katrina: lessons learned. Available online at: http://library.stmarytx.edu/acadlib/edocs/katrinawh.pdf. Accessed 4 Dec 2015
42. Turner B (1979) The social aetiology of disasters. Disasters 3(1):53–59
43. Van Heerden I (2007) The failure of the New Orleans levee system following Hurricane Katrina and the pathway forward. Public Adm Rev 67:24–35
44. Wisner B, Blaikie P, Cannon T, Davis I (1994) At risk: natural hazards, people's vulnerabilities and disasters. Routledge, Oxon

Disasters and Mishaps: The Merits of Taking a Global View

Simon Bennett

Abstract On 24 March, 2015, Andreas Lubitz, the First Officer of Germanwings Flight 4U9525, committed suicide by aircraft. Following the disaster there was a 'rush to blame', with Lubitz painted as the sole villain. Few reviewed the wider circumstances. While accepting the primacy of Lubitz's actions in the destruction of Germanwings Flight 4U9525, this chapter scans the horizon for contributory factors. In doing so it demonstrates the contribution systems-thinking can make to understanding failure in complex, transnational socio-technical systems (such as commercial aviation). The chapter offers a counterweight to the fundamental attribution error. It proffers an antidote to blamism. It references the work of Ross, Reason, Turner and Fiske and Taylor. While blame and punishment satisfy our baser instincts (the urge to hurt those who have hurt us is hard to resist), they generally undermine safety. The chapter argues that, from a safety standpoint, blamism is an inappropriate response to mishap and disaster.

Keywords Reductionism · Systems-thinking · Actor-network-theory · Holism · Disasters · Germanwings flight 4U9525

1 Introduction

There is an understandable human need to identify those responsible for death and injury [16, 33]. Too often, however, that need manifests as an ugly and ill-informed witch-hunt. While a witch-hunt may satisfy our desire to discern agency, it often undermines efforts to avoid a repeat. Focusing on a single action may obscure the underlying causes of mishap and disaster. Blamism may lead to the misdirection or inadequate resourcing of remediations, making it more likely that the failure will reoccur. Reductionism (simplification) is the enemy of safety.

S. Bennett (✉)
Civil Safety and Security Unit, University of Leicester, Leicester, UK
e-mail: sab22@leicester.ac.uk

A.J. Masys (ed.), *Disaster Forensics*, Advanced Sciences and Technologies for Security Applications, DOI 10.1007/978-3-319-41849-0_7

For safety's sake, it is wise to identify not only a disaster's immediate causes (like a flight crew's failure to de-ice), but also any factors that may have contributed to the undesired outcome (like a lack of ground-support equipment, schedule pressures, poor training or flawed CRM). Pioneered by the Honourable Mr Justice Virgil P. Moshansky in his ground-breaking investigation (Fig. 1) into the 1989 Dryden Fokker F-28 accident—where a crew's failure to de-ice cost the lives of 24 people—the holistic, inclusive, systems-focused approach to accident investigation ensures that nothing is missed. Speaking in 1992, Moshansky claimed that Dryden "… was the result of a failure in the air transportation system". In Professor James Reason's [56–58] argot, Dryden was a 'systems accident' that originated in the spaces and interactions between the aviation system's myriad social, economic and political components (personnel, equipment, training regimes, rules, laws, free-market competition, shareholder and investor agendas, etc.).

Systems-thinking recognises the interactive complexity and opacity of modern socio-technical systems (like nuclear power generation, global finance and

Fig. 1 The Moshansky Report [50]: genesis of the holistic, systems-thinking-informed approach to air accident investigation

aviation). It posits that output-oriented systems (like the socio-technical system that delivers air service across international borders) are purposive aggregations of actants (rules, regulations, beliefs, cultures, financing, workers, managers, training, tools, machines, computer software, etc.). In the argot of actor-network theory (ANT), output-oriented systems are 'hybrid-collectifs', the product of 'heterogeneous engineering' [11]. Actor-network theory reveals how actants are bent to a purpose—for example, the provision of loan finance or cancer care, creation of National Parks, or delivery of air service across international borders. It reveals how network authors (animate and inanimate) 'bend space around themselves' [12].

Crucially, from a safety-assurance standpoint, systems-thinking recognises that complex, output-oriented systems behave in unexpected ways. Non-linear interactions—where large inputs generate unexpectedly small outputs, small inputs generate unexpectedly large outputs and, through time, identical inputs generate qualitatively different outputs—render the behaviour of complex output-oriented systems unpredictable: "In complex systems, outcomes are often emergent, and not simply a result of the performance of individual system components. Hence system behaviour is hard to understand and often not as expected" [60]. Emergence obtains when a system "exhibits behaviours that cannot be identified through functional decomposition" [34].

The risk inherent in a complex system "arises not from a singular cause but from … interactions at the systemic level" [48]. Hollnagel [32] concludes: "[A]ccidents [should be] seen as emerging phenomena in complex systems … the result of an aggregation of conditions … ". Dekker [19] observes: "[I]t is critical to capture the relational dynamics and longer-term socio-organisational trends behind system failure". Unpredictable (emergent) behaviour may accompany a system modification/upgrade: "In general, emergence is problematic because its behaviours are unexpected. In telecommunications this is called the feature-interaction problem, in which adding features to a system produces new behaviours not included in any single feature. In system safety, emergence causes two types of problems. The first occurs when interaction among components produces hazardous behaviours that are previously unidentified …. The second problem occurs when interaction among components produces emergent behaviours that violate a known system safety goal … " [8].

2 What Does a Reductionist Analysis Look like?

A reductionist analysis explains failure in terms of discrete actions (like a flight crew's decision to depart without de-icing their aircraft). According to Horlick-Jones [33], society's desire or, as he puts it, "need" to allocate blame, encourages reductionism. According to Cook [16], the tendency to look for a 'root-cause' speaks to "… the social [and] cultural need to blame specific, localised forces or events for outcomes". In the aftermath of disaster, the social and cultural need to individuate responsibility may give rise to an unedifying, demeaning and cruel 'blame-game'.

2.1 Reductionism in the 1950s—the Munich Air Disaster

In 1958 a British European Airways (BEA) Airspeed Ambassador (Flight 609) crashed on take-off at Munich Airport killing twenty-three and injuring nineteen (Fig. 2). Some of the dead and injured played for Manchester United Association Football Club. In a tense post-disaster atmosphere the German authorities alleged that the crew had failed to clear the wings of ice. It was subsequently found that the Ambassador's failure to achieve take-off speed was caused by runway contamination (slush). As the only surviving pilot, Captain James Thain bore the full weight of German opprobrium: "The politically inspired scapegoating of Captain Thain following the 1958 Munich air disaster served no constructive purpose. Indeed, because it obscured one of the underlying causes of the disaster (the impact of runway contamination on take-off rolls), it subsequently exposed passengers and employees to extra risk .... It was only through the efforts of Thain's own family ... that the risk was eventually understood. It was too late for Thain, however, who never flew again ... " [3]. Even though the British government believed Thain's account, he had to wait until 1969 to be publicly exonerated. The desire to secure a post-World War Two political and cultural rapprochement between Britain and Germany made Thain a pawn of international diplomacy: "Recently declassified British files show that, while the authorities privately took Thain's side all along, they did not exert more pressure in order to avoid embarrassing the Germans in the fraught postwar atmosphere" [40]. Anglo-German politicking destroyed Thain's career. What does a reductionist analysis look like? In the case of the victimisation of BEA's Captain James Thain, ugly.

Fig. 2 The hulk of Flight 609 in which several Manchester United players died

2.2 Reductionism in the 1980s—Kegworth

On the evening of 8 January, 1989, at an altitude of around 28,000 ft, the number one (left-side) engine of British Midland Boeing 737-400 G-OBME began to disintegrate. The outer nine inches of the number one engine's fan blade number 17 broke away, lodging itself shrapnel-like in the acoustic panel of the intake casing. Out of balance, the engine began to vibrate. At the same time, frequent, high-speed contacts between the remaining blades and the engine's abradable seals caused smoke and fumes to enter the aircraft cabin. Subsequent compressor stalls produced loud thumping noises and flaming that could be seen from the cabin. According to Brookes [9], "No. 1 engine vibration indicator [reached] the top of its [five point] scale within two seconds of the onset of vibration, and [remained] there for about three minutes (until after beginning the descent)".

The Boeing 737-400's engine vibration indicators (Fig. 3) are small. According to Brookes [9] they are about the same size as a British twenty pence piece. Further, unlike those installed in the earlier Boeing 737-300, they use electro-luminous [29] display technology: "Criticisms would later be made of the dials' … edge-lit symbology, rather than old-fashioned electro-mechanical pointers" [9]. Believing the vibration and smoke to be emanating from the number two engine, the crew performed an in-flight shut-down. Purely by chance, at the moment the number two engine was shut down, the number one engine momentarily stabilised, reinforcing the crew's belief that the problem lay with the number two engine. (A false positive reinforced the crew's misperception). Some of the passengers and crew knew otherwise, having seen fire and sparks emanating from the number one engine. Unfortunately, this information was kept from the flight crew. Today, intra-crew communication is a major theme of airline training [6].

With the aircraft flying on a disintegrating engine, all on board G-OBME were in jeopardy. The number one engine stopped generating thrust at 900 ft, causing the aircraft to crash-land short of Runway 27 at East Midlands Airport. The 737 came to rest on the western embankment of the M1 motorway, the fuselage splitting into three. Thirty-nine passengers died on the night. A further eight died later. Many of those who survived suffered broken backs, fractured skulls, brain damage and post-traumatic stress. In 1998 one traumatised survivor successfully sued British Midland for £57,000. Captain Kevin Hunt was trapped in his seat for more than two hours. His spinal and leg injuries consigned him to a wheelchair for several years. First officer David McClelland spent several months in hospital (Fig. 4).

After it was revealed that Hunt and McClelland had shut down the working engine, both men were effectively sacked: in October 1990, British Midland told Hunt that he was being compulsorily retired; at about the same time, McClelland's contract was not renewed [9, 64]. Captain Hunt made a number of observations in his defence. Hunt claimed, for example, that experience had taught him to treat engine vibration readings with scepticism. In his opinion engine vibration sensors were unreliable.

Fig. 3 The 737-400's primary and secondary engine displays. The engine vibration indicators are *arrowed*

2.2.1 Proximate Causes

It was suggested that airframe vibration would have made it difficult to read the twenty pence-sized engine vibration dials: "[A] high level of vibration in the aircraft … would have hindered reading the secondary engine instruments" [29]. It was further suggested that the use of 'edge-lit symbology' (where a small needle described an arc around the dial) made the vibration dials difficult to read. The 737-400's predecessor, the venerable 737-300, shipped electro-mechanical engine vibration dials, whose relatively long, white-painted needles described an arc *within* the dial. Brookes [9] observes: "Arguably, little yellow cursors one-third the length of old white needles were too small and discrete to warn effectively".

The layout of the 737-400's head-down instrumentation may have contributed to the flight-crew's misreading of powerplant serviceability: "The number 2 … engine was throttled back …. This was possibly a result of the layout of the primary and secondary engine instrumentation screens …. [T]he power levers [engine throttle

Fig. 4 The broken 737 lies on the western embankment of the M1 motorway. Note how the tail section jack-knifed up the embankment

levers] were … located between the two screens, thus the number 1 engine vibration gauge was directly above the number 2 engine power [throttle] lever" [29]. The under-pressure crew may have associated the readings from the secondary instruments column for the number 1 engine, with the throttle lever for the number 2 engine.

Finally, it was pointed out that throughout the emergency the two-man flight crew (engineers having been automated off the flight-deck) was deluged with requests and tasks. The relationship between workload and task performance is unclear. As Hockey [31] explains: "[S]erious effects of both workload and stress on task performance have been difficult to demonstrate". Importantly, the relationship is conditioned by factors like knowledge, experience, innate ability, environmental conditions, and required standards [13, 20, 44]. Nevertheless, the possibility exists that overburdened pilots *may* lose situation awareness. They *may* focus obsessively on a small number of issues and/or inputs (coning of attention). In a laboratory experiment with young, low-hours (10–300 h flight-time) student pilots, Morris and Leung [49] noted "high prioritisation error rates associated with increased mental workloads". Green et al. [27] note: "At extremely high levels of workload (overload), important information may be missed, due to the narrowing or focusing of attention onto only one aspect of the task". Those subjected to high workloads may experience temporal distortion (a few minutes may seem like hours and an hour may seem like a few minutes) [63]. They may revert to learned behaviours inappropriate to their current operating environment: "[U]nder stress, behaviour may regress to the earliest learnt, such as operating a control or selector in a manner which would have been appropriate to the previous type of aircraft flown, but not the current one" [13]. While pilots regularly train for high-workload, high-stress emergency situations, simulation can only ever be an approximation of reality. Dekker [20] argues that pilots' foreknowledge of the simulation exercise's adverse event "[makes] the unexpected less unexpected". How good a test is an expected unexpected event?

2.2.2 Pilots' Sensemaking

Reacting to their treatment by British Midland and elements of the Fourth Estate, G-OBME's pilots posited a link between their misidentification of the problem engine and the physical and political context of their labour. First Officer David McClelland (cited in [51]) observed: "Pilot error is a very neat term. What they're saying is that the people who designed it, manufactured it and carried out the specifications all got it right, but the two chaps at the front got it wrong. Straight away it sweeps all the problems below the mat". Captain Kevin Hunt (cited in [51]) noted: "We were the easy option, the cheap option if you wish. We made a mistake—we both made mistakes—but the question we would like answered is why we made those mistakes".

Scapegoating is a means of distracting attention from resident pathogens (see Reason [56–58] for a definition of 'resident pathogen') [7] like poor flight-deck ergonomics or unreliable mechanicals. Reason and Hobbs [59] observe: "Blaming people for their errors is emotionally satisfying but remedially useless. Moral judgments are only appropriate when the actions go as intended and the intention is reprehensible. Blame and punishment make no sense at all when the intention is a good one, but the actions do not go as planned". The Head of Psychology at the Royal Air Force's Institute of Aviation Medicine passed this judgment on Kegworth: "Any pilot's performance is a product of the training he has had, and the equipment he has been given to operate .... [I]f you give him equipment which isn't as easy to use as it could have been, are you going to blame the pilot who maybe gets it wrong when he doesn't get time to sort it out?" (Green cited in [9]).

3 Reductionism—A Disease of Contemporary Discourse?

The fundamental attribution error is not only made in relation to aviation disasters, but in relation to other losses, too. The 2015 death of a former British politician, the ex-Liberal Democrat leader Charles Kennedy, is a case in point. Kennedy had a drink problem. Doctors attributed the 'massive haemorrhage' that killed him to chronic alcohol abuse. Unfortunately, in doing so they committed the fundamental attribution error. While the proximate cause of Kennedy's death was an alcohol-induced haemorrhage, there were broader, contextual reasons for his demise. For example:

- a British popular culture that trivialises alcohol abuse (the phenomenon of alcohol abuse is relentlessly mined by self-interested, money-grubbing British comedians)
- government policies that pander to the interests of the drinks industry [62].

In 2005, Tony Blair's Labour government introduced 24-h drinking on the assumption that revellers would drink at a more sensible pace. Revellers did not behave as expected. Many drank more. The anticipated European-style 'café culture' failed to materialise. Alcohol consumption generates too much tax revenue for *any* government to intervene in a meaningful way in the *modus operandi* of the drinks industry. Political self-interest has led successive governments to gloss over the fact that alcohol is a poison that destroys lives, families and communities. Political self-interest has helped make alcohol a socially acceptable drug.

As a drug, alcohol is addictive and destructive: "Today [in Britain] around nine-million adults drink at levels that pose some risk to their health with 2.2 million drinking at higher-risk of harm. An estimated 1.6 million may have some degree of alcohol dependence .... The total annual cost to society of alcohol-related harm is estimated to be £21bn. The NHS incurs £3.5bn a year in costs related to alcohol. Few other health harms have such high overall costs when the impact on productivity and crime are included" [55]. It is also antisocial: "Acpo [the Association of Chief Police Officers] estimates that half of all violent crime is alcohol-related, while offenders are thought to be under the influence of drink in nearly half of domestic abuse cases" [62]. The 'shared enemy' of alcohol forged a friendship between Charles Kennedy and Alistair Campbell, Tony Blair's former spin doctor. Campbell was one of the few public figures to discuss the wider societal origins of the disease (alcoholism) that killed his friend. His was an all-too-rare discourse on the *systemic* origins of the disease.

3.1 Blamism Creates Comprehension Lacunae

By individuating responsibility for failure, blamism denies the contribution of wider political, societal and organisational factors, like geopolitical calculation, political self-interest, bureaucratic incompetence, under-funding, cost-cutting, inadequate training, unrealistic schedules, equipment shortages and bad design. Blamism produces what Fiske and Taylor [21] call the fundamental attribution error—an analysis that overlooks the possibility of a link between circumstance and mistake. As Glendon, Clarke and McKenna [26] explain in their book *Human Safety and*

Risk Management, those who make the fundamental attribution error assume that employees act in a vacuum:

> There is a strong tendency, called the fundamental attribution error, to overemphasise the influence of internal causes, as opposed to external ones, when judging other people's behaviour. Thus, a manager looking for a causal explanation for a worker's behaviour will tend to focus on cues related to the actor, rather than the situation .... The fundamental attribution error is only made when judging other people's behaviour [like that of Lubitz], not our own.

Horlick-Jones [33] offers this interpretation of Fiske and Taylor's [21] fundamental attribution error theory:

> The 'fundamental attribution error' ... is a tendency to blame undesirable events on individuals ... without taking into account situational factors beyond these agents' control.

3.2 *Blamism and Victimisation Are Normalcy*

Glendon, Clarke and McKenna's fundamental attribution error is a commonplace. Events with complex aetiologies and subtle impacts are frequently the subject of simplistic analyses and inappropriate and ineffectual knee-jerk policy responses (often because under-pressure civil servants and politicians believe they have to be seen to be doing something. The media has a lot to answer for). In the United Kingdom, health-care sees a good deal of naming, shaming and ill-informed blaming, as the following case studies demonstrate:

3.2.1 Case Study 1: Misprescription

In February 2012, a National Health Service nurse, Amanda Young, administered the wrong dose of an anti-psychotic drug to paranoid schizophrenic Joshua Gafney: "Amanda Young told Joshua Gafney to drink six bottles of anti-psychotic medicine clozapine—instead of the required dose of just one teaspoon" [46]. Mr Gafney did as he was told, and died. Young was sacked. In 2015 Amanda Young was found not guilty of manslaughter by gross negligence. It was judged that Young's actions, while misconceived, were not intended to harm. In other words, the court found that Nurse Young had made a mistake. The court heard that Young "had been a kind and dedicated nurse" [46]. The dead man's family issued a statement: "This verdict means that no-one has been held accountable for [Joshua's] death" (cited in [46]). The Gaffney family's framing of a criminal conviction as the *sine qua non* of accountability and justice supports Horlick-Jones's [33] contention that blamism is a commonplace of social discourse: "The need to blame seems to be fundamental in a wide range of cultures and societies .... Blaming ... seeks to make some sense of the world, and to defend [citizens] from future harm". The tendency to blame, victimise, prosecute and imprison honourable people whose well-intentioned actions cause harm serves no useful purpose [59].

3.2.2 Case Study 2: Teamworkers

In 2013, Dr Louise Schodlok was found guilty of serious misconduct after she lost her temper with a junior colleague at London's Queen Elizabeth Hospital. In a 2015 systems-thinking-informed examination of the facts, the Court of Appeal judged Schodlok's behaviour to be the product of a dysfunctional National Health Service culture: "A surgeon accused of being 'rude and aggressive' was really a victim of the 'shocking and old-fashioned culture' at [the Queen Elizabeth Hospital, Woolwich, south-east London] .... Lord Justice Vos at the Court of Appeal said ... the 'difficult atmosphere' at work 'contributed to events'" [47]. Note how Schodlok's employers were content to discipline her without exploring wider contextual factors, like the dysfunctional health service culture they helped create and sustain. Lord Justice Vos (cited in [17]) observed in his summing-up: "I should not leave this case without mentioning something upon which Dr Schodlok placed great reliance, and which was acknowledged by the Panel, namely the difficult atmosphere and the shocking and old-fashioned culture at the hospital at the time of these events (see the evidence of Dr Ian Stell, who was clinical director for the emergency department at the hospital at the relevant time). This climate plainly contributed to the events that have now been all too extensively litigated. The numerous references and testimonials that Dr Schodlok has adduced demonstrate that she has been a satisfactory, or even a good, doctor for many years both before and after her year at the hospital. The incidents of serious misconduct wrongly found proved against her related to only two days even in the unsettled period she spent at the hospital".

3.2.3 Case Study 3: Victims

At 10:30 on February 18, 2011, Jack Adcock, a boy with Down's Syndrome, was admitted to the Children's' Assessment Unit (CAU) of the Leicester Royal Infirmary (LRI) with severe vomiting, diarrhoea and breathing difficulties. Attended to by an agency Staff Nurse, Jack's care was overseen by the CAU's Dr Hadiza Bawa-Garba. The child was X-rayed. Bloods were taken. At 21:00 Jack was transferred to the children's ward. He went into cardiac arrest. At 19:45 he stopped breathing. He died at about 21:20, despite resuscitation attempts. When Dr Bawa-Garba saw staff trying to resuscitate him she told them to stop, believing he was under a "do not resuscitate" order (that actually applied to another child). On realising her error, she participated in efforts to resuscitate him [23, 52]. Following a Police investigation the Crown Prosecution Service (CPS) decided to prosecute Dr Bawa-Garba, the agency Staff Nurse, and the CAU's Ward Sister for gross negligence manslaughter. Specialist prosecutor Fiona Morrison claimed that a prosecution would be in the public interest. In November 2015, Dr Bawa-Garba and agency Staff Nurse Isabel Amaro were found guilty of gross negligence manslaughter [18]. Nicola Adcock (cited in [2]) commented: "We have always believed that someone needed to be held accountable for what happened to our son. The guilty verdicts will bring us some closure". Prosecutor Andrew Thomas QC (cited in [2]) said: "[Dr Bawa-Garba] neglected her duty to care for Jack …. It was not just a momentary lapse …. Jack's care was neglected over a protracted period of time …. These were not just simple breaches of duty, but really serious breaches amounting to gross negligence".

A systems analysis suggests that organisational factors contributed to Jack Adcock's death. For example:

(A) Bawa-Garba was routinely responsible for 30–40 patients
(B) Bawa-Garba had worked twelve hours without a break. Hersman [30] notes: "The NTSB's 1994 study of flight crew-related major aviation accidents found that captains who had been awake for more than about 12 h made significantly more errors than those who had been awake fewer than 12 h". Caruso and Hitchcock [14] note: "Fatigue-related impairments can lead to reduced performance on the job …. [T]he odds for a nurse making an error at work increased by three times when work shifts lasted 12.5 h or longer, compared with 8.5-h shifts"
(C) Bawa-Garba had been immediately put to work after returning from a 13-month maternity leave. She observed: "My skills were probably not at the level they should have been"
(D) According to Bawa-Garba it was normal practice for the nurse allocated to the child to inform the doctor that the X-rays were ready [52]. Jack had been allocated an agency nurse who failed to inform Dr Bawa-Garba that the X-rays were ready
(E) A malfunctioning computer system caused a delay in publishing the results of a comprehensive blood test.

The 'person model' of mishap recognises that error is a commonplace, and that anyone, however talented or well-trained, can make a mistake [36]. It suggests that the public good is best served by learning from, rather than punishing, those who err. The moral precepts of the 'legal model' of mishap take us down a different path: "Central to [the legal model of mishap] is the belief that responsible and highly trained professionals should not make errors. They have a duty of care …. Errors with bad consequences are assumed to be negligent or even reckless and thus deserve deterrent sanctions" [58]. Prosecuting those who make unintentional errors serves no rational purpose. It merely satisfies the irrational desire (fuelled by religious belief) to hurt those who hurt us. Staff at the Leicester Royal Infirmary did not intend Jack Adcock to die. Malice of forethought played no part in his death. Human error and circumstance did. Prosecuting carers for being human is nonsensical. There were six victims in the Adcock case: Jack Adcock; Jack's father and mother; Dr Hadiza Bawa-Garba; Staff Nurse Isabel Amaro; and British justice.

3.3 The Universality of the Fundamental Attribution Error

The fundamental attribution error is committed in every culture and under every circumstance. In September 2015, at Saudi Arabia's annual Hajj pilgrimage, over 1400 worshippers were killed (and hundreds injured) in a stampede. The blame-game started almost immediately, with some attributing the disaster to poor infrastructure and organisation, and others attributing it to the pilgrims themselves. One official blamed a specific racial minority: "Prince Khaled al-Faisal, head of the central Hajj committee, was criticised for saying 'some pilgrims from African nationalities' were responsible for the stampede" [41]. It took several days to establish exactly how many pilgrims had been killed and injured [22]. In the uncertainty following the disaster how could al-Faisal have discerned who caused the stampede? Did al-Faisal's analysis reflect the facts, or a political agenda? Denialism and scapegoating distract attention from institutional weaknesses. The rich and powerful may see scapegoating as a relatively risk-free damage-limitation strategy—risk-free because those whom the rich and powerful blame are often the poor and powerless. Responding to al-Faisal's claim, Tarek Fatah, journalist and fellow of the Middle-East Forum think tank, described the Saudi official's remarks as a "display of Arab anti-black racism" (Fatah cited in [22]).

4 The Germanwings Disaster

On 24 March 2015, at Prads-Haute-Bléone, Alpes-de-Haute-Provence, France, Germanwings Airbus A320-211 D-AIPX flew into a mountain killing all on board (154 passengers and six crew) (Fig. 5). Germanwings is a subsidiary of German airline Lufthansa. According to the Bureau d'Enquêtes et d'Analyses pour la sécurité de l'aviation civile's (BEA's) [10] May 2015 *Preliminary Report*, the primary cause was the suicide of D-AIPX's First Officer, Andreas Lubitz: "[Lubitz] intentionally modified the autopilot instructions to order the aeroplane to descend until it collided with the terrain. He did not open the cockpit door during the descent, despite requests for access made via the keypad, the cabin interphone and knocks on the door".

In its *Preliminary Report* the BEA (2015) promised a systems-theory-informed investigation that would take in two major systemic factors: "The investigation will also study the systemic failings that may have led to this accident or to similar events, with two main investigative orientations: **Medical aspects**: the investigation will seek to understand the current balance between medical confidentiality and flight safety. It will specifically aim to explain how and why pilots can be in a cockpit with the intention of causing the loss of the aircraft and its occupants, despite the existence of: regulations setting mandatory medical criteria for flight crews, especially in the areas of psychiatry, psychology and behavioural problems; recruitment policies, as well as the initial and recurrent training processes within airlines. **Cockpit security**: the investigation will seek to understand the

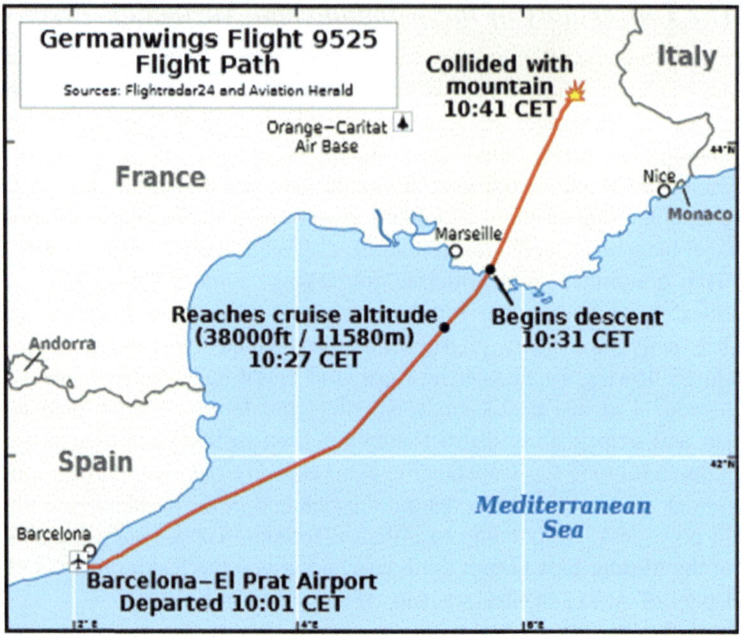

Fig. 5 Lubitz locked his colleagues out of the flight-deck as he flew the airbus to its destruction

compromises that were made between the requirements of security, specifically those that followed the attacks on 11 September 2001, and the requirements of flight safety. In this context, the investigation will include a focus on cockpit door locking systems and cockpit access and exit procedures".

4.1 The Germanwings Disaster Through a Systems-Thinking Lens

The BEA's initial finding of 'suicide by commercial aircraft' is a reasonable and evidence-based conclusion (Fig. 6).[1] But what of the proximate causes? In the aftermath of the 24 March, 2015 Germanwings disaster, numerous institutions and parties committed Fiske and Taylor's [21] fundamental attribution error. Instead of

[1]Other pilots have used large passenger aircraft to commit suicide. According to the United States's National Transportation Safety Board the 'probable cause' of the 1999 loss of an EgyptAir Boeing 767 some 60 miles south of Nantucket Island was pilot suicide by commercial aircraft. Like Lubitz, EgyptAir Flight 990's First Officer, Gameel al-Batouti, waited until he was alone on the flight-deck before diving the aircraft: "The flight data recorder showed that he waited for the captain to leave the cockpit and then disengaged the autopilot. As the plane descended, he could be heard saying in Arabic, 'I rely on God', over and over" [61].

Fig. 6 Lubitz enrolled D-AIPX as an actant in his suicide actor-network

The Guardian	A picture emerges of a man disturbed and ill. Yet allowed to fly.
The Times	Killer pilot 'had made plans to go down in history'. Girlfriend was scared of his erratic behaviour.
The Daily Telegraph	Doctor had ordered killer pilot to stay off work on day of disaster
The Independent	Pilot 'had a sick note' for day he killed 149 people
Financial Times Weekend	Co-pilot destroyed sick note declaring him unfit to fly
i on Saturday	Co-pilot hid illness from his employers
Daily Express	Death crash pilot was depressed and ripped up his sick notes
The Sun	Kill pilot tore up flight-day sicknote
Daily Star	Killer pilot's secret gay torment

Fig. 7 How British newspapers framed the Germanwings disaster

considering wider organisational factors, politicians, independent commentators and much of the press rounded on the pilot. The ferocity with which most newspapers condemned Lubitz—and he alone—merits review. The following is a sample of the front-page headlines that appeared in Britain on 28 March, 2015 (Fig. 7):

The attributional myopia of the Fourth Estate is obvious. Wider organisational and human factors that were overlooked or ignored by the Press included:

– the inability of Lufthansa's in-house medical department to communicate medical data directly to the airline. As a Lufthansa spokesperson explained: "It would be against the law for our, or any other, aeromedical centre to share information with the company, or any other third parties …. According to German Federal Law, under no circumstances is medical information to be shared with the company. Only between medical practitioners, aeromedical centres and/or the medical service of the German Aviation Authority (LBA) is this allowed" (cited in [15])

- the possibility that the authorities failed to maintain a coherent, long-term picture of Lubitz's physical and mental health (such that a timely medical intervention could have been made)
- the possible impact of careerism and peer-pressure on Lubitz's decision to report for duty against medical advice. Pilots don't like letting colleagues or employers down [4]
- the possibility that Lubitz feared disciplinary action if he did not report. As Bennett [4] notes in *How Pilots Live—An Examination of the Lifestyle of Commercial Pilots*, such fears are not unknown amongst flight crew
- the possibility that Lubitz was in such a confused state on the day he committed suicide that he was incapable of reflecting on his fitness to fly
- the possibility that chronic fatigue induced by high-frequency operations in busy airspace (typical fare for short-haul European carriers like Germanwings) worsened Lubitz's physical and mental state, and impaired his judgment. During the sometimes acrimonious debate over the European Aviation Safety Agency's new Europe-wide flight and duty-time limitations regulations, concerns were expressed about the possible impact of the new regulations on the prevalence of acute and chronic fatigue amongst flight-crew [4].

Politicians were no more circumspect than journalists. France's Prime Minister, Manuel Valls, claimed that all the evidence pointed to an act that was 'crazy'. "Everything is pointing towards an act that we can't describe: criminal, crazy, suicidal" exclaimed Valls (cited in [1]). In failing to reference the broader context to the disaster, politicians, commentators and the Fourth Estate distracted attention from potentially relevant factors like:

- the possibility that moral imperatives like medical confidentiality (see above) may have created a window of opportunity for Lubitz. Viewed through Merton's [45] theory of the unintended consequences of purposive social action, Germany's medical confidentiality statute created an opportunity or affordance for suicide-by-commercial-aircraft
- the possibility that the capacity of an airline's human resources (HR) department to effectively support its employees may be compromised by cost-cutting. According to Lawton [39], cost-cutting is a prominent feature of airline praxis (airlines' operating philosophies and methods).

4.2 *Limitations of the Bureau d'Enquêtes et d'Analyses pour la sécurité de l'aviation civile's Investigation*

The scope or reach of the BEA's systems-thinking-informed investigation was too narrow. While the Bureau committed to investigate "… the systemic failings that may have led to this disaster", it considered just two potential contributory factors/proximate causes: the balance between medical confidentiality and flight safety; and the impact on flight safety of armoured, lockable flight-deck doors. Apparently, other systemic factors, like the possible impact of cost-cutting on the long-term health monitoring of flight-crew, or the prevalence of chronic fatigue amongst flight-crew, were not considered worthy of investigation. If one accepts that system safety is an emergent property of the interactions between system components (employees, customers, equipment, design philosophies, laws, rules, regulations, procedures, etc.) [8, 16, 24], it follows that arbitrarily excluding some components will produce a false or skewed analysis. A partial investigation may create more problems than it solves—because it may cause attention and resources to be focused on irrelevancies. A partial investigation may be thought of as a latent error/resident pathogen in our understanding of a complex socio-technical system like air service (or nuclear power generation or global finance or health-care provision, etc.). Viewed through Merton's [45] unintended-consequences-of-purposive-social-action prism, a partial investigation may increase rather than reduce the chances of a complex socio-technical system being afflicted by the same (or a similar) problem.

Reflecting in the *Foreword* to the book *Beyond Aviation Human Factors—Safety in High-Technology Systems* on a clash with Transport Canada's legal counsel, the Honourable Mr Justice Virgil P. Moshansky (the originator of the systems-thinking-informed approach to aviation accident investigation) warned against the arbitrary circumscription of incident and accident investigations: "During the early stages of the [Dryden] Inquiry, counsel for the Regulator attempted to limit the scope of the Inquiry with threats to limit my mandate by seeking an order in the Federal Court of Canada. When it became clear that intimidation would not succeed, these attempts were abandoned … " (Moshansky cited in Maurino et al. [44]). Reasons for limiting the scope of an inquiry include fear of what an investigation that follows the evidence might reveal. It is wise to always be suspicious of institutional defensiveness. Alarm bells should ring when terms of reference are constrained.

5 Reductionism—Leveraged by Third-Party Claims?

On 31 October, 2015, an eighteen year-old Airbus A321 carrying 224 passengers and crew fell from circa 31,000 ft onto the Sinai desert. The aircraft, operated by Metrojet, a small Russian carrier, was en-route from the Egyptian holiday resort of Sharm el-Sheikh to St Petersburg. A month before the disaster, Russia began military operations against those fighting to overthrow Syria's President Bashr al-Assad. Russian ground-attack aircraft struck at both the Free Syrian Army, and fundamentalist groups like Jabhat al-Nusra and the Islamic State of Iraq and the Levant (ISIL) (also known as Daesh) [25]. In response, radical Islamist groups, including ISIL/Daesh, called for attacks on Russian interests. Not unsurprisingly, ISIL/Daesh immediately claimed responsibility for the Metrojet loss, stating online:

"Soldiers of the caliphate were able to bring down a Russian plane above Sinai province" (cited in [42]). Few believed that ISIL/Daesh had the hardware and military intelligence to down an aircraft cruising at 31,000 ft, although the possibility of an improvised explosive device could not be discounted.

Efforts to secure political capital from a disaster like the loss of Metrojet Flight 9268 impede rational, detached investigation. However ill-founded or bizarre, claims of responsibility feed the media rumour-mill and divert resources from the investigation proper (because it is necessary to refute even the most outlandish claims). Malcontents understand that news reporting amplifies statements and claims (see, for example, Kasperson et al's. [35], and Pidgeon et al's. [54] work on the social amplification of risk), which is why media-savvy terrorist groups like ISIL/Daesh engage with the Fourth Estate: "Signals about risk are processed by … social amplification stations, including … the news media, cultural groups, interpersonal networks, and others. Key steps of amplifications can be identified at each stage. The amplified risk leads to behavioural responses, which, in turn, result in secondary impacts" [35]. The tight-coupling that results from increased media connectivity and the widespread adoption of a 24-h rolling news format (that creates a voracious appetite for news [5]) ensures the almost instant, global reporting of claims.

Faced with a maelstrom of speculation, accident investigators must guard against reacting in such a way that jeopardises their work. In the aftermath of the Metrojet Flight 9268 disaster, several parties, including a Russian investigation team conscious of the need to avoid committing Fiske and Taylor's [21] fundamental attribution error, patiently made the case for a measured examination of the facts. Responding to the ISIL/Daesh claim, the Russian aviation regulator, Rosaviatsia, warned against speculation: "Until there is reliable evidence about the circumstances of what happened, there is no sense in putting forward and discussing any versions" (cited in [42]). Determined not to be panicked or undermined by the ISIl/Daesh claim, a spokesperson for the Egyptian Army announced: "[The Islamic State of Iraq and the Levant] can put out whatever statements they want, but there is no proof at this point that terrorists were responsible for this plane crash. *We will know the true reasons when the Civil Aviation Authority in coordination with Russian authorities completes its investigation* [my emphasis]" (Samir cited in [42]). The Metrojet Flight 9268 war of words that erupted in the hours following the loss shows that pressures to commit the fundamental attribution error originate both within, and without organisations. There is always the possibility that malcontents will seek to capitalise on misfortune.[2]

[2]On November 17, 2015, the Russian security service, the FSB, claimed that the Metrojet Airbus had been downed by an improvised explosive device. The loss marked the start of a bloody ISIL/Daesh campaign against Western interests that included the murder of 130 Parisians on the night of 13 November, 2015. Over 360 persons were injured in the Paris attacks. The hunt for the terrorists who escaped saw Brussels locked-down for several days. The French government declared a three-month state-of-emergency. Further major ISIL/Daesh-inspired terrorist attacks followed in 2016.

6 The Importance of Systems-Thinking to Risk Managers

Only systems investigations like that conducted by the Honourable Mr Justice Virgil P. Moshansky [50] into the 1989 Dryden accident, or that conducted by Mr Charles Haddon-Cave QC [28] into the 2006 RAF Nimrod loss over Afghanistan, are of use to risk managers and legislators, because only this type of investigation can identify the conditions that incubate failure. As Professor Barry Turner [65, 66] demonstrated in his Six Stage Model of Failure, disasters often 'cook' unseen over months or years until a trigger (like a budget cut, overly-tight schedule or poorly thought-out rule or regulation) provokes a catastrophic failure. Echoing Turner's work, Professor James Reason [56–58] talked about the latent conditions or resident pathogens that spark failure. As demonstrated by the Fukushima, Chernobyl and Three Mile Island nuclear disasters [37, 43, 53], complex socio-technical systems may harbour myriad weak points, both human and non-human. The creation of vast 'systems-of-systems' amplifies problems of incomprehensibility, emergence and latent error:

> [T]he risk of crisis is … becoming structural as large networks become more complex, more vulnerable …. [C]rises continue to become more frequent and destabilising [38]

We deceive ourselves if we believe we can make the world a safer place by victimising prime-movers like Germanwings First Officer Andreas Lubitz. By focusing on unfortunates like Lubitz we risk overlooking failure's messy, convoluted origins. In time, the loss of Germanwings Flight 4U9525 may be shown to have been a preventable systems accident (Figs. 8 and 9).

Fig. 8 Dusseldorf airport's temporary memorial to the victims of Germanwings Flight 4U9525

Fig. 9 Germanwings flags flying at half-mast as a sign of respect

References

1. Beck J (2015) Depressed doesn't mean dangerous. News reports are focusing on the Germanwings pilot's possible depression, following a familiar script in the wake of mass killings. But the evidence shows violence is extremely rare among the mentally ill. http://www.theatlantic.com/health/archive/2015/03/depresseddoesntmeandangerous/388922/. Accessed 21 Sept 2015
2. Belfast Telegraph (2015) Jack Adcock parents: convictions bring some closure but void remains. http://www.belfasttelegraph.co.uk/. Accessed 14 Nov 2015
3. Bennett SA (2010) Human factors for maintenance engineers and others—a prerequisite for success. In: Blockley R, Shyy W (eds) Encyclopaedia of Aerospace Engineering. Wiley, Chichester
4. Bennett SA (2014) How pilots live: an examination of the lifestyle of commercial pilots. Peter Lang International Academic Publishers, Oxford
5. Bennett SA (2015) Airshows are risky—that's why we like them—but they also have a strong safety record. theconversation.com/. Accessed 12 Nov 2015

6. Bennett SA (2015) Aviation safety and security: the importance of teamwork, leadership, creative thinking and active learning. Libri Publishing Ltd., Faringdon
7. Bennett SA (2015) Framing the MH17 disaster—more heat than light? Int J Aviat Aeronaut Aerosp 2(4):1–26
8. Black J, Koopman P (2009) System safety as an emergent property in composite systems. Carnegie Mellon University, Pittsburgh, PA
9. Brookes A (1992) Disaster in the air. Ian Allan, Shepperton
10. Bureau d'Enquêtes et d'Analyses pour la sécurité de l'aviation civile (2015) Preliminary report. Accident on 24 March 2015 at Prads-Haute-Bléone (Alpes-de-Haute-Provence, France) to the Airbus A320-211 registered D-AIPX operated by Germanwings. Bureau d'Enquêtes et d'Analyses pour la sécurité de l'aviation civile, Paris
11. Callon M (1986) The sociology of an actor-network: the case of the electric vehicle. In: Callon M, Law J, Rip A (eds) Mapping the dynamics of science and technology: sociology of science in the real world. Macmillan, London
12. Callon M, Latour B (1981) Unscrewing the big leviathan: how actors macro-structure reality and how sociologists help them to do so. In: Knorr-Cetina K, Cicourel AV (eds) Advances in social theory and methodology: towards an integration of micro and macro sociologies. Routledge and Kegan Paul, Boston, pp 277–303
13. Campbell RD, Bagshaw M (1999) Human performance and limitations in aviation. Blackwell Science, Oxford
14. Caruso CC, Hitchcock EM (2010) Strategies for nurses to prevent sleep-related injuries and errors. Rehabilitation Nursing. 35(5):192–197
15. Chandler J, Thomas G (2015) Germanwings 9525: a matter of safety vs privacy? http://www.airlineratings.com/news.php?id=471/. Accessed 24 Sep 2015
16. Cook RI (2000) How complex systems fail (being a short treatise on the nature of failure; how failure is evaluated; how failure is attributed to proximate cause; and the resulting new understanding of patient safety). Cognitive Technologies Laboratory, University of Chicago, Chicago
17. Court of Appeal Civil Division (2015) England and Wales Court of Appeal (Civil Division): decisions: Schodlok v the General Medical Council [2015]: case no: C1/2013/1887: EWCA Civ 769 (21 July 2015). http://www.bailii.org/ew/cases/EWCA/Civ/2015/769.html. Accessed 19 Sep 2015
18. Crown Prosecution Service (2015) Two Leicester Royal Infirmary medical staff found guilty of manslaughter following the death of six year old boy. http://www.cps.gov.uk/news/latest_news/leicester_royal_infirmary/ Accessed 12 Nov 2015
19. Dekker S (2006) Resilience engineering: chronicling the emergence of confused consensus. In: Hollnagel E, Woods DD, Leveson N (eds) Resilience engineering: concepts and precepts. Ashgate Publishing Ltd., Aldershot
20. Dekker S (2007) Workload: A Strange Concept. Hindsight, July
21. Fiske S, Taylor S (1984) Social cognition. Random House, New York
22. Gallagher P, Johnston I (2015) Blame-game begins as death toll reaches 719 in worst Hajj stampede for 25 years. The Independent, 25 Sept 2015
23. Gardner B (2014) Doctor and two nurses charged with manslaughter over death of six-year-old boy. http://www.telegraph.co.uk/. Accessed 10 Nov 2015
24. Gherardi S, Nicolini D (2000) The organizational learning of safety in communities of practice. Journal of Management Inquiry 9(1):7–18
25. Gilsinan K (2015) The confused person's guide to the Syrian civil war. *The Atlantic*, Oct 29
26. Glendon IA, Clarke SG, McKenna EF (2006) Human safety and risk management. Taylor and Francis, London
27. Green R, Muir H, James M, Gradwell D, Green RL (1996) Human factors for pilots. Ashgate Publishing Ltd., Aldershot
28. Haddon-Cave C (2009) The Nimrod review. An independent review into the broader issues surrounding the loss of the RAF Nimrod MR2 aircraft XV230 in Afghanistan in 2006. HC 1025. Her Majesty's Stationery Office, London

29. Harris D (2004) Head-down flight-deck display design. In: Harris D (ed) Human factors for civil flight deck design. Ashgate Publishing Ltd., Aldershot
30. Hersman DAP (2009) Fighting fatigue: a team effort. Aerosafety World, June
31. Hockey R (2002) Human performance in the working environment. In: Warr P (ed) Psychology at work. Penguin, London
32. Hollnagel E (2004) Barriers and accident prevention. Ashgate Publishing Ltd., Aldershot
33. Horlick-Jones T (1996) The problem of blame. In: Hood C, Horlick-Jones T (eds) Accident and Design. Contemporary Debates in Risk Management. UCL Press, London
34. Johnson CW (2005) What are emergent properties and how do they affect the engineering of complex systems? Department of Computing Science, University of Glasgow, Glasgow
35. Kasperson RE, Renn O, Slovic P, Brown HS, Emel J, Goble R, Kasperson JX, Ratick S (1988) The social amplification of risk: a conceptual framework. Risk Anal 8(2):177–187
36. Kohn LT, Corrigan J, Donaldson MS (2000) To err is human: building a safer health system. National Academy Press, Washington D.C.
37. Ladkin P (2012) The Fukushima accident. In: Dale C, Anderson T (eds) Achieving systems safety: proceedings of the twentieth safety-critical systems symposium. Bristol, UK, Springer, London, 7–9 Feb 2012
38. Lagadec P (1993) Ounce of prevention worth a pound in cure. Management Consultancy. June
39. Lawton TC (2002) Cleared for take-off. Structure and strategy in the low-fare airline business. Ashgate Publishing Ltd., Aldershot
40. Leroux M (2008) Captain James Thain cleared of blame after the thawing of hostilities. The Times, Jan 30
41. Logan R (2015) 700 pilgrims die in Hajj stampede. People were climbing over each other to breathe, say witnesses. Metro, 25 Sept 2015
42. Luhn A, Khalil J (2015) Russian plane crash: investigation begins into cause of Sinai crash. http://www.theguardian.com/world/2015/oct/31/russianplanecrashinvestigatorsseekcauseofa321crash. Accessed 1 Nov 2015
43. Mackay L, Thompson M (1988) Something in the wind: politics after Chernobyl. Pluto Press, London
44. Maurino DE, Reason J, Johnston N, Lee RB (1998) Beyond aviation human factors—safety in high-technology systems. Ashgate Publishing Ltd., Aldershot
45. Merton RK (1936) The unanticipated consequences of purposive social action. Am Sociol Rev 1(6):894–904
46. Metro (2015) Nurse goes free after fatal medicine slip killed patient. Metro, Jun 16
47. Metro (2015) 'Rude' surgeon is cleared on appeal. Metro, Jul 24
48. Miller KD (2009) Organisational risk after modernism. Organ Stud 30(2–3):157–180
49. Morris CH, Leung YK (2006) Pilot mental workload: how well do pilots really perform? Ergonomics 49(15):1581–1596
50. Moshansky VP (1992) Moshansky, Commission of Inquiry into the Air Ontario accident at Dryden, Ontario: final report, vol 1–4. Minister of Supply and Services, Ottawa
51. Nottingham Post (2009) Kegworth: Sacked pilots claim they were 'scapegoats'. http://www.nottinghampost.com/KegworthSackedpilotsclaimscapegoats/story12186326detail/story.html. Accessed 19 Oct 2015
52. Owen DW (2015) Jack Adcock manslaughter trial: doctor defends her actions in court. http://www.leicestermercury.co.uk/ Accessed 14 Nov 2015
53. Perrow C (1982) The President's Commission and the normal accident. In: Sils D, Wolf C, Shelanski V (eds) Accident at Three Mile Island: the human dimensions. Westview, Boulder
54. Pidgeon N, Kasperson RE, Slovic P (2003) (eds) The social amplification of risk. Cambridge University Press, Cambridge
55. Public Health England (2014) Alcohol treatment in England 2013–2014. Public Health England, London
56. Reason J (1990) Human error. Cambridge University Press, Cambridge
57. Reason J (1997) Managing the risks of organisational accidents. Ashgate Publishing Ltd., Aldershot

58. Reason J (2013) A life in error. Ashgate Publishing Ltd., Aldershot
59. Reason J, Hobbs A (2003) Managing maintenance error: a practical guide. Ashgate Publishing Ltd., Aldershot
60. Shorrock S, Leonhardt J, Licu T, Peters C (2014) Systems thinking for safety: ten principles. Eurocontrol, Brussels
61. Sinha S (2015) A history of crashes caused by pilots' intentional acts. The New York Times, 26 Mar 2015
62. Slack J (2013) 24 hour drink law disaster: police chiefs brand round-the-clock pubs a terrible mistake as they call for 'drunk tanks' to cope with late-night mayhem. http://www.dailymail.co.uk/news/article2424718. Accessed 15 Oct 2015
63. Smith D (2012) CRM series: workload management. http://www.cfidarren.com/crmworkload.htm. Accessed 31 May 2015
64. Truslove B (2014) Kegworth air disaster: plane crash survivors' stories. http://www.bbc.co.uk/news/ukenglandleicestershire25548016. Accessed 19 Oct 2015
65. Turner BA (1978) Man-made disasters. Wykeham, London
66. Turner BA (1994) Causes of disaster: sloppy management. Br J Manag 5(3):215–219

Dynamics of Information Flow Before Major Crises: Lessons from the Collapse of Enron, the Subprime Mortgage Crisis and Other High Impact Disasters in the Industrial Sector

Dmitry Chernov and Didier Sornette

Abstract The analysis of the two largest financial disasters in the USA so far in the first decade of this century—the collapse of Enron in 2001 and the subprime mortgage crisis of 2007–2008—suggests that the huge scale of these disasters stemmed from a lack of timely information. We present extensive evidence that regulators, investors and associates were not informed of the conditions and risks associated with the activities of Enron management in the first case, or with the assessment and underwriting of collateralized debt obligations (CDOs) in the second; and with little understanding of the "whole picture" of risks, they could not intervene decisively to prevent or minimize disaster. Moreover, we identify similar obstacles to the transmission of reliable risk information in past cases such as the Barings Bank crash, the Deepwater Horizon oil spill, the nuclear accidents at Chernobyl and Fukushima-Daiichi as well as in the current development of the US shale energy industry. Based on the careful observation of events before the moment of collapse in three financial events (Barings, Enron and subprime crisis), one mixed financial-industrial case and three industrial catastrophes, we document and discuss how the inadequate transmission or outright concealment of risk information constitutes a powerful engine of disasters.

Keywords Financial crises · Information concealment · Early warning · Complexity · Resilience · Risk management

D. Chernov · D. Sornette (✉)
ETH Zürich, Zurich, Switzerland
e-mail: dsornette@ethz.ch; dsornette@gmail.com

A.J. Masys (ed.), *Disaster Forensics*, Advanced Sciences and Technologies for Security Applications, DOI 10.1007/978-3-319-41849-0_8

1 Introduction

The private sector actors and policy makers are all interested in developing and fostering innovations and industry developments that can provide higher profits, growth and employment. But there is always a trade-off between unbridled innovations that may lead to serious negative externalities (pollution, accidents, crises) and full-fledged regulatory control that can stifle innovation. This belongs to the general principle-agent issues, where the size of the subsequent impacts can reflect amplifying externalities.

In this article, we review three financial crises (Barings, Enron and subprime crisis), one mixed financial-industrial disaster and three industrial catastrophes, and identify the role of seriously missing information in generating the accident. The lacking knowledge resulted from both inadequate transmission between involved actors and direct concealment of risk information. We find that regulators and representatives of these industries had a mutual interest in the weakening of any existing regulation to facilitate the launching of innovative development. So the budgets of government oversight bodies were reduced, preventing them from hiring qualified and experienced inspectors who understood the innovations; the authorities demanded less reporting from industries as they brought in the now deregulated innovations; and ultimately regulators lost the comprehensive understanding of the challenges involved. In the absence of strict government control to protect the long-term interests of society, private industries were free to choose the most effective way of implementing innovations to maximize dividends to shareholders, growth of capitalization and bonus plans to motivate executives—all which served short-term interests. Their solutions for introducing innovation, while serving their own interests very well, were less effective in protecting the interests of society.

During the early stages of a new development, it seems to executives and regulators that the expansion of innovations is going well because, in the wake of deregulation measures, nobody fully understands all aspects of the development. Sometimes, they realize some of the shortcomings of the innovations. But previous poor decisions, an unwillingness to believe that the worst could happen, and an industry culture of risk concealment, which prevents the transmission of timely information about existing risks and the adequate assessment of potential ones—all of these lead to a misplaced confidence among executives that the present state of the innovation process is sound, when in reality it is moving towards catastrophe.

Given the evidence, the question arises whether risk-mitigation could have been undertaken in the seven reviewed cases. Were the risks in the pre-crisis event activities observable? Could ex-ante regulatory action have been taken to control or even prevent the crises? A paradox of risk management is that its value (and in fact failure) is revealed mostly when a crisis occurs while, in the absence of observable problems, it seems that risk management is redundant and even constitute an obstacle to sound business and growth. It is also often debatable whether regulatory actions would be welcome and not actually worsen the conditions with unintended consequences. It may seem unrealistic to assume that regulation applied prior to the crisis event would have been effective in addressing the risks prior to its occurrence. In the presence of information problems (and possible externalities) that we document below, what regulation would necessarily be effective? One possible approach would be to assume that well-functioning markets provide best or most efficient outcomes (abstracting from less well-defined welfare issues). In this space, regulation would be most effective only when it addresses market failures, providing direct economic benefits. In this framework, any policy is, by definition, net costly and leads to inefficient outcomes. However, the extensive pieces of evidence we provide below suggest that starting from the reference point of well-functioning markets and studying possible deviations from it may be misconceived. The evidence suggests instead that another anchor might be more representative: markets seem to be endowed with intrinsic negative externalities, pitting the short-term private interests of actors against those of the largest society, in a ubiquitous principle-agent context.

Our purpose is modest. We do not attempt to list and rank the dominant factors leading to crises but rather document what we diagnose as an under-recognized dimension, that of the failure of risk information transmission and risk information concealment. Our extensive case studies suggest that much of the hidden information was actually available, and in fact discussed by some involved actors prior to the occurrence of the crisis. Our contribution is mainly to raise the awareness of risk managers and decision makers by providing vivid instances of risk information transmission failure and concealment that can be analogously identified in other situations, so that counter-measures could be developed on a case-to-case basis. We focus on two distinct meanings of the behaviors uncovered in our investigations: (i) the condition that facts and knowledge about an organization and its functioning are hidden from those that should use them; the concealment can be due to many causes, including complexity, miscommunication, and so on; (ii) the conscious and deliberate action of keeping important information secret or of misrepresenting it; this second meaning is a surprisingly important part of the pieces of evidence that we present. We do not believe in a "one size fits all" solution, in particular in the regulation space, given the complexity and large variety of circumstances, constraints and cultures. However, we do believe in the existence of robust patterns of information gaps, as documented below, which can therefore be targeted systematically by responsible and attentive risk managers and regulators.

Our examination of seven cases shows that risk information concealment played an influential role in creating or aggravating a catastrophe. One could argue that this does not prove anything, except showing that we fall in the standard "confirmation bias" that consists in looking for examples than confirm our prior belief. In principle, one would like to have a selection process of case studies that is independent of the hypothesis being considered. This requires using a standard database of crises and studying to what extent the failure of information flow is a determinant of the crises. While laudable, this program is beyond the scope of the present more modest article, as it falls within a much larger research agenda. In our defense, let us point out that we just selected the three most important financial crises in the last decade. We thus argue that we do not have any selection bias other than immediacy and recent relevance. For the industry catastrophes, we choose the two largest nuclear disasters. Overall, this suggests that an accusation of "confirmation bias" would not be reasonable. But, of course, this remains to be demonstrated rigorously by adding up other case studies in a systematic way.

In Sect. 2, we present the evidence showing the impact of deregulation and/or insufficient control over innovative development in the creation of the conditions for future disasters. Section 3 dissects the internal organization environments that promote the large level of risk concealment boosting the likelihood and severity of future crises. Section 4 discusses the influence of the absence of reliable information and of adequate processes for sound risk assessment in the formation of the conditions of catastrophic crises. In each of the Sects. 2–4, we review the Barings Bank crash, the Enron collapse, the subprime crisis, the nuclear accidents at Chernobyl and Fukushima-Daiichi, the Deepwater Horizon oil spill, as they demonstrate that the problems are not specific to the financial industry but represent a generic structure in business development in the presence of innovations and novel opportunities. We also end each of the Sects. 2–4 with an analysis along these three dimensions of what we consider a crisis in the making, that of the US shale energy industry. This on-going bubble is characterized by huge debt, great hype together with low or even negative real return on investment and poor long-term prospects of the productivity of shale wells. Section 5 concludes.

A final word of caution is in order before presenting the case studies. Notwithstanding our attempt to sample a representative set of crises, one should be careful before generalizing to every organization in the world. Because of the high complexity of modern technical and organizational systems and multi-cultural differences, one should consider the existence of other mechanisms in addition to the one described here, before claiming the existence of an "universal theory risk obfuscation". With this caveat in mind, this article presents our effort towards the goal of developing a best practice approach for the management of sensitive organizations.

2 Deregulation or Absence of Sufficient Control over Innovative Development

2.1 *The General Picture*

It is remarkable that the largest disasters in recent decades in terms of damage and casualties have one important similarity: government oversight over the innovative industries involved was weak. This applies to the Enron bankruptcy with losses of up to US$63 billion in assets; the collapse of Lehman Brothers, which resulted directly in more than US$600 billion losses and triggered the global financial and economic crisis in 2008–2009, causing the loss of over US$30 trillion worldwide in stock market capitalization [1]; the collapse of Barings, one of the oldest merchant banks in the world, which began to work with innovative securities; the accident at the Chernobyl nuclear power plant, which caused the largest civil nuclear disaster in world history in term of the radioactive release; the largest maritime oil spill in the world, after the Macondo well blowout and the collapse of the Deepwater Horizon platform; and the meltdown of the Fukushima-Daiichi nuclear power plant's reactors after the Tohoku earthquake and tsunami, the largest ever nuclear accident in terms of damage costs; and finally, as we shall show, the development of the highly hazardous shale industry in the US, whose environmental and economic consequences remain to be assessed.

In all these cases, national governments were interested in launching innovative development in order to gain a competitive advantage over other countries and promote economic growth. Governments preferred to rely on the experience of industry to accept the implementation of new technologies and issue permits for their development without a proper assessment of the potential impact of these technologies on society in the long term, and without a rigid system of governmental control over their testing and implementation.

The fake growth of Enron's revenue started when the SEC allowed the mark-to-market accounting method and deregulated over-the-counter (OTC) derivatives in order to increase liquidity of the American stock market and the nominal size of American GDP, and massage the figures on national economic growth. The Bush administration supported the development of subprime mortgage lending, and hesitated to regulate the market in collateralized debt obligations and other derivatives, because it sought to trigger a real estate boom through permanent economic growth after 9/11, encourage foreign investment in the US stock market, reduce unemployment, and raise revenues from individual and property taxes. The collapse of Barings was partly caused by the British government allowing British merchant banks to work with securities in order to increase the competitive advantage of British banks on the international markets. Decades before the accident at Chernobyl, the Politburo (the executive committee for the Communist Party of the Soviet Union) had decided to transfer full responsibility for design and construction in the civil nuclear industry to the developers of the Soviet nuclear weapons—all nationally respected and honored scientists. They had a monopoly on decision-making regarding reactor types and on any technical solutions on nuclear plants in the USSR. Because nobody in the Academy of Sciences of the USSR or the Soviet government was more qualified than these experts, the Politburo had to rely on their experience in nuclear science. So, for two decades, they made technical decisions unchallenged by any effective government oversight; during this period, several minor errors were introduced in the design of Chernobyl-type reactors, and these had still not been eliminated when the accident took place. Although they did not monitor the safety of the nuclear program, the Politburo put constant pressure on the program's directors to increase the rate, and reduce the cost, of nuclear plant construction to ensure a cheap electricity supply for domestic needs. In the case of the Deepwater Horizon disaster, innovation in deepwater drilling promised to increase domestic oil and gas output and hasten the energy independence of the United States. Because of a prevailing belief in reducing government oversight of private enterprise, there were ongoing cuts in the budget and authority of the Minerals Management Service, the government body overseeing deepwater drilling; so the regulator could not hire qualified staff and had to rely on the expertise of private deepwater operators and contractors while innovative drilling methods were being developed [2]. The Nuclear and Industrial Safety Agency (NISA), under the authority of the Ministry of Economy, Trade and Industry of Japan, was intended not only to ensure that Japanese nuclear power plants were safely run but also to achieve energy independence for Japan by supporting low-cost electricity production from more than 50 Japanese nuclear reactors; to do this, they created stable financial conditions for operators by balancing safety issues and spending [3]. And the US government allowed private shale operators to pump billions of cubic meters of water mixed with harmful chemicals into the ground in order to stimulate a backflow of hydrocarbons from domestic shale formations—again in pursuit of energy independence from imported oil and gas. Alan Greenspan, Chairman of the US Federal Reserve during the presidencies of Reagan, George H.W. Bush, Clinton and George W. Bush, summarized the argument for deregulation: "*Those of us who support market capitalism in its more competitive forms might argue that unfettered markets create a degree of wealth that fosters a more civilized existence. I have always found that insight compelling... The market-stabilizing private regulatory forces should gradually displace many cumbersome, increasingly ineffective government structures*" [4, 5].

All these assumptions were disproved by reality. Nevertheless, politicians informally gave industry executives carte blanche to conduct high-risk operations, the true details of which were disclosed neither to regulators, employees or external audiences such as investors and contractors. And the dominance of democratic electoral procedures in many countries also forces politicians to promise voters that they will achieve visible results in a very short period of time. This affects their choice of foreground national projects: short-term high-impact projects, which can immediately revive economic growth and provide jobs, often win out over longer-term strategic initiatives whose results are only seen over decades. Since politicians often begin to prepare for the next election campaign immediately after the start of a new term, they generally favor populist measures to fulfill the immediate desires of their voters; understandably, they are reluctant to handle painful and unpopular (but necessary) reforms, or to develop strategic programs that are important for the survival and resilience of a society. Since people would rather not think about difficult times and decades of hard work, politicians and the economic elite proclaim what voters want to hear—they promise specific and tangible results in the near future, but keep quiet about the fact that these short-term results may be harmful to the survival of a state in the long term. It is in this spirit that the deregulation policies of many governments did not allow them to compel industry to give them the full picture about the risks involved in developing an innovation.

During the testing and implementation of innovations, industry obviously prefers to focus on short-term profitability, on cheap and effective solutions, regardless of the long-term national interests that regulators have to defend. Moreover, globalization has expanded opportunities for investment and increased competition for investment resources. In this fight, the winners are those companies that can show high profitability in a very short period, and keep production costs low. This forces companies to focus on continuous cost reduction at the expense of long-term business (and national) interests, and to avoid investing in long-term projects due to the unpredictability of global developments. Industries typically declare, when introducing such innovative technologies, that they are doing so in the national interest. But economic feasibility, and the pressure to demonstrate the effectiveness of a technology to society and the authorities, force industries to make ill-considered decisions and implement new technology as quickly and cheaply as possible. Companies prefer to hide the full costs that society will pay for the use of such unproven technology over the whole life-cycle of a project. Any disadvantage of a new technology is actively hidden by its developers and promoters, because if deficiencies were recognized early the entire industry would become unattractive to investors, generating a wave of lawsuits with multi-billion dollar compensations.

Collaboration between governments and private industries in the development of state economic policy, the widespread practice of getting a job in the private sector after a career in government, and corporate financing of election campaigns, whether legitimate or outright corrupt—all these contribute to a convergence of interests between the political and business elites. When business has poured vast sums of money into cultivating and supporting politicians, government all too readily agrees to promote the deregulation of markets and industries. The experience of the disasters we will elaborate below shows that corporations, having actively lobbied for deregulation measures and weakened public control over their activities, lost powerful and objective external controllers. Before deregulation, such controllers could have prevented the implementation of risky and reckless management decisions by strict legislation, and through the continuous monitoring of business practice by highly skilled government representatives. But politicians, often elected with a mandate to cut back on bureaucracy, and subjected to determined lobbying by private industry, cut the salaries of government representatives. So regulators can no longer hire highly educated and experienced staff, and pay levels in private industry are several times higher than for government jobs, leading to a decline in the prestige of public service and a dearth of skilled government officials. Faced with executives who have a strong interest in pulling the wool over their eyes, those officials lack the education and experience to understand new technologies being implemented in the industries they are overseeing, or identify major risks that these industries want to hide from regulators and the public. Moreover, the loyalty of regulatory representatives can either be bought by private business—through employment guarantees after the end of the government job, or just simple bribery—or extorted by threats of dismissal following a word in the ear of the right government official.

The examples that follow clearly confirm that constant deregulation of industry leads to the disappearance of compulsory reporting to the authorities, and to a fragmented and inadequate perception of risk among the regulators who are primarily responsible for disaster prevention within any industry.

2.2 *Deregulation Measures in the Financial Industry*

2.2.1 Deregulation in the British Financial Sector and the Collapse of Barings Bank

In February 1995, Barings PLC—the oldest and most reputable bank in Britain and one of the oldest merchant banks in the world—collapsed through the unauthorized trading of Nick Leeson, a Singapore-based trader at the bank, who single-handedly lost about US$1.4 billion (£827 million). Barings began to work in the futures market after Margaret Thatcher's Conservative government deregulated the British financial sector in 1986, allowing traditional commercial banks to provide investment bank services such as securities brokerage and underwriting. The government wanted to give British banks an edge in the international markets and ensure the status of London as one of the world's financial centers. Following the deregulation, Barings Bank made more than 50 % of its total profits from securities [6]. Whenever Leeson asked the London headquarters to fund his speculations between the Japanese and Singapore stock exchanges, he always received money, because he manipulated the accounts to make the Singapore subsidiary seem highly profitable to Barings management. None of Barings' executives could see the potential problem with its Singaporean branch in time because they lacked comprehensive knowledge of futures trading and this large financial institution lacked adequate internal controls. Top managers of Barings were blinded by falsified reports from Singapore and transmitted themselves these inadequate assessments to the regulator. In 1993, the bank's chairman Peter Baring commented to Brian Quinn, Director of the Bank of England: "*The recovery in profitability has been amazing following the reorganization, leaving Barings to conclude that it was not actually terribly difficult to make money in the securities markets*" [7]. Barings had an exclusive relationship with the Bank of England: according to Lord Hollick, the British central bank had an "*informal regulatory regime*" concerning Barings [8]. This cozy relationship with the authorities allowed Barings to violate restrictions on capital adequacy in order to increase their profits in Singapore. According to Leeson: "*[Barings bank's] capital base was only $250 million, [but] at the end of 1994 I had $500 million in Singapore, so twice the capital base of the bank. I think it was 10 times the legal limit that [a bank] could lend to a subsidiary, which the Bank of England had allowed to happen*" [9]. The regulator must have been delighted that its efforts towards deregulation seemed to be leading to greater profitability in the British banking sector.

2.2.2 Accounting, Electricity and Energy Futures Deregulation in the United States, and the Rise of Enron

Close and corrupting relationships between Enron executives and the US political elite played an important role in Enron's growth. Kenneth Lay had a cozy relationship with the Bush family as a devoted friend and major contributor to the election campaigns of George H.W. Bush, George W. Bush and other Republicans [10–15]. This familiarity helped Lay and Enron to benefit from the easing of government control in several spheres.

Firstly, George H.W. Bush was Vice President of the USA during the eight-year presidency of Ronald Reagan, an apologist for deregulation in many spheres, including finance, transport and energy. Reagan made his position very clear: "*Government is not the solution to our problem; government is the problem*" [16]; "*We who live in free market societies believe that growth, prosperity and, ultimately, human fulfillment are created from the bottom up, not the government down... [We] believe in the magic of the marketplace*" [17]. From 1989 to 1993, Bush continued Reagan's deregulation strategy as President in his own right. In the dozen years of Republican power, new principles were established for the federal deregulation of the American wholesale and retail electricity markets. In the 1990s, this energy deregulation continued on a state level. For example in California, the Republican Pete Wilson, state governor from 1991 to 1999, deregulated electricity supply in 1996: state power plants were sold off and electricity bought from a single wholesale pool [18]. But in 2000–2001, an electricity crisis erupted. Enron energy traders manipulated electricity supplies, creating an artificial power shortage and causing blackouts by shutting down Californian power plants, and raised state wholesale prices by 1000 %; the price of natural gas, also provided by Enron, jumped by the same amount [19]. Enron earned billions on overpriced electricity and natural gas.

When George W. Bush won the US presidential elections in 2000, he appointed a Secretary of Energy who had previously received campaign contributions from Enron as Republican senator for Michigan [20]; and the new Chairman of the Federal Energy Regulatory Commission—which regulates the transmission and sale of electricity, natural gas and oil in interstate commerce—was Lay's recommended candidate. In spring 2001, the new Governor of California Gray Davis asked Bush's Republican administration for a federal response to the state's electricity crisis, but Bush refused any federal government intervention or price controls. He maintained that there were still too many state regulatory restrictions, and that federal government had nothing to do with energy companies manipulating the market; and he personally did not see Enron's role in the California crisis [21–23]. This passive attitude probably reflected the wider political context: California had voted for Democratic candidate Al Gore in the recent presidential elections, and Democrat Gray Davis had presidential ambitions for the 2004 election cycle [24]. Moreover, Davis had signed the nation's first state law requiring car makers to limit auto emissions—damaging the interests of oil companies and car manufacturers, both heavyweight supporters of the Republican Party through campaign contributions.

Secondly, Enron benefited from the deregulation of energy futures trading. In 1989, early in George H.W. Bush's presidency, Enron started trading natural gas commodities and commodity derivative financial contracts. From this time, along with the investment banks, Enron lobbied for the removal of regulatory restriction on over the counter derivatives—and particularly energy derivatives—from the Commodity Futures Trading Commission. In 1989, the Securities and Exchange Commission (SEC) "*began requiring that managers make specific disclosures of financial contingencies and off-balance-sheet arrangements when a particular 'trend, demand, commitment, event or uncertainty' was 'reasonably likely'. [However], if management determined that the contingency was not reasonably likely to occur, no disclosure was required*" [25]. And on January 30 1992, the SEC accepted the mark-to-market accounting method for the energy contracts of Enron Gas Services group, which allowed Enron to calculate its revenue by the market value of derivative trading, creating the illusion that they were "*larger*" than General Electric, Citigroup, or IBM [26]. Lay was co-chairman of George H.W. Bush's re-election committee for his second presidential race in 1992 —Bush lost, but Enron continued lobbying. Derivative traders also found support from Alan Greenspan, Chairman of the U.S. Federal Reserve during four US presidencies (Reagan, Bush, Clinton and Bush), and from Robert Rubin and Lawrence Summers, Secretaries of the US Treasury during Clinton's terms—all ardent apologists for deregulation in the financial sector [27]. This deregulation would ultimately be a significant catalyst for the global financial and economic crisis in 2008–2009. And in 2001, Harvey Pitt, the private lawyer of the "Big Five" accounting firms, including Arthur Andersen, was appointed as SEC Chairman in George W. Bush's administration [20]. Over decades of lobbying, the SEC budget was consciously reduced, even though derivatives trading was becoming more complex [28].

Thirdly, US accounting practice is based on state regulation, and both the Texas-registered Enron and the Houston office of Arthur Andersen, which approved "creative accounting" at Enron, were under the jurisdiction of the Texas State Board of Public Accountancy (TSBPA) [29]. Mike Conaway was appointed as TSBPA chairman until 2004 during George W. Bush's term as state governor in 1995–2000. The worst accounting falsifications at Enron occurred while Conaway was at the TSBPA. In the 1980s, Conaway had been chief financial officer of Arbusto and Bush Exploration [30]. These were among several small oil companies particularly owned by George W. Bush; they were drilling in Texas in the 1980s–1990s, but the wells ran at a loss because of high production costs during a decade of low energy prices [31]. According to Paul Krugman, a Nobel laureate in economic sciences and columnist for The New York Times, Harken Energy—a merger of Bush Exploration/Arbusto and a third Bush-owned company—tried to falsely inflate their revenues in order not to go bankrupt: "*Mr. Bush [who was on the board of directors and head of the finance audit committee] profited personally from aggressive accounting identical to recent scams that have shocked the nation*" [32].

Once Enron's activities were deregulated and the company was no longer legally required to disclose them, Enron management began years of systematic accounting falsifications, which were revealed only after the company's bankruptcy.

2.2.3 Deregulation of the American Financial Sector and the Subprime Mortgage Bubble

The American financial lobby looked to the academic world for theoretical credibility for deregulation, and engaged prominent academics to study the possible advantages of deregulated markets. They offered millions of dollars in funding and grants, generous speaking fees and salaries for membership on the boards of financial institutions [33]. Naturally, this led to the dominance of a free-market theory, supported by apparently solid academic studies arguing for a reduced government role in the economy. This gave financial lobbyists a legal justification for deregulation, and convinced politicians to dismantle the legal framework that had been in place since the Great Depression: from 1999 to 2008, the financial sector spent US$2.7 billion on federal lobbying. The sector also contributed more than US$1 billion to political campaigns during this period [34].

In 1998, during the Clinton administration, Citibank took over Travelers Insurance Group—which owned Salomon Brothers investment bank—to establish the largest financial institution in the world, Citigroup Inc. This deal violated the Glass–Steagall Act of 1933, but the Federal Reserve made an exception for the merger. The Glass–Steagall Act restricted securities activity by commercial banks, and affiliations between commercial banks and securities firms, to avoid conflicts of interests. The creation of a single financial institution combining an investment bank, a commercial bank and an insurance company was prohibited. At the time of the deal, the Secretary of the Treasury was former Goldman Sachs executive Robert Rubin; after the merger he joined the board of directors of Citigroup Inc., becoming chairman of the executive committee and chairman of the board (1999–2009). Citigroup Inc. paid him up to US$126 million [35]. And in 1999, after lobbying from the financial sector, Congress passed a new act to lift all restrictions against the combination of banking, securities and insurance operations within a single financial institution, paving the way for further mergers [36]. By 2005, the ten largest US commercial banks held 55 % of the industry's assets—twice the share held by the top ten in 1990 [37]. Lawrence Summers, Rubin's successor as Secretary of the Treasury and a former academic economist, said on the passing of the new act: "*Today, Congress voted to update the rules that have governed financial services since the Great Depression and replace them with a system for the 21st century. This historic legislation will better enable American companies to compete in the new economy*" [38]. But the Financial Crisis Inquiry Commission (FCIC) convened after the crisis described the new arrangements as "*a 21st-century financial system with 19th-century safeguards*" [39].

As well as continued lobbying from the financial sector, the inherent complexity of creating and calculating derivatives helped to impede serious government regulation over innovative financial instruments. When the Commodity Futures Trading Commission expressed their intention to regulate OTC derivatives, their attempts to do so were suspended by Alan Greenspan, Robert Rubin and Lawrence Summers [27]. Greenspan advised: "*Regulation of derivatives transactions that are privately negotiated by professionals is unnecessary*" [40]. In the 20 years from early 1990 to 2009, the unregulated global derivatives market—which dealt almost entirely in OTC derivatives—grew from US$10 trillion to US$605 trillion [41], or nearly ten times the world GDP at the time [42]. As we have already seen the SEC, far from expanding to deal with this huge market, was continually cut back over the same period.

Furthermore, the banking, securities and insurance operations of the new merged financial institutions were still overseen by separate bodies: there was no unified regulator building up a holistic picture of the risks involved in the housing bubble and the securitization pipeline. The development of these gigantic multi-industry mergers was happening too fast for the government regulatory framework to keep up, and there was no parallel development of a "mega-regulator". According to John Snow, US Secretary of the Treasury from 2003 to 2006, regulators tended not to see problems at their own institutions: "*Nobody had a full 360-degree view. The basic reaction from financial regulators was, 'Well, there may be a problem. But it's not in my field of view*'" [43]. The CEO of Citigroup told the FCIC commission that US$40 billion invested in highly rated mortgage securities would "*not in any way have excited my attention*", and the commission reported the co-head of Citigroup's investment bank saying that he spent "*a small fraction of 1 % of his time on those securities*". The commission summed up: "*too big to fail meant too big to manage. We conclude a combination of excessive borrowing, risky investments, and lack of transparency put the financial system on a collision course with crisis*" [44]. So neither government nor financial executives had the whole picture of the risks involved in a complex combination of businesses with different interests, and especially in the widening distribution of derivatives.

2.3 Deregulation Measures in Different Industrial Sectors

2.3.1 Chernobyl Nuclear Disaster

After the 1973 oil crisis, the Soviet Union began to receive tremendous export revenue from hydrocarbons, so it was rational to try and shift domestic energy production towards the active development of cheap nuclear energy rather than burning valued hydrocarbons. As we have seen, the Politburo invited the developers of the Soviet nuclear arsenal to oversee the construction and development of civil nuclear energy. These brilliant minds developed the RBMK reactor, which was highly economical, fast and easy to construct. The reactor core design contained several minor theoretical and technical mistakes, which together cause what is known as the "positive SCRAM effect" [45], a complication only occurring in practice under a rare combination of circumstances. The reactor's designers and developers were distinguished members of the Academy of Sciences of the USSR, and had close and friendly relations with senior ministers in all the government departments responsible for the Soviet civil nuclear industry. This made it almost impossible to create an independent government body to oversee the industry and spot potential deficiencies in the reactor design. The developers convinced everybody—especially the Politburo and the plant operators—that the RBMK reactor was absolutely safe [46–48]. Among Soviet decision makers and the industry alike, there was widespread wishful thinking and overconfidence about the development of the civil nuclear industry.

2.3.2 Deepwater Horizon Oil Spill

After the dramatic increase in oil prices following the 1973 oil crisis, the industry turned to offshore drilling—especially in the Gulf of Mexico. In 1978, Shell Oil launched drilling there at a depth of 1000 ft (304 m) underwater. In 2006 Chevron, Devon Energy and Statoil drilled an exploratory well 7000 ft (2133 m) underwater [49], reaching a total depth of 28,125 ft (8572 m). And in 2009, working from the Deepwater Horizon platform, BP discovered the gigantic Tiber Oil Field, with resources of between 4 and 6 billion barrels at a total depth of 35,056 ft (10,685 m) and under 4130 ft (1258 m) of water [50]. In 2011, 30 % of U.S. crude oil was extracted in the Gulf of Mexico [51].

In 1982, the U.S. Department of the Interior established the Minerals Management Service (MMS) to regulate such intensive offshore drilling. But with the rise of the doctrine that government oversight of private enterprise should be kept to a minimum, and with active lobbying from the industry, the budget of MMS was cut from US$250 million in 1984 to less than US$200 million ($100 million at 1984 dollar values) in 2009. Meanwhile oil companies progressed considerably in the development of deepwater drilling over this period [52]. The regulator could not afford to hire specialists who understood innovations in the field, and instead had to rely on the expertise of deepwater operators and contractors. Moreover, by 2009, there were far fewer unannounced MMS inspections of offshore oil infrastructure than there had been in the 1980s [2]. This regulatory impotence led to a situation where US offshore drilling operators were free to implement or reject innovations in the safety requirements for offshore drilling, even when other countries had brought them in as compulsory measures after accidents [53]. One such innovation is the acoustics trigger now required in all deepwater blowout preventers in Norway and Brazil, enabling a well to be shut down remotely in an emergency. These triggers cost over US$0.5 million apiece—and in the USA the use of such devices was optional [54]. Worse still, BP had no contingency plan for emergencies arising while drilling the Macondo well: again, this was not required under US deepwater drilling legislation [55].

2.3.3 Fukushima-Daiichi Nuclear Accident

Japan started developing civil nuclear energy in the mid 1960s. It has been a national strategic priority since the oil crisis in 1973 because Japan depends heavily on imported fuel, which provided 84 % of its energy needs in the 2010s [56].

Before the accident at Fukushima-Daiichi, the Nuclear and Industrial Safety Agency (NISA) worked under the authority of the Ministry of Economy, Trade & Industry (METI) to both promote and regulate nuclear energy. Clearly, this was a conflict of interests: NISA's primary goal was to protect society from radiation threat, but the agency also sought the energy independence of Japan. This involved supporting low-cost electricity production from a large number of nuclear plants, and maintaining a stable financial climate for the further development of nuclear technology by the nuclear industry. So a cozy relationship developed between operators, regulators and academics, leading to a situation where "*the regulators and the operators prioritized the interests of their organizations over the public's safety, and decided that Japanese nuclear power plant reactor operations* '*will not be stopped*'. *Because the regulators and operators have consistently and loudly maintained that* '*the safety of nuclear power is guaranteed*', *they had a mutual interest in averting the risk of existing reactors being shut down due to safety issues, or of lawsuits filed by anti-nuclear activists. They repeatedly avoided, compromised or postponed any course of action, and any regulation or finding that threatened the continued operation of nuclear reactors*" [3].

Neither the industry nor the regulators felt any need to implement safety improvements learnt from the experience of nuclear accidents elsewhere, because nuclear plants in Japan were already designed to cope with severe disasters such as high-magnitude earthquakes. This was part of a pervading sense of superiority of the Japanese technology and know-how over those of foreigners. Within this Japanese Zeitgeist, a sense of superiority was difficult to avoid, with the extraordinary success of Japan Inc. to export worldwide all kinds of high-tech products, translating in the massive growth of Japanese stock market valuation and trading volumes, which towards the end of the Japanese bubble in 1989, topped that of the U.S. market! "*What is there to learn from accidents in foreign nuclear plants, given that the Japanese way is so much better?*" was a common thinking. In 1991, this complacency led to a situation where operators on Japanese nuclear stations were left free to apply the safety measures they saw fit, independent from the control of regulators: "*the accident management, including expedient and flexible measures that might be required under actual situations, shall be considered and implemented by the operators based on their* '*technical competency*' *and* '*expertise,*' *but [it] shall not require authority to regulate the specific details of measures*" [57]. Investigations after the accident revealed that many IAEA safety recommendations and guidelines generated by nuclear accidents elsewhere in the world had been ignored, or their implementation postponed, by Japanese nuclear operators [58–60].

2.3.4 Shale Energy Development

The United States represent just 4.4 % of the world's population but consume more than 26 % of the world's energy. So the country is constantly seeking new ways to increase energy independence. One of these is the development of unconventional oil and gas resources within the US. In the late 1990s and early 2000s, Halliburton, a leading service company in the petroleum industry, was developing methods for the extraction of gas and oil from shale formations based on hydraulic fracturing or "fracking". Halliburton's CEO from 1995 to 2000 was Dick Cheney, who had been US Secretary of Defense under George H.W. Bush from 1989 to 1993, including the first Iraqi campaign of 1990–1991. He would subsequently become Vice-President of the US during the two terms of George W. Bush from 2001 to 2009. As CEO of Halliburton, Cheney pushed to improve the technology of hydraulic fracturing. Then, during the eight years of his vice-presidency, Cheney directed the Energy Task Force. One goal of this group was to promote the development of domestic unconventional gas and oil resources, based on the technology of fracking. In 2001, the group described the technology: "*This is a common procedure used by producers to complete gas wells by stimulating the well's ability to flow increased volumes of gas from the reservoir rock into the wellbore. During a fracture procedure, fluid and a propping agent (usually sand) are pumped into the reservoir rock, widening natural fractures to provide paths for the gas to migrate to the wellbore. In certain formations, it has been demonstrated that the gas flow rate may be increased as much as 20-fold by hydraulic fracturing*" [61]. However, the report did not mention that the "fluid" involved contains a mixture of 500 chemicals, added to the water to allow it to permeate more effectively through the rock. These chemicals are extremely harmful to the environment [62–67]. Some of these substances are recognized as extremely hazardous for human health and the environment [68]. Water, at just US$0.8 a barrel, was the obvious choice as the cheapest basis for fracking fluids. Even so, more than a quarter of operating costs are water-related expenses [69]. A shale gas well requires an average of 15,000 tons of water in its lifetime: 10–100 times the amount consumed by a conventional gas well [70, 71]. Cost considerations have prevented the development of water-free fracking using a thick propane gel in place of water—to reduce damage to the environment. This alternative is at least 20–40 % more expensive than hydraulic fracking [72, 73] but uses no water at all.

Not surprisingly, hydraulic fracturing technology contravened strict American standards on environmental pollution. Nevertheless, widespread shale oil and gas drilling has been carried out since 2005, when the Bush administration made changes to most of the relevant legislation in favor of the shale industry, including the Safe Drinking Water Act, the Clean Air Act, the National Environmental Policy Act, and the Clean Water Act. By 2013 more than 82,000 wells had been drilled and fracked on 31 shale plays in the United States; around 1 billion tons of water were contaminated and more than 1400 km^2 of land damaged [74].

A whistleblower at the US Environmental Protection Agency (EPA) revealed that during the Bush administration top EPA officials had conflicting interests with shale gas companies, and allowed them to continue fracking despite the possible risks of water contamination [75]. During the Obama administration, the EPA has continued to underplay the environmental damage of the shale industry in pursuit of national energy independence by any means. The US has also undertaken a massive promotional campaign for the development of shale gas in Europe, but an independent study undertaken by the European Parliament, "*Impacts of shale gas and shale oil extraction on the environment and on human health*", revealed the real environmental threat of fracking to highly populated territories there [76].

Because each of the 30-odd shale deposits being "played" in the US is geologically unique, operators persuaded the government to implement state level regulation of shale exploration, instead of an integrated federal oversight. So there has been little or no interaction between the various regulators involved—which enables energy companies to hide the whole picture of risks, and the authorities to keep the dangers of fracking out of the public eye at a national level. For instance, the EPA only announced that fracking may be to blame for causing groundwater pollution in 2011, after fracking chemicals were detected in groundwater beneath the Pavillion field in Wyoming. The regulator emphasized that the findings were specific to the Pavillion area [77]—but in fact shale operators generally use similar exploration technologies across different shale plays. Consequently regulators have a fragmented picture of risks: they cannot fully understand the risks associated with the large-scale exploitation of shale formations using innovative, but complex and unproven technology. This situation is reminiscent of the regulation of subprime mortgage deals leading up to the financial crisis in 2008–2009. In what became known as the "securitization pipeline", mortgages were sold to people who were highly unlikely to keep up their repayments, re-packaged into bundles and presented to investors as safe investments—but these secondary investors had no idea of the true instability of that they were buying. And because there was no mega-regulator to oversee the whole picture "*nobody had a 360-degree view*" [43]. Like the financial sector just five years earlier, the shale industry was left with toothless regulation: massive cuts in government spending at both federal and state level led to the budgets of most EPA departments being reduced, and many representatives of the agency were fired.

To convince investors and other countries that a "shale revolution" could deliver resources for centuries to come, the US government even changed the rules for evaluating unconventional reserves and resources. In the last weeks of its term in early 2009, Bush's administration implemented new SEC rules. Domestic energy companies could now book unconventional oil and gas reserves more generously, based on "in-house" estimates of the amount of hydrocarbons recoverable from a proposed oil or gas field—estimates made without test drilling, and ignoring the economic viability or technical feasibility of extracting these reserves. A New York Times investigation, after the implementation of the new SEC rules, revealed that at least seven of the largest 19 shale operators increased their estimated reserves—some by more than 200 % [78]. On the basis of data filed in various states on actual well production, some analysts estimate that operators now overbook their shale reserves by 400–500 % [79].

Resource manipulation also allows the US government to predict a century of natural gas abundance—see for example Barack Obama's declaration that "*We have a supply of natural gas that can last America nearly 100 years*" [80]. With such a discrepancy between estimates on shale gas resources and real production data, the US Government Accountability Office was reduced to informing the main governmental comptroller: "*the amount of domestic technically recoverable shale gas could provide enough natural gas to supply the nation for the next 14 to 100 years*" [81]. Again, these changes to the SEC resource accounting rules resemble those in 1992 under George H.W. Bush, when the SEC accepted the mark-to-market accounting method for the energy contracts of Enron Gas Services. As we have seen, this allowed Enron to calculate its own revenue by the market value of derivatives trading—in other words, on the basis of an "in house" estimate of how deals would perform in the future—and to create the illusion of being "larger" than General Electric or IBM [82].

3 Internal Organizational Environment Promoting Risk Concealment

3.1 Main Mechanisms

In "Das Kapital", Karl Marx quoted T.J. Dunning, an English trade unionist: "*Capital is said ... to fly turbulence and strife, and to be timid ... but ... with adequate profit, capital is very bold. A certain 10 % will ensure its employment anywhere; 20 % ... will produce eagerness; 50 %, positive audacity; 100 % will make it ready to trample on all human laws; 300 %, and there is not a crime at which it will scruple, nor a risk it will not run ... If turbulence and strife will bring a profit, it will freely encourage both. Smuggling and the slave-trade have amply proved all that is here stated*" [83].

In all the cases we have mentioned, managers and owners prioritized their company interest or personal income over the interests of investors or the public, because of weak public oversight, public ignorance about their activities, regulators inadequately qualified to assess the risks, and cozy relationships—if not outright collusion in corruption—between business and government. After the financial meltdown in 2008, Fed Chairman Greenspan confessed: "*Those of us who have looked to the self-interest of lending institutions to protect shareholders' equity (myself especially) are in a state of shocked disbelief*" [84].

Investors focused on quick turnover projects, the unpredictability of a globalized world market, politicians looking for a quick fix to please voters, all these factors encourage companies to develop short-term strategies. Shareholders and politicians want short-term business development—and thus economic progress through revenue growth and tax payments—and this translates into ambitious and often unrealistic business strategies. So companies develop operational plans that create enormous stress in the workforce, who have to ignore risks to achieve unrealistic results, and distort information about risks to show their superiors that they can deliver: if managers judge them unable to handle a challenging environment their jobs may be on the line. Those who disagree with such practice will not last long in the organization. To meet unrealistic targets, managers promote a risk-taking approach, facilities are operated close to their tolerance limits while cutting back on costs and labor… and the whole situation may be heading for disaster. And managerial compensation usually involves the payment of annual bonuses, which encourages short-term business strategies because it motivates managers to show short-term results by any means.

Unrealistic targets, set under pressure for short-term results in a competitive market, also force managers to demand constant haste from their subordinates. There is no time for rigorous evaluation of potential risks, or for communication with other units, so employees have to rely on their own experience to assess risks. And the rush to complete projects forces workers to compromise on quality: experience shows that having to settle for half-baked and barely tested solutions motivate organizations to withhold information about shortcomings, to avoid causing the outrage of customers and regulators.

Unrealistic expectations and the push to increase productivity compel managers to shift responsibility for the implementation of plans onto their subordinates; a culture of "*no bad news*" develops, where only those who produce clear short-term results will survive. Workers have to find "*their own solution*", and take the initiative without bothering the management. They are afraid of layoffs, afraid of being written off as incompetent if they fail, and under pressure to keep setting unrealistic goals. This fear culture obliges employees to distort information about their own success, to falsify records, to tell managers what they want to hear and withhold information about risks or shortcomings until they have no choice, and to deny personal responsibility.

When executives are under pressure to cause impressive results quickly, the weakening of internal control seems to serve their interests. A professional, thorough and independent audit department, collecting information about both staff and managerial activity and making impartial assessments, seems a dangerous witness that could be used by regulators and investigators in the event of disaster. So in many of the cases we have investigated, in-house regulatory departments were axed—or if not they were deliberately staffed with incompetent employees, unable or unwilling to work with integrity.

Maximizing revenue and reducing costs dictates significant wage cuts, so the most competent staff looks elsewhere. Thus, the organization can lose those who understand the complicated or risky aspects of its work. On complex technological sites, this loss of expertise is dangerous and always leads to accidents. But in the financial sector, companies like Enron and those implicated in the subprime mortgage bubble deliberately encouraged a high staff turnover precisely to stop employees grasping the true scale of the risks the organizations were taking. These companies looked for young people with little experience—they needed loyal, ambitious staff who would do whatever was required to achieve short-term results at any cost.

Neither a government keen to support the development of risky new technologies, nor an industry which is implementing them, has any interest in encouraging whistleblowers—active citizens with the courage to inform regulators and the public about illegal or dangerous practice. In the background to many disasters, there have been close ties between regulators and executives who had a common interest in the concealment of risky working practice in an industry. Regulators often have explicit or implicit orders from politicians to overlook risks in order to maintain economic growth; so whistleblowers threaten not just the industries who are cutting corners, but the regulators who tacitly approve of dangerous practices. Most employees would rather keep quiet about risks at work than be condemned or even ostracized by colleagues and lose their income. Experience shows that most whistleblowers have acted solely out of personal moral principles—but in spite of public approval, their careers were ruined and they paid for their integrity in their personal and professional lives.

3.2 *Internal Culture of Risk Concealment in the Financial Industry*

3.2.1 The Collapse of Barings

Nick Leeson admitted that his main motivation for concealing his losses and falsifying his profits was "*fear of failure*". The work culture, both on the Singapore stock exchange and within Barings bank, respected success and profit and despised failure and loss: if Leeson's true losses were revealed, his "*incompetence, negligence and failure*" would be exposed [6]. Barings managers were dazzled by the apparent profits from Singapore, which would directly affect their annual bonuses; they assumed Leeson was making fully matched trades at no real risk to Barings [85]. So they sent more money to Singapore to cover his losses, apparently convinced that he would make them millions: "*[Barings was] driven to make profits, profits, and more profits…*" [86]. "*[I]t was their greed that lay at the root of the whole problem. They did not want to know about the internal structure of the firm*" [8].

3.2.2 The Enron Case

Enron "achieved" the tremendous annual growth of its revenues—from about US$13 billion in 1996 to US$138.7 billion for the first 9 months of 2001—by using the aforementioned mark-to-market accounting method, and pioneering the "*merchant model*" to set up Enron Online, one of the first online trading platforms. Mark-to-market accounting was based on reporting, for both online and "traditional" deals, "*the entire value of each trade on which it was a counterparty as its revenue, rather than reporting as revenues only its trading or brokerage fees*" [87], while investment banks used the "*agent model*", a more conservative approach based on brokerage fees alone [87]. Enron's business model aimed to continuously raise executive earnings by maintaining constant growth in the company's market value.

To achieve this, the company needed to continually increase short-term revenue figures while keeping debts low. So Enron's executives bribed their auditors and several investment banks, offering lucrative secure contracts if they produced the required figures. By falsifying the accounts, Enron was able to declare a market value growth of more than 450 % between 1996 and 2000, to over US$60 billion—70 times their income and six times book value [88]. Over those five years, Enron paid its top five executives more than US$500 million in options, bonuses and salaries [89].

Under the leadership of CEO Jeffrey Skilling, Enron developed a "*cut-throat*" corporate culture, unusual for an energy company: it would have been more appropriate for an investment bank. Because mark-to-market accounting allowed profits from long-term deals to be recorded in the current year, traders were under enormous pressure to maintain the growth of company revenue and market capitalization by delivering ever more gigantic new deals. At Enron, it was not quality that mattered, but the size of deals and the maintenance of a constant 'deal flow' [90]: "*Good deal versus bad deal? Didn't matter. If you could give it a positive Net Present Value it got done*" [91]. And deals were no sooner done than forgotten, since the trader received compensation immediately; so the entire staff of Enron was focused on short-term output [90]. The company hired "*the best and the brightest*" young MBA school graduates: too inexperienced to immediately grasp the flaws of the Enron corporate system, but very smart, ambitious, and hungry for short-term money. Rewards for traders who met their earnings targets were huge: some common traders could earn up to US$15 million a year [90]. The message was simple: "*If you were smart enough and tough enough to work at Enron, you deserved to live like last year's Oscar winner*" [92]. In 2000, base salaries exceeded the peer group average by 51 %, bonus payments by 382 % and stock options by 484 % [89]. And because employee pension funds were invested in Enron stock, and stock options formed a significant slice of their compensation, employees pushed to increase Enron's capitalization by any means. In return for such large compensation, the company demanded high productivity, unquestioning loyalty and faith in "*the Enron way*": employees were even nicknamed "*Enronians*" and "*believers*" [93]. Furthermore, Skilling set up a system of selection and ranking of personnel unparalleled in corporate America for its ruthlessness. This was called the Performance Review Committee (PRC), and it was a six-monthly audit of the number, profitability and permanency of the deals an employee had brought into Enron. Every six months, staff dreaded finding themselves among the bottom 15 % in the PRC rating; if they were still there in the following review, they would be fired [91, 94]. The system strengthened competition between traders and alienated them from each other. Nobody in the company could afford to be honest with anyone else about the risks they were taking: "*Clearly, the switch from affirmation to punishment within Enron meant that employees regularly received mixed messages. On the one hand, they were the cleverest and best in the world—a form of positive reinforcement, or love bombing, that it would be hard to better. On the other, they could be branded as 'losers', and fired at any time. Consistent with general cultic norms, the overall effect was disorientation, an erosion of one's confidence in one's own perceptions and, most crucially, a further compliance with the group's leaders that strengthened conformist behavior in general… It is clear that Enron management regarded kindness as a show of weakness. The same rigors that Enron faced in the marketplace were brought into the company in a way that destroyed morale and internal cohesion. In the process of trying to quickly and efficiently separate from the company those employees who were not carrying their weight, Enron created an environment where employees were afraid to express their opinions or to question unethical and potentially illegal business practices. Because the rank-and-yank system was both arbitrary and subjective, it was easily used by managers to reward blind loyalty and quash brewing dissent… [There was a] prevailing culture [of] 'the undiscussability of the undiscussable also undiscussable'… [A] former senior manager's summary of the internal culture: 'There was an unwritten rule… a rule of 'no bad news.' If I came to them with bad news, it would only hurt my career'*" [93]. "*Paranoia flourished and trading contracts began to contain highly restrictive confidentiality clauses. Secrecy became the order of the day for many of the company's trading contracts, as well as its disclosures*" [94]. "*Enron Gas Services was developing a reputation as a predatory place where people would sell each other out to survive*" [95]. This internal climate of fear and concealment soon distorted communication with external audiences too. Mark Koenig, Enron's former head of investor relations, testified: "*I wish I knew why I did it. I did it to keep my job, to keep the value that I had in the company, to keep working for the company. I didn't have a good reason*" [90].

In the 1990s, Enron's auditors Arthur Andersen were actively expanding their operations into accounting consulting. Similar to what was happening at Enron, different units at Arthur Andersen competed with each other, avoiding open communication about the problems of their clients and pursuing continuous growth—regardless of the source of new revenue, the quality of clients or even the legality of their recommendations [96]. The largest of Arthur Andersen's clients worldwide was WorldCom, which was to file for bankruptcy in 2002; Andersen's second largest client worldwide, and the largest at their Houston office, was Enron [97]. The Houston office provided both auditing and the new consulting service to Enron: more than 70 % of the fees that Andersen received from Enron came from consulting. Andersen consultants helped to bring in more aggressive accounting and oversaw the creation of special purpose entities (SPEs)—limited liability companies formed solely in order to separate profit, debts or risks from the main company and keep them off the books. Meanwhile the Andersen audit unit earned US$1 million a week for internal and external auditing [98]. With no fraud examiners and no internal audit department [99], Enron outsourced their own "internal audit" to Arthur Andersen—and over time, many of Enron's own accountants and controllers were recruited as former Andersen executives [94]. The bonuses of staff at the accountants' Houston office depended on the stable growth at Enron, and many Andersen employees, "*[l]ured by promises of undreamt-of-wealth… aspired to work for Enron and were therefore very reluctant to 'rock the boat' with the company*" [100]. It was only a matter of time before auditors were approving falsified accounting reports to increase their own bonuses. Andersen auditor Carl

Bass, amongst others, voiced his concern about the way Enron was using mark-to-market accounting and special purpose entities—but David Duncan, Andersen senior executive at the Houston office, simply responded by removing Bass from the Enron account [101]. If Bass had gone directly to the Texas State Board of Public Accountancy—which as we have noted was under the control of a friend of George W. Bush and Ken Lay—he could have lost not just the account, but his job or even his career as an auditor in the state of Texas, with no assurance that the case would even be properly investigated by the TSBPA. In the end, although he was eventually recognized as an accounting hero, Bass lost his license along with many of Enron's other former auditors [102]. Moreover, the Houston's irregularities were unlikely to come to the notice of Andersen international headquarters: the company had a weak system of internal control over its regional offices, and senior managers were delighted by the continuous growth of the Houston office's revenue, so they avoided asking awkward questions about the details of consulting and audit practice.

Investment banks were also more interested in Enron's results than in their methods. A substantial part of their income came from underwriting merger deals, whereas broker fees brought relatively insignificant profits. They took generous fees from Enron transactions, invested in Enron's off-balance-sheet SPEs and therefore had credit exposure to Enron [103]. So, investment bank analysts had no interest in publishing negative reports about Enron, and sell-side bank traders recommended Enron to their clients—though Enron stocks, with an average annual growth rate of over 65 %, more or less sold well by themselves [104]. When analysts like Merrill Lynch's John Olson made a "*sell*" recommendation on Enron stocks or published a "*neutral*" report, they were simply fired, since their employers had a close relationship with Enron's management. Merrill Lynch was rewarded handsomely for the dismissal of John Olson, with at least US$45 million in fees from Enron deals [105]. Enron's claim—apparently backed by the figures—that they would become "*the world's leading company*" attracted worldwide investment. The company stated confidently: "*We believe wholesale gas and power in North America, Europe and Japan will grow from a US$660 billion market to a US$1.7 trillion market over the next several years. Retail energy services in the United States and Europe have the potential to grow from US$180 billion to $765 billion in the not-so-distant future. Broadband's prospective global growth is huge – it should increase from just US$17 billion today to $1.4 trillion within five years. Taken together, these markets present [a several] trillion [dollar] opportunity for Enron... Our stock price is going to go to $120 per share*" [93, 106]. Enron shares were selling at a registered maximum of US$90 in August 2000; by late November 2001, the value of a share was less than $1. Goldman Sachs extolled Enron in an analytic report: "*Enron has built unique and, in our view, extraordinary franchises in several business units in very large markets*" [107]. According to Thomson First Call, 13 of Enron's 18 analysts were still recommending to buy Enron stocks in early 2001 [107]. Incredibly, 10 out of 15 analysts who followed Enron were even rating the stock as a "*buy*" or a "*strong buy*" on November 8, 2001—when Enron finally confessed to accounting fraud [108].

3.2.3 Subprime Mortgage Crisis

In the case of the mortgage bubble, the real estate boom had become a major source of revenue for the American financial sector. Between 1978 and 2008, the total debt held by the financial sector exploded from US$3 trillion to US$36 trillion; financial institutions were generating more than 33 % of all corporate income in the United States by 2003, when in 1980 they had accounted for only 15 % [37]. Before the deregulation of the early 1980s, lenders selected borrowers carefully, because they needed, for their own sake, to ensure that a borrower could pay a 30-year fixed-rate mortgage. The stability of financial institutions depended on the reliability of their debtors. Even in the 1990s, only the highest quality clients who could comply to tough requirements—known as "prime" borrowers—were eligible. For example, one requirement was that first-time homebuyers should be able to make a 20 % down payment. However, deregulation and active encouragement from the government allowed lenders to lower the acceptable standard for borrowers, and provide credit for people with no credit history or proof of income—and the "subprime" market was born.

Deregulation allowed the creation of the "securitization pipeline": lenders packaged subprime loans into residential mortgage–backed securities, and investment banks like Goldman Sachs, Merrill Lynch, Bear Stearns or Lehman Brothers repackaged these securities into collateralized debt obligations (CDOs). In their turn, CDOs were promoted among more conservative American investors like retirement systems, hospitals, endowment funds, and global investors such as pension funds and sovereign funds, as a "*super-senior*" and "*super-safe*" alternative to US Treasuries—with the same AAA-rating but a higher yield [109, 110]. Collateralized debt obligations were bundles, or "tranches", of mortgage–backed securities from a range of different quality debtors. Economist James Grant described the "*mysterious alchemical processes [by which] Wall Street transforms BBB-minus-rated mortgages into AAA-rated tranches of mortgage securities*" [111]. The banks insured themselves against potential default by setting up "credit default swaps" (CDSs) with companies like American International Group (AIG), the largest insurance company in the world. By 2007, AIG had issued CDSs with a total underlying value of $379 billion [112]. Having sold a mortgage securities package, lenders did not need to monitor the financial situation of debtors: payments—or defaults—from borrowers went to the owners of mortgage securities.

CDOs received their AAA rating from such respected rating agencies as Moody's, Standard & Poor's and Fitch. Just as Enron's executives had corrupted their auditors at Arthur Andersen and their underwriters among the investment banks a few years earlier, the investment banks now bribed rating agencies by paying "*handsome fees to the rating agencies to obtain the desired ratings*" [113]—between US$0.5 million and $0.85 million for every mortgage-related security. In the 1990s, the obligations were still of reliable quality, but as lending became more widespread it was harder to track the quality of borrowers. The rating agencies were perfectly aware of what they were doing. One employee at S&P wrote: "*Rating agencies continue to create an even bigger monster—the C.D.O. market. Let's hope we are all wealthy and retired by the time this house of cards falters*". Another wrote in an email: "*We rate every deal. It could be structured by cows and we would rate it*" [114]. And executives at Moody's testified: "*We had almost no ability to do meaningful research... The threat of losing business to a competitor* [Standard & Poor's or Fitch], *even if not realized, absolutely tilted the balance away from an independent arbiter of risk towards a captive facilitator of risk transfer... Bankers were pushing more aggressively, so that it became from a quiet little group to more of a machine... Subprime [residential mortgage–backed securities] and their offshoots offer little transparency around composition and characteristics of the loan collateral... Loan-by-loan data, the highest level of detail, is generally not available to investors*" [115]. In their standard disclaimer, Moody's stated that "*The ratings ... are, and must be construed solely as, statements of opinion and not statements of fact or recommendations to purchase, sell, or hold any securities*", thereby protecting the rating agency against lawsuits from misled investors. Nevertheless, this appropriation of fictitious ratings resembles the corruption at the Houston office of Arthur Andersen during the Enron case. Between 2000 and 2007, Moody's gave AAA ratings to nearly 45,000 mortgage-related securities. In 2006 alone, earnings on mortgage ratings reached US$887 million, or 44 % of overall corporate revenue, and Moody's was approving 30 mortgage-related securities as AAA every day. Ultimately, during the crash in 2007–2008, 83 % of the mortgage securities rated AAA in 2006 would be downgraded [116].

Thousands of new young people, with no mortgage experience, were hired to sell credit products "*to, in some cases, frankly unsophisticated and unsuspecting borrowers*" [117]. Lenders offered low monthly payments in the first few months of a loan and delayed bigger fees in later payments, which were seldom disclosed to borrowers. Executives at Countrywide—which was financing up to 20 % of all mortgages in the United States, around 25 million homebuyers—recognized even during the boom that many of the loans they were selling could cause "*catastrophic consequences*" to buyers and "*financial and reputational catastrophe*" to the firm. In an internal e-mail, the company's proprietor wrote: "*In all my years in the business, I have never seen a more toxic [product]*" [118]. But Countrywide and the investment banks continued to sell to investors nonetheless, and insurance companies continued to insure the sellers against default. According to the FCIC commission after the crisis, other lenders withheld critical information from investors too: while in Countrywide's portfolio, 59 % of its loans were "non-traditional" loans, Wells Fargo had 58 %, Washington Mutual 31 %, CitiFinancial 26.5 %, and the Bank of America 18 % [119].

In June 2006, Citi's chief business underwriter Richard Bowen discovered that up to "*60 % of the loans that [were bought] and packaged into obligations were defective. If the borrowers were to default on their loans, the investors could force Citi to buy them back. He tried to alert top managers at the firm by 'email, weekly reports, committee presentations, and discussions'; but though they expressed concern, it 'never translated into any action'. He finally took his warnings to the highest level he could reach—Robert Rubin, the chairman of the Executive*

Committee of the Board of Directors and a former US treasury secretary. He sent Rubin and the others a memo with the words 'URGENT—READ IMMEDIATELY' in the subject line. Sharing his concerns, he stressed to top managers that Citi faced billions of dollars in losses if investors were to demand that Citi repurchase the defective loans. Rubin told the Commission in a public hearing in April 2010 that 'I do recollect this and that either I or somebody else, and I truly do not remember who, but either I or somebody else sent it to the appropriate people, and I do know factually that that was acted on promptly and actions were taken in response to it'. According to Citigroup, the bank undertook an investigation and the system of underwriting reviews was revised... There was no disclosure made to the investors with regard to the quality of the files they were purchasing... Bowen told the Commission that after he alerted management by sending emails, he went from supervising 220 people to supervising only 2, his bonus was reduced, and he was downgraded in his performance review" [120]. Such practice was common not only in Citi, but also among other players of the securitization pipeline: Richard Fuld, CEO at Lehman Brothers, was eliminating internal critics who realized early that Lehman was heading for serious trouble. Warnings from researchers and even from managing directors were ignored.

Until the 1980s, most investment banks were private companies; a loyal employee would work for the same bank for decades and receive a bonus upon retirement. But the compensation model completely changed when investment banks became public companies, and staff began to trade with shareholders' money. Huge annual bonuses focused executives and managers' attention to short-term financial results: they pursued current capitalization growth and short-term profitability regardless of the possible risk in the long-term. In 2007, Wall Street executives received roughly US$33 billion in year-end bonuses [121]. Nobody wanted to overturn the teetering mortgage market by exposing the flaws in the business model they had created. After the crisis, JP Morgan CEO Jamie Dimon testified: "*I blame the management teams 100 % and ... no one else*" [122].

By December 2006, executives at Goldman Sachs had recognized "*the major risk in the mortgage business*". Ignoring their own rule that "*clients' interests always come first*", they secretly decided to sell mortgage securities only to their own clients. Comments like these make the prevailing attitude only too clear: "*Distribute junk that nobody was dumb enough to take first time around*"; "*[They] structured like mad and traveled the world, and worked their tails off to make some lemonade from some big old lemons*"; "*How much of that sh—deal did you sell?*" [123, 124]. The FCIC found that "*the firm targeted less-sophisticated customers in its efforts to reduce subprime*" [123]. In July 2007, Goldman Sachs withheld vital information from investors about the low quality of ABACUS 2007-AC1 [125], a CDO on which those investors would lose most of their $150 million investment only months later [126]. After the crisis Goldman Sachs, JP Morgan Chase, the Bank of America and other institutions were heavily fined by the SEC for overstating the quality of the mortgages they had been selling to investors.

3.3 Internal Environment of Risk Concealment in the Industrial Sector

3.3.1 Chernobyl Nuclear Disaster

The constant growth of domestic electricity needs put pressure on the Soviet nuclear industry to organize the rapid construction of a number of RBMK reactors. Moreover, the Soviet Planning Commission was against building containment domes over nuclear reactors as protection against the release of radioactivity in the event of a reactor accident, because this would raise the cost of the plant by 30 %. There was in fact an accident in 1975—at the prototype RBMK reactor, commissioned in 1973 on Leningrad NPP under oversight of the military—which revealed the "positive SCRAM effect" for the first time. Despite this clear warning, the reactor design was not revised or improved for the RBMK series [127]: by April 1986, when disaster struck at Reactor #4 of Chernobyl NPP, 14 reactors with the defective design had already been installed across the USSR, and nine more were under construction.

Because the reactor had initially been developed in a Soviet military context, information about shortcomings in the design was not transferred to the civil Ministry of Energy and Electrification of the USSR, which was responsible for operating the nuclear plants. The developers simply assumed that the conditions required for the positive SCRAM effect to come into play would never occur [128]: they were confident that, with the right organizational measures (clear and exhaustive instructions, staff training, etc), operators would be able to prevent any incident escalating to a dangerous level [129]. Nevertheless—in addition to the incident mentioned above at the prototype RBMK reactor—the positive SCRAM effect was registered during the launch of both Reactor #1 at Ignalina NPP and Reactor #4 at Chernobyl NPP in 1983 [130, 131]. The developers again discussed the defects [132] but decided to implement the required changes during the planned reconstruction of the existing reactors [133, 134]. The operating staff of the NPPs—civilians who worked for the Ministry of Energy and Electrification—were informed neither of discussions within the development team, nor of near-misses having occurred on other plants. A state regulator for the nuclear power industry was in fact established three years before the ultimate accident; but it had no legal basis, and lacked the human and financial resources even to understand the physics of the deficiencies of RBMK reactor design, let alone to grasp the degree of danger and monitor operations accordingly [135–137]. And as we have remarked already, wishful thinking was endemic among the Politburo and the Ministry of Energy and Electrification: they were so sure of the infallibility of Soviet nuclear technology that executive positions on Soviet nuclear plants were often given to managers with neither education in nuclear science nor experience of running nuclear power plants. As an illustration, the director of the Chernobyl plant had worked in the past on coal power stations and was a turbine specialist, while the chief engineer at the plant was an electrician with experience on thermal stations and the electric grid.

With hindsight, this may seem incredible—but nobody among the personnel of the RBMK plant was aware that this type of reactor was unsafe under certain conditions. So, the Chief Engineer at the Chernobyl NPP had no idea of what he was unleashing when he decided to conduct an experiment with the emergency power supply system on Reactor #4—an experiment stipulated by the reactor project as part of compulsory measures [138, 139]. To supervise what he assumed was to be an electro-technical experiment, he brought in a service contractor from the Ministry of Energy and Electrification who specialized in electrical equipment, but had no nuclear experience. There was no requirement in the project guidelines that the program should be approved by the reactor developers [140], so nobody informed the developers of the plans to implement the test. Consequently, the test violated twelve different sections of the operating instructions for an RBMK reactor [141]. As operators were starting to shut the reactor down during the experiment on the night of the accident—causing an uncontrollable power excursion, a reactor explosion and the largest ever release of radioactivity in an industrial accident—they were confident that the reactor was absolutely safe, because nobody had given them specific instructions on how to handle RBMK reactors to avoid the positive SCRAM effect [142–144].

3.3.2 The Deepwater Horizon Oil Spill

The Deepwater Horizon platform started drilling the Macondo well in February 2010, expecting to finish the job in 51 days on a budget of US$96.2 million [145]. But following delays and over-expenditure, drilling was still incomplete by the deadline, and BP managers urged Transocean and Halliburton staff to work faster [146]: their expenses for leasing the platform were over US$1 million a day. By the disaster date, there was a delay of 43 days and BP were more than US$58 million over budget [145]. BP engineers described Macondo as "*[a] nightmare well, which has everyone all over the place*" [145]; even so, the well was successfully drilled by the middle of April 2010.

On the day of the accident, Halliburton's cementing engineer sent an email to a colleague in Houston: "*We have completed the job and it went well*" [147]. A BP engineer informed onshore colleagues: "*just wanted to let everyone know the cement job went well. Pressures stayed low… The Halliburton cement team … did a great job*". BP executives responded: "*Great job guys*!" [148]. The cementing of the walls is vital to the safe exploitation of deepwater wells: an MMS study concluded that cementing was the single most significant factor in 18 of 39 well blowouts in the Gulf of Mexico over a 14-year period [149]. In this case, BP managers had reduced the number of centralizers, which distribute cement evenly in a well, from 21 to 6 in order to save money and time. Neither Transocean's rig crew nor several BP representatives were aware that between February and April 2010, Halliburton had run three laboratory tests on cement stability for the well, all of which had failed [150]. BP were relying on the good quality of Halliburton's cement to compensate for cost-reduction measures they had already taken: on the morning of the disaster day, BP managers even canceled the final acoustic test of the cement job, assuming they had saved $128,000 by doing so [148].

The National Commission on the disaster found that managers at Halliburton "*did not comment on the evidence of the cement slurry's instability, and there is no evidence that BP examined the foam stability data in the report at all… BP, Transocean, and Halliburton failed to communicate adequately. Information appears to have been excessively compartmentalized at Macondo as a result of poor communication. BP did not share important information with its contractors, or sometimes internally even with members of its own team. Contractors did not share important information with BP or each other. As a result, individuals often found themselves making critical decisions without a full appreciation for the context in which they were being made (or even without recognition that the decisions were critical)*" [151].

3.3.3 The Fukushima-Daiichi Nuclear Disaster

In 2002, the Japanese government launched an investigation into the widespread practice of falsifying routine safety inspection data at NPPs run by the Tokyo Electric Power Company (TEPCO) when the true data had been deleted. TEPCO eventually confirmed 200 cases of data falsification between 1977 and 2002. Tsunehisa Katsumata, who was appointed as president of TEPCO after the scandal, revealed "*serious cases of inappropriate conduct in which employees should have reported cracks in the shroud to the national government [and] failure to keep records of problems. The engineers involved were afraid that, if they notified the national government of the problem, they would have to shut down the plant for a longer period of time than planned. This fear resulted in a conservative mentality that led them to avoid reporting problems to the national government as long as they believed that safety was secured. Engineers, who were reluctant to report problems, therefore eventually came to believe that they would be allowed not to report faults if the faults did not pose an immediate threat to safety and, as a result, they went as far as to delete factual data and falsify inspection and repair records*" [152]. Other nuclear operators followed the same practice—for instance, in 2007 Hokuriku Electric Power admitted that they had hidden a nuclear incident at the Shika NPP in 1999 [153]. Nevertheless, according to research by James Acton and Mark Hibbs, "*the relationship between NISA and the Japanese government, on the one hand, and that between NISA and industry, on the other, was not fundamentally challenged*" by the falsification scandal [154].

In 2003, when operation had resumed at nuclear plants suspended in the falsification scandal, TEPCO "*implemented a [c]ompany-wide program to reduce cost, including measures to curb maintenance expenditures* [155]. To help operators reduce costs on safety installations, NISA ruled that "*actions should be taken autonomously by the operator*". Furthermore, "*Since 2006, the regulators and TEPCO were aware of the risk that a total outage of electricity at the Fukushima Daiichi plant might occur if a tsunami were to reach the level of the site… NISA knew that TEPCO had not prepared any measures to lessen or eliminate the risk, but failed to provide specific instructions to remedy the situation... NISA informed the operators that they did not need to consider a possible station blackout because the probability was small and other measures were in place. It then asked the operators to write a report that would give the appropriate rationale for why this consideration was unnecessary*" [156].

The regulators and TEPCO underestimated the risk of a high-wave tsunami partly because the Japanese nuclear industry had focused so much on the possibility of earthquakes. They felt confident that the comprehensive calculations they had made would guarantee safety from beyond-design accidents. The Fukushima-Daiichi plant was designed by two American companies, General Electric (who designed the boiling water or BWR reactors) and EBASCO (who designed the plant) in the 1960s. Its foundations were on a bluff at a height of 35 m above sea level, but TEPCO civil engineering staff lowered the bluff by 25 m to reduce the threat posed by earthquakes and cut the cost of running the seawater pumps [157]. The maximum expected height of a tsunami wave near Fukushima-Daiichi NPP was 3.1 m above sea level: this figure was based on 13 earthquake tsunami statistics dating from 1611, within which the largest tsunami to hit the Fukushima coastline was the 1960 Chilean Earthquake tsunami, at 3.122 m [157]. However, since 1498 there had been 12 tsunamis off the coast of Japan and the Russian Kuril Islands with maximum amplitudes of more than 10 m, generated by earthquakes with magnitudes between 7.4 and 9.2—and half of these had maximum amplitudes over 20 m [158]. The BWR reactors on the ocean coastline of Japan were similar in design to American General Electric reactors sited near rivers, which had never been intended to withstand sudden high-level waves or flash flooding. American engineers had housed backup emergency diesel generators and DC batteries in turbine buildings around 4 m above sea level, and TEPCO had agreed with this because nobody was expecting a tsunami wave of more than 3.1 m [159]. NISA had also accepted this solution because the regulator had focused for decades on earthquake-resistant solutions, not on the possible threat of a tsunami. Accordingly, they had funded academic grants for research on earthquake safety, and marginalized tsunami safety [160]. Toshiba engineers wanted to improve on the General Electric design during construction of the Fukushima-Daiichi NPP, but TEPCO blocked any major changes: "*TEPCO, conservative by nature, didn't allow the Japanese companies building the plant to make any alterations to GE's basic design... [TEPCO] told the Japanese makers to build the plants exactly in the same way as those of foreign makers... TEPCO was very bureaucratic*" [159]. Once the Fukushima-Daiichi plant was operational, engineers there were concerned about the placement of the generators: "*If an earthquake hits and destroys some of the pipes above, water could come down and hit the generators. DC batteries were also located too close to the diesel generators. It's not at all good in terms of safety. Many of the middle-ranking engineers at the plant shared the same concern*" [159].

The NAIIC commission concluded: "*this was a disaster "Made in Japan"... Its fundamental causes are to be found in the ingrained conventions of Japanese culture: our reflexive obedience; our reluctance to question authority; our devotion to 'sticking with the program'; our groupism; and our insularity... This conceit [disregard for anything 'not invented here'] was reinforced by the collective mindset of Japanese bureaucracy, by which the first duty of any individual bureaucrat is to defend the interests of his organization. Carried to an extreme, this led bureaucrats to put organizational interests ahead of their paramount duty to protect public safety. Only by grasping this mindset can one understand how Japan's nuclear industry managed to avoid absorbing the critical lessons learned from Three Mile Island and Chernobyl; and how it became accepted practice to resist regulatory pressure and cover up small-scale accidents. It was this mindset that led to the disaster at the Fukushima Daiichi Nuclear Plant*" [161].

3.3.4 Shale Energy Development

As the SEC was changing its rules on the estimation of resources, shale energy operators were telling investors and the public that shale wells would have a similar output to conventional gas and oil wells. An average conventional gas well, where intra-formational pressure helps to push natural gas out of the rock, lasts 20–30 years. But in a shale gas well, fluid is forced into the rock by tremendous artificial pressure during the fracking process. The natural intra-formational pressure of a typical shale well is far weaker that in a conventional well over the long term; this is why fracking and pressurization are needed in the first place. Shale wells therefore produce 74–82 % of their lifetime output in the first three years of operation [162]: the productive life of a shale well is only around three years. However, shale operators claimed a higher estimated ultimate recovery (EUR) for each of their wells by multiplying the high production rate during the first weeks after fracking by the three-decade lifetime predictable for a conventional well, rather than the three-year lifetime typical for a shale well. Moreover, producers initially search and drill "sweet spots", so production levels and rates of return in the first months of operation in any shale play are higher than those from average wells on the play, which are only drilled later. This creative accounting enabled them to claim a tremendously high EUR for their wells, and overstate the reserves of whole shale plays. Thus, in presentations to investors, Chesapeake Energy Corp declared average EURs of 4.2 billion cubic feet (Bcf) for their wells in the Marcellus region of Pennsylvania, Range Resources claimed 5.7 Bcf, and Cabot Oil and Gas Corporation 15 Bcf. In fact, according to the impartial US Geological Survey (USGS), the average EUR for the Marcellus wells is only 1.1 Bcf [163]. In 2009, a typical unconventional oil well in Oklahoma produced about 1200 barrels per day during the first weeks after fracking, but after four years output was down by more than 90 %, at just 100 barrels per day [164]. Overstating yields also allowed operators to declare lower "official" production costs for exploiting shale formations: they would divide current expenses for drilling any shale well by their generously-evaluated EUR for the well. If operators declared "tremendous" EURs for their wells, then production costs could be calculated as very low, when in reality unconventional energy production is tens times more expensive than conventional production.

Within the United States, there is still no unified nationwide government database giving details of well productivity, gas content, geophysical features, injected fracking fluids and development activity for every unconventional well in the US, and including capital and operating costs for leasehold, drilling and completion, maintenance expenses, and re-fracturing. Only the shale energy companies have such information, and they reveal it to the authorities and the public on a voluntary basis. So national regulators have received incomplete data, which in its turn has led to inadequate decision-making on a national scale. Thus the US Energy Information Administration (EIA)—the statistics department of the US Department of Energy—promoted overstated forecasts for the total unconventional reserves and likely rates of shale gas and oil production in the United States, in order to attract investment into the industry and increase domestic energy production. Some EIA forecasts were recognized as too optimistic [165], because it was private companies affiliated with the shale industry that provided the data about resources and drilling progress on shale plays. An investigation by the New York Times stated: "*The Energy Information Administration's annual reports are widely followed by investors, companies and policy makers because they are considered scientifically rigorous and independent from industry… The Energy Information Administration, for example, relies on research from outside consultants with ties to the industry. And some of those consultants pull the data they supply to the government from energy company news releases... Projections about future supplies of natural gas are based not just on science but also some guesswork and modeling... Two of the primary contractors, Intek and Advanced Resources International, provided shale gas estimates and data for the Energy Information Administration's major annual forecasting reports on domestic and foreign oil and gas resources. Both of them have major clients in the oil and gas industry, according to corporate tax records from the contractors. 'E.I.A.'s heavy reliance on industry for their analysis fundamentally undermines the agency's mission to provide independent expertise'... a senior petroleum geologist who works for the Energy Information Administration wrote that upper management relied too heavily on outside contractors and used 'incomplete/selective and all too often unreal data*" [166]. In the report "*Drilling Deeper: A Reality Check on U.S. Government Forecasts for a Lasting Shale Boom*" in October 2014, J. David Hughes concluded: "*[Shale] oil production from major plays will peak ... by 2017 and the remaining plays will make up considerably less of future production than has been forecast by the EIA... By 2040, production rates from the Bakken and Eagle Ford will be less than a tenth of that projected by EIA... [The] forecast by the EIA ... is in most cases highly optimistic and unlikely to be realized at the rates projected... Conventional wisdom holds that the shale boom will last for decades, leaving the U.S. woefully unprepared for a painful, costly, and unexpected shock when the shale boom winds down sooner than expected*" [162]. And the US Government Accountability Office points out: "*[The] EIA reports that experience to date shows production rates from neighboring shale gas wells can vary by as much as a factor of 3 and that production rates for different wells in the same formation can vary by as much as a factor of 10*" [167]. Moreover, the EIA's overstated forecasts of abundant shale gas resources around the world focused the attention of governments in Poland, Romania and Ukraine on the "wealth" beneath their feet, motivating them to raise domestic investment to bring American drilling and service companies to their countries: these were the only companies in the world with experience in and technologies for shale exploration and production. Most of these forecasts overstated real shale resources by at least ten times [168–172].

Exaggerating the figures on resources helped operators to conceal their huge production costs, and increase the perceived value of their companies, in order to convince backers to continue investing in the "shale revolution". Investment banks on Wall Street make billions of dollars in transaction fees from shale mergers and acquisitions. The total value of deals in 2009 was US$50 billion; in 2010 $38 billion; in 2011 $47 billion [173]; and since 2006, when the shale boom began, more than US$200 billion [174]. Because their fees depend on the value of merger and acquisition (M&A) deals, banks do not reveal the financial problems of shale operators in their reports and presentations. Once again we are reminded of the behavior of these same banks in the Enron case between 1996 and 2001, when they made serious money from underwriting or merger deals, whereas brokerage fees brought insignificant profits. A similar situation is being played out again with the shale industry: Wall Street investment banks have an interest in publishing positive or neutral reports on the financial conditions of shale operators. Deborah Rogers, financial consultant for several major Wall Street firms, states that "*analysts and investment bankers ... emerged as some of the most vocal proponents of shale exploitation*". She cites the assessment of Wood Mackenzie's head of consulting Neal Anderson on the investment community and shale exploration: "*It seems the equity analyst community has played a key role in helping to fuel the shale gas M&A market, acting as chief cheerleaders for shale gas plays*" [175]. Conventional producers in America and worldwide, and unsophisticated international investors from Asian countries like Japan, China and South Korea, all bought American shale assets; and the Wall Street investment banks made substantial transaction fees from all of these deals. One should be concerned with the significant risk of future large losses for the concerned investors.

4 The Absence of Reliable Information and of an Adequate System for Risk Assessment

4.1 General Picture

When risky and innovative solutions are being implemented in pursuit of short-term benefits, the systematic distortion of information leads to a dangerous situation: even management have a limited understanding of the complex consequences of the new developments, a fragmented or unbalanced picture of the real situation in an organization and the condition of its different units.

The deliberate concealment of risks in the cases we have discussed also leads to the danger of self-deception—managers start to believe in a distorted picture of the situation, assuming that it reflects "reality". Rather than tackling the situation realistically, studying the facts, going to primary sources and independently assessing information, people tend to convince themselves of what they want to believe. Wishful thinking is a sure path to an inaccurate perception of reality, and thus to misguided action, or disastrous inaction, when essential changes are overdue.

4.2 Examples of Lack of Reliable Information for Adequate Risk Assessment in the Financial Industry

The senior management at Barings, including those responsible for auditing and supervision, had a merchant banking background: they were unfamiliar with derivatives and associated them with tremendous risks. This is why they were blinded by Nick Leeson's reports of "profits" from the Singapore office: as we have already noted, they believed that he was making fully matched trades which represented no unusual risk to Barings [85]. Later, Leeson described the Barings management as "*idiots*" [176] who did not even grasp the basics of the futures trading business: "*How little did the management of Barings know about what was going on? They had no clue. In 1994 [they] came from London, New York, and Tokyo to receive an award from SIMEX for the 'Highest Customer Volume*'" [9]. A year later, Leeson's unauthorized trading had left them bankrupt.

Similarly, in the case of the mortgage bubble, it was not just external observers, but also investment bank executives themselves, who failed to understand the real impact of the new over-the-counter derivatives on their business. The FCIC commission after the crisis declared: "*The mortgage pipeline... introduced leverage at every step. High leverage, inadequate capital, and short-term funding made many financial institutions extraordinarily vulnerable to the downturn in the market in 2007*" [177]. Through OTC derivatives, traders at the five major investment banks (Bear Stearns, Goldman Sachs, Lehman Brothers, Merrill Lynch, and Morgan Stanley) could operate with leverage ratios as high as 40 to 1 on their capital. So for every US$40 in assets, they had only US$1 in capital to cover losses: a drop in asset values of less than 3 % would be enough to bankrupt any major investment bank [178]. Not even senior managers at the financial institutions had a sense of "the whole picture" of the risks; yet they continued to assure investors, partners, competitors and the authorities that their organizations were financially stable. For instance, Richard Fuld, CEO of Lehman Brothers, assured shareholders at a meeting in April 2008—just after the failure of Bear Stearns—that "*the worst ... [is] ... behind us*" [179]. Some sources considered Fuld to be mainly a bond trader, with little technical understanding of new financial instruments like CDOs and CDSs. In fact, the majority of Lehman's board of directors lacked specialized financial expertise: nine of them were retired, four of them over 75 years old, one was a theater producer, another a former Navy admiral… only two actually had direct experience in the financial services industry [180, 181]. Even after Lehman Brothers filed for bankruptcy—a step for which Fuld voted—he insisted: "*There was no capital hole at Lehman Brothers. At the end of Lehman's third quarter [of 2008], we had US$28.4 billion of equity capital*" [123]. The insurance firm AIG was caught unaware in a similar way: executives there told the FCIC commission that "*they did not even know about these terms of the [credit default] swaps until the collateral calls started rolling*" in July 2007 [182]. Even the unfortunately named Thrift Supervision, the regulators who supervised AIG on a consolidated basis, had not grasped the true level of risk the company was underwriting [182]. A former Bear Stearns executive told the commission how a Federal Reserve representative, hearing that the housing securitization market was on shaky ground, said: "*We don't see what you're talking about because incomes are still growing and jobs are still growing*". With such a superficial understanding of a bewilderingly complex market, regulators "*relied extensively on banks' own internal risk management systems*", and clung blindly to the dogma that "*markets will always self-correct*" [183]. One FCIC commissioner observed: "*it appears that market participants were unprepared for the destructiveness of this bubble's collapse because of a chronic lack of information about the composition of the mortgage market. Information about the composition of the mortgage market was simply not known when the bubble began to deflate*" [182]. After the crash, Ben Bernanke, Chairman of the Federal Reserve, admitted that he had missed the systemic risks: "*Prospective subprime losses were clearly not large enough on their own to account for the magnitude of the crisis*" [184]. Property prices peaked in 2006, and Bear Stearns investment bank was judged to be problematic the following year, but regulators maintained that it was a "*relatively unique*" case. They continued to assure the financial world that there was "*comfort about the capital cushions*" at the big investment banks until Bear Stearns actually collapsed in March 2008 [185]. The US Secretary of the Treasury during the crisis was Henry Paulson, who had been CEO at Goldman Sachs, one of the key players of the securitization pipeline, between 1999 and 2006. In October 2007, he warned that the burst of the housing bubble was "*the most significant risk to our economy*" [186]. Despite his warning—in a year which saw US$100 billion in mortgage-related losses—the government still took no decisive action to assess the real state of the financial institutions, or to try and avert impending crisis, until the autumn of 2008. In the words of the Financial Crisis Inquiry Commission: "*The captains of finance and the public stewards of our financial system ignored warnings and failed to question, understand, and manage evolving risks within a system essential to the well-being of the American public. Despite the expressed view of many on Wall Street and in Washington that the crisis could not have been foreseen or avoided, there were warning signs. The tragedy was that they were ignored or discounted... Little meaningful action was taken to quell the threats in a timely manner*" [37]. Because nobody really saw the whole picture, few could guess the full magnitude of the approaching calamity.

According to some researchers on the Enron case, "*The board of directors simply did not understand what was going on; they trusted that Jeffrey Skilling's and Andrew Fastow's labyrinthine special purpose entities made sound financial sense; after all, both Skilling and Fastow had graduated from top MBA programs. Thus, neither the auditors nor the Board of Directors performed effectively their function of monitoring the activities of insiders for the benefit of outsiders* [100] … *The Auditing Committee of the Board of Directors continued to rely on its public auditing firm, Arthur Andersen, who continued to write favorable opinion letters that ENRON's accounting was 'adequate to provide reasonable assurance as to the reliability of financial statements*'" [187]. Others consider that Enron's board of directors may have kept silent for financial reasons: "*Each director received nearly $350,000 per year for serving on Enron's board. That amount was double the high end of normal large public company director fees. The board routinely bragged about Enron's management team. One may ask how much of their 'Enron can do no wrong' attitude was impacted by the fees they received*?" [99]. Such board could not perform its primary function—to control the top management. Impressed by the company's exceptional growth, Harvard University prepared a case study on Enron for its MBA students; Business Week, Forbes, Fortune and other business magazines and newspapers were likewise dazzled by the "Enron Miracle", and portrayed the company in a very favorable light [99]. For example, Fortune listed Enron stocks among its "*10 stocks to last the decade... that should put your retirement account in good stead and protect you from those recurring nightmares about stocks that got away*" [188]; Skilling was hailed as "*The #1 CEO in the USA*" for pioneering radical new theories of business and making enormous profits from these innovations [90]. Ultimately however, this was a cautionary tale of "*individual and collective greed born in an atmosphere of market euphoria and corporate arrogance. Hardly anyone ... wanted to believe the company was too good to be true... Many kept on buying the stock, the corporate mantra and the dream*" [94]. The dream started to unravel on March 5, 2001, when Fortune magazine published the first serious investigation into Enron's accounting practices—an article by Bethany McLean, simply entitled "*Is Enron Overpriced?*"—which finally brought the company's problems to the attention of its shocked investors [107].

4.3 Examples of the Absence of Reliable Information for Adequate Risk Assessment in the Industrial Sector

During the first hours after the explosion at Chernobyl NPP, the staff could not believe that the reactor had been completely destroyed because nobody thought a beyond-design accident could ever happen [189]. Reassured by reports from the operators that the reactor was intact, the director of the plant ignored warnings from the Civil Defense Service that the radiation level near the plant was 80,000 times the maximum acceptable level; in fact he flatly contradicted these warnings in an update to the Politburo in Moscow: "*The reactor is intact, continuing to pump water into the reactor, the radiation level is within the normal range*" [190]. Years after the disaster, he admitted: "*People [were] doing this [misrepresentation] with no malice. There was such practice within the industry: nothing bad to report. We always had to say—everything is going well*" [191]. So reliable scientifically verified information about the real scale of the disaster at the plant simply did not reach the Politburo in the first few days [192, 193]. As a result, no immediate measures were taken to evacuate the residents of Pripyat', a town of 47,000 inhabitants located near the nuclear power plant; and the Politburo adopted a policy of silence or understatement of the possible threat. The inability of Mikhail Gorbachev and other Politburo members to deal adequately with the situation caused huge disappointment in the communist leaders among the Soviet people, and the Chernobyl disaster was one of the triggers for the collapse of the Soviet Union [194]. At a Politburo meeting in July 1986, Mikhail Gorbachev berated the nuclear industry: '*Over the last 30 years, we hear from you [the developers of RBMK reactors] that everything here [in the nuclear industry] is reliable. In addition, you expect that we will look at you as on gods. From this, all went [wrong]. It occurred because all the ministries and research centers were out of control [of the Politburo and the Soviet government]. Finally it ended [in failure]... The accident could have been prevented. If there had been proper and timely information [about the defects of RBMK reactors], then [the Politburo] could have taken action and we would have avoided this accident*" [135, 195].

After the Macondo well blowout, BP's new CEO Robert Dudley claimed that BP had never anticipated such a huge spill: "*We've been drilling in the Gulf of Mexico, in the deep water for 20 years now. You just never see an accident like this*" [135]. But according to U.S. officials, there have been 948 fires or explosions on offshore oil platforms in the Gulf of Mexico since 2001, many of which occurred during the drilling of exploratory wells, where there is an extremely high risk of blowouts [196, 197]. Furthermore, in 1979, there was a blowout on the Mexican Ixtoc I oil rig, at a depth of just 50 m in the south-western part of the Gulf of Mexico: the flow could not be shut down for 10 months, and three million barrels of oil were discharged (more than half the total estimated amount of oil discharged during the BP deep horizon disaster).

The Japanese "*reluctance to question authority*" [198]—combined with an administrative system that was slow, bureaucratic, and geared to communicating only good news—led to a situation when managers similar to other cases had little understanding of the real condition of their plants and were fully satisfied with reassuring news from the stations. Moreover, TEPCO's corporate system "*tolerated or encouraged the practice of covering up problems*" [161]; this meant that "*utilization of risk information was insufficient, and the risk of [a station blackout] was not widely recognized by the management*" [199]. Any issue of operating risk or nuclear safety was considered a matter for the on-site plant department, and would never have been raised at central risk management meetings [200]. Masatoshi

Toyota, a former senior vice president of TEPCO and one of the directors of the construction of the Fukushima plant, later admitted: "*I didn't know until March 11 that the diesel generators were placed in the turbine buildings. If I had known, I would have definitely changed that*" [201].

The investment boom in the shale industry was achieved through concerted actions, by US government and the industry itself, to promote an American "shale revolution" among institutional investors while concealing unpalatable features of the technology. Even experienced traditional energy producers were deceived with respect to the resource estimates for shale assets and the manipulated figures for potential production from American shale plays made by the US government. For instance, in 2010, Exxon acquired XTO Energy, one of the leaders in unconventional production in the US, with an estimated value of between US$26 billion and $41 billion. With this deal, Exxon became the largest natural gas producer in the US; but in 2012, Rex Tillerson Exxon's CEO, complained to investors: "*We are all losing our shirts today... We're making no money [on shale gas]. It's all in the red*" [159]. Exxon were not alone: in 2012, BP declared write-downs of US$4.8 billion, the British BG Group debited $1.3 billion on shale investments, and the Canadian EnCana lost $1.7 billion and informed shareholders that losses would increase if gas prices did not return to an "acceptable" level [202]. Shell has spent about $30 billion on shale plays in the US and Canada [203] but, in September 2013, they put land in Texas, Kansas and Colorado up for sale, including their largest field Eagle Ford. They admitted that 192 wells "*are not in a position to reach the planned volume production*," and they announced debts of US$2.1 billion and began a strategic reassessment of investment in oil shale deposits in the United States. The following year, Shell announced a new "fix or divest" strategy for its Marcellus shale assets—for which they had invested a total of US$4.7 billion in the Warrendale-based East Resources—after underwhelming results over several years [204]. However, in March 2014, when oil prices were high, Ben van Beurden, Shell's CEO, was forced to admit: "*Financial performance there is frankly not acceptable [for US onshore assets] ... some of our exploration bets have simply not worked out*" [205]. The Australian BHP Billiton joined the shale race in 2011 and had a similar experience, investing billions in US shale assets only to report a devaluation of at least US$2.8 billion just a year later. By October 2014, BHP Billiton was looking for a buyer for its "assets" in North Carolina [206, 207]. The Bloomberg News surveyed the whole industry in April 2014: "*...It's an expensive boom. ... The spending never stops ... Since output from shale wells drops sharply in the first year, producers have to keep drilling more and more wells to maintain production. That means selling off assets and borrowing more money... [In the words of] Tim Gramatovich... of Peritus Asset Management: 'People lose their discipline. They stop doing the math. They stop doing the accounting. They're just dreaming the dream, and that's what's happening with the shale boom*'" [208]. In the following issue, the author focused particularly on debt: "*Shale debt has almost doubled over the last four years [2010–2013] while revenue has gained just 5.6 percent... A measure of the shale industry's financial burden, debt hit $163.6 billion in the first quarter [2014], according to company records compiled by*

Bloomberg on 61 [shale] exploration and production companies" [209]. And in August 2014, Tim Morgan, former head of research at Tullett Prebon, concluded: "*We now have more than enough data to know what has really happened in America. Shale has been hyped ... and investors have poured hundreds of billions of dollars into the shale sector. If you invest this much, you get a lot of wells... If a huge number of wells come on stream in a short time, you get a lot of initial production. This is exactly what has happened in the US. The keyword here, though, is "initial". The big snag with shale wells is that output falls away very quickly indeed after production begins... [So] the only way to keep production rates up (and to keep investors on side) is to drill yet more wells. This puts operators on a "drilling treadmill", which should worry local residents just as much as investors. Net cash flow from US shale has been negative year after year, and some of the industry's biggest names have already walked away... In the future, shale will be recognised as this decade's version of the dotcom bubble*" [208].

Citigroup estimated the following relationship between oil price and profitability of US shale oil production. Above $70 per barrel, almost all US shale oil production is profitable; at $60 per barrel, about 40 % of unconventional oil production becomes uneconomical; at $50 per barrel, almost 90 % of US shale oil production is unprofitable [Andrews R., Oil Wars: Why OPEC Will Win, Oil Price. December 11, 2014, http://oilprice.com/Energy/Oil-Prices/Oil-Wars-Why-OPEC-Will-Win.html]. By not reducing their productions during 2014–2016, notwithstanding the evidence of over-supply, in order to keep the largest share of the world oil market, Saudi Arabia and other OPEC members oversaw the dramatic falling of oil prices by more than 70 %, from $95 per barrel in September 2014 to $26 per barrel in January 2016 [51]. Due to the fall in oil prices, the U.S. rigs count dropped by 78 % from its peak: from 1930 oil rigs in September 2014 to 421 in June 2016 [162]. As a consequence, oil production in the United States was reduced from 9,6 million barrels per day in July 2015 to 8,6 million barrels per day in June 2016, with forecast of further reduction due to low oil prices [51].

5 Conclusion

Disasters and crises rarely come out of the blue. There are often significant early warnings and near-misses but, unfortunately, they are generally ignored. The information is present in some form but the firm is not structured to use it. The managers have often other short-term goals and construct incentives for their collaborators that are not conducive to resilience.

A general efficient risk management system, which can provide the minimum information to avoid as much as possible the kind of disasters that we have discussed here, should include (i) effective monitoring processes, (ii) relevant risk metrics, (iii) adequate tools to analyse the drift and time evolution of these risk metrics and (iv) a communication system that favors the transmission of information bottom-up with the right incentives. Reliable and sustainable operations of sensitive financial and industrial systems require a communication process to share information between teams, a management system of resources and risks, a verification and validation of hypotheses of the causes of looming risks, risk identification and tracking, and questioning of assumptions. Moreover, the awareness of past cases, as provided in the present article, should be continuously in the consciousness of the decision marker and manager, who should always worry whether any of the weaknesses documented here in previous developments leading to disasters are at any time present in their own structure.

In this article, we have stressed that blatant information concealment developed before promoted the likelihood and severity of major crises and disasters. We have dissected in details how this proceeded in the seven analyzed case studies. Arguably, a core mechanism is captured by the adage: "*No one sees any pressing need to ask hard questions about the sources of profits when things are doing well*". This is fundamentally associated with the problem of incentives, and often, but not always, of moral hazard, i.e., not having "*skin in the game*" (no supporting sufficiently the consequences of the risks taken by the enterprise as a consequence of one's decisions). Allowing the concealment or misrepresentation of information on the scale shown in our case studies is in fine a choice of society, since it touches many of its levels and is tolerated or promoted by many of its prominent actors.

Finally, the concealment or misrepresentation of information is greatly facilitated by the mismatch between the inherent complexity of our society based on technology and finance on the one hand and the human capacity to comprehend this complexity. With our limited and biased cognitive abilities, we have intrinsic weaknesses against developing the correct insights into complex system behavior, against using the correct models and developing the adequate regulatory responses. And this is generally exploited to hide or misrepresent the developing risks of complex human activities. We argue that this can be addressed by a focus on developing suitable metrics, applying them to measurements, recording, analysing trends and developing the corresponding responses in a perpetual virtuous circle.

There have been calls for greater sophistication of models to cope with the booming complexity [210–212]. But, the solutions are often known and turn out to be quite simple, but forgotten on the basis of modernism (that what is new is always better), the argument that "this time, it is different", the belief in a "new economy" to which previous methods of valuation do not apply and so on. We argue that simple "satisfying" solutions exist [213] (to use the neologism created by H. Simon) but are ignored on the altar of optimization and the search for ever greater yield, at the cost of increasing fragility and loss of resilience. We should continuously keep in mind the tension between (i) short-term growth maximization and systemic (in-)stability, (ii) the bearers of costs and putative beneficiaries, (iii) the ethos of individual gratification, and (iv) conflict between social and eco-systemic optimization.

References

1. Lin JY (2009) Policy responses to the global economic crisis, development outreach. World Bank Inst 11(3):29–33
2. National commission on the BP deepwater horizon oil spill and offshore drilling. Report to the President, Jan 2011, p 75
3. The official report of The Fukushima nuclear accident independent investigation commission. The National Diet of Japan, Executive summary, 5 July 2012, p 43
4. Remarks by Chairman Alan Greenspan, before the council on foreign relations. Washington, D.C., 19 Nov 2002
5. Financial crisis inquiry report, the financial crisis inquiry commission. Washington, D.C., Jan 2011, p 28
6. Documentary "going rogue", journeyman pictures, Dec 2011
7. Heffernan S (2005) Modern banking. Wiley, p 381
8. Lords Hansard entry for 21 Jul 1995 (150721-14). http://www.parliament.the-stationery-office.co.uk/pa/ld199495/ldhansrd/vo950721/text/50721-14.htm
9. Hutchens G (2012) Barings wake up call unheeded: Leeson, Sydney Morning Herald, 20 Oct 2012
10. Documentary "enron: the smartest guys in the room", Director Alex Gibney, 2005
11. Enron's Kenneth Lay is a bush family friend. The Los Angeles Times, 02 Dec 2001
12. Nichols J, Lay K (2006) Guilty. George bush—guilty. The Nation, 25 May 2006
13. Bush-Lay letters suggest close relationship CNN, 17 Feb 2002
14. Gerstein J (2001) Enron's close ties to Bush. ABC News, 10 Dec 2001
15. Huffington A (2006) Ken Lay on trial: why are the media forgetting the Bush/Cheney connection? The Huffington Post, 26 Apr 2006
16. Inaugural address of Ronald Reagan, 20 Jan 1981
17. Remarks at the annual meeting of the boards of governors of the world bank group and international monetary fund, 29 Sept 1981
18. Dembinski PH, Lager C, Cornford A, Bonvin JM, Enron and world finance. A case study in ethics, p 26
19. Palast G (2001) California Reamin, the guardian, 22 May 2001
20. O'Harrow R Jr, Shackelford L, Enron case. Political players. Washington post. http://www.washingtonpost.com/wp-srv/business/enron/4a.html
21. Van Natta D (2002) Enron's many strands: the administration; Bush's California energy stance faulted. New York Times, 08 May 2002
22. Leopold J (2006) Lay and skilling in the dock. Enron and the Bush Administration, Counterpunch, 01 Feb 2006
23. Pellegrini F (2002) Bush's Enron problem. Time Magazine, 10 Jan 2002
24. Sigal C (2002) Notes from Los Angeles; a gray future for California voters. The New York Times, 17 Oct 2002
25. Dembinski PH, Lager C, Cornford A, Bonvin J-M, Enron and world finance. A case study in ethics. p 69
26. Dharan BG, Bufkins WR (2008) Red flags in enron's reporting of revenues & key financial measures. Soc Sci Res Netw 3
27. Schmitt R 2009) Prophet and loss. Stanford Magazine, March/April
28. Gudikunst A, Enron: a study of failures, who, how and why, Bryant college working paper series. Faculty Newsletter, Sept, 2002, p 10
29. Cunningham GM, Harris JE (2006) Enron and arthur andersen: the case of the crooked e and the fallen A, global perspectives on accounting education, vol 3, p 33
30. Mike Conaway's profile of The Washington Post, 2004. http://www.washingtonpost.com/wp-srv/elections/2004/candidates/24150/
31. Unger C (2004) House of Bush, house of Saud: the secret relationship between the world's two most powerful dynasties. Simon and Schuster, New York, p 117

32. Krugman P (2002) Succeeding in business. The New York Times, 7 July 2002
33. Documentary film "inside job", director Charles Ferguson, 2010
34. The financial crisis inquiry report, the financial crisis inquiry commission. Washington, D.C., Jan 2011, p xviii
35. Cohan WD (2012) Rethinking Bob Rubin from Goldman Sachs star to crisis scapegoat. Bloomberg, 20 Sept 2012
36. Sherman M (2009) A short history of financial deregulation in the United States. Center for economic and policy research, July 2009, p 10
37. The financial crisis inquiry report, The Financial Crisis Inquiry Commission, Washington, D. C., January 2011, p. xvii
38. Stephen Labaton, Congress Passes Wide-Ranging Bill Easing Bank Laws, The New York Times, 5 Nov 1999
39. The Financial Crisis Inquiry report, the financial crisis inquiry commission. Washington, D. C., January 2011, p xx
40. Testimony of chairman Alan Greenspan, the regulation of OTC derivatives, before the committee on banking and financial services. U.S. House of Representatives, 24 July 1998
41. Top 10 challenges for investment banks 2011. Chapter challenge 2: dealing with OTC derivatives reform". Accenture, 2010
42. In search of growth, the economist online, 25 May 2011
43. The financial crisis inquiry report, the financial crisis inquiry commission. Washington, D.C., Jan 2011, p 172
44. The financial crisis inquiry report, the financial crisis inquiry commission. Washington, D.C., January 2011, p xix
45. The SCRAM system refers to the control rods that are inserted into a nuclear reactor core to suppress nuclear fission. A "positive SCRAM effect" is a localized increase of activity in the bottom of the core of a reactor during emergency shutdown, when introducing graphite rods leads to decreased absorption of neutrons by the xenon in the core ("xenon poisoning") and accelerates the nuclear reaction
46. Dyachenko A (2004) Experience of liquidation of Chernobyl disaster, Federal State unitary enterprise "institute of strategic stability" of Rosatom, Moscow, 2004, http://www.iss-atom.ru/book-7/glav-2-3.htm
47. Interview with Mikhail Gorbachev, documentary "the battle of Chernobyl. Director: Thomas Johnson, 2006
48. Karpan N (2006) Vengeance of peaceful atom. Dnepropetrovsk, p 401
49. Chevron announces record setting well test at Jack, Press Release, Chevron, San Ramon, California, 5 Sept 2006
50. BP announces giant oil discovery in The Gulf Of Mexico, Press release, BP, 2 Sept 2009
51. The U.S. Energy Information Administration, 2011
52. National commission on the bp deepwater horizon oil spill and offshore drilling. Report to the President, Jan 2011, p 73
53. National commission on the BP deepwater horizon oil spill and offshore drilling. Report to the President, Jan 2011, p. 72
54. Doggett T, Gardner T (2010) U.S. mulls requiring remote shutoffs for oil rigs. Reuters, 3 May 2010
55. Milov V (2010) Oversept the accident, Gazeta. RU, 16 June 2010, http://www.gazeta.ru/column/milov/3385462.shtml
56. Nuclear power in Japan, World nuclear association, updated February 2014
57. The official report of The Fukushima nuclear accident independent investigation commission, the national diet of Japan, executive summary, 5 July 2012, p 28
58. The official report of The Fukushima nuclear accident independent investigation commission, the national diet of japan, executive summary, 5 July 2012, p 75
59. The official report of The Fukushima nuclear accident independent investigation commission, the national diet of japan, chapter 1. Was the accident preventable? 5 July 2012, pp 53–57

60. Acton JM, Hibbs M (2012) Why Fukushima was preventable, Carnegie endowment for international peace, Mar 2012, p 2
61. National energy policy, report of the national energy policy development group, May 2001, pp 5–6
62. Miller G (2009) Review of the revised draft supplemental generic environmental impact statement on the oil, gas and solution mining regulatory program, toxicity and exposure to substances in fracturing fluids and in the waste water associated with the hydrocarbon bearing shale. Consulting environmental toxicologist to the natural resources defense council, 29 Dec 2009
63. Miller G (2012) Review of the revised draft supplemental generic environmental impact statement on the oil, gas and solution mining regulatory program well permit issuance for horizontal drilling and high-volume hydraulic fracturing to develop the marcellus shale and other low-permeability gas reservoirs. Prepared for natural resources defense council, 6 Jan 2012
64. Finkel M, Hays J, Law A (2013) The shale gas boom and the need for rational policy. Am J Publ Health 103(7):1161–1163
65. Colborna T, Kwiatkowskia C, Schultza K, Bachran M (2011) Natural gas operations from a public health perspective. Hum Ecol Risk Assess 17:1039–1056
66. Witter R, McKenzie L, Towle M, Stinson K, Scott K, Newman L, Adgate J (2010) Health impact assessment for battlement mesa. Garfield County Colorado, Colorado School of Public Health, University of Colorado Denver, Sept 2010
67. Linnitt C (2010) Report "fracking the future—how unconventional gas threatens our water, health and climate. DeSmogBlog Project
68. Ohio: shale drillers must report chemicals locally. Associated Press, 1 Oct 2013
69. Tomson R (2013) Water use in fracking needs to be refined. Houston Bus J
70. Standing committee on natural resources, number 040, 3rd session, 40th parliament, Parliament of Canada, 1 Feb 2011
71. Stokstad E (2014) Will fracking put too much fizz in your water? Science 344(6191):1468
72. Goodman M (2012) Waterless fracking method targets natural gas industry's gaze. CBS News, 12 Jan 2012
73. Milmo S (2011) Fracking with propane gel, royal society of chemistry, 15 Nov 2011
74. Ridlington E, Rumpler J (2013) Fracking by the numbers, key impacts of dirty drilling at the state and national level. Environ Am Res Policy Cent 4
75. Documentary film "Gasland" by Josh Fox. URL: www.gaslandthemovie.com (evidence that EPA's top-level executives had close relation with shale gas lobby on 30th minute of the film)
76. Impacts of shale gas and shale oil extraction on the environment and on human health, European Parliament, Directorate General for Internal Policies, Policy Department A, IP/A/ENVI/ST/2011-07, PE 464.425, June 2011
77. Gruver M (2011) EPA: Fracking may cause groundwater pollution. Associated Press, 8 Dec 2011
78. Ian Urbina S.E.C. shift leads to worries of overestimation of reserves. The New York Times, 27 June 2011
79. Ahmed N (2013) Shale gas won't stop peak oil, but could create an economic crisis. The Guardian, 21 June 2013
80. Remarks by the president in state of the union address. The United States Capitol, 24 Jan 2012
81. Oil and gas: information on shale resources, development, and environmental and public health risks. U.S. Government Accountability Office, Sept 2012, p 19
82. Dharan BG, Bufkins WR (2008) Red flags in enron's reporting of revenues & key financial measures. Soc Sci Res Netw 3
83. Marx 1887, p 748, with reference to Dunning, 1860, p 35
84. Scannell K, Reddy S (2008) Greenspan admits errors to hostile house panel. Wall Street J

85. Report to the board of banking supervision inquiry into the circumstances of the collapse of Barings. Bank of England, 18 July 1995, Conclusion chapter, subsection: "why was the true position not noticed earlier?"
86. How Leeson broke the bank. BBC, 22 June 1999
87. Dharan BG, Bufkins WR (2008) Red flags in enron's reporting of revenues & key financial measures. Soc Sci Res Netw 7
88. Healy PM, Palepu KG (2003) The fall of Enron. J Econ Perspect 17(2):3
89. Ackman D (2002) Pay madness at Enron. Forbes Magazine, 22 Mar 2002
90. Free C, Stein M, Macintosh N (2007) Management controls: the organizational fraud triangle of leadership, culture and control in Enron. The Organization, July/Aug 2007
91. Dembinski PH, Lager C, Cornford A, Bonvin J-M, Enron and world finance. A case study in ethics, p 196
92. Cruver B (2003) Anatomy of greed: telling the unshredded truth from inside Enron. Basic books, p 191
93. Tourish D, Tourish N (2005) Charismatic leadership and corporate cultism at Enron: the elimination of dissent, the promotion of conformity and organizational collapse. Leadership, Nov 2005
94. Thomas CW (2002) The rise and fall of Enron. When a company looks too good to be true, it usually is. J Accountancy
95. Dembinski PH, Lager C, Cornford A, Bonvin J-M, Enron and world finance. A case study in ethics, p 28
96. Sawayda J, Andersen A, An accounting confidence crisis, Daniels fund ethics initiative. University of New Mexico, p 2, 6
97. Cunningham GM, Harris JE (2006) Enron And Arthur Andersen: the case of the crooked e and the fallen A. Glob Perspect Account Educ 3:31
98. Cunningham GM, Harris JE (2006) Enron And Arthur Andersen: The case of the crooked E and the fallen A. Glob Perspect Account Educ 3:43
99. Carozza D (2007) Interview with Sherron Watkins. Constant warning, Fraud Magazine, Jan/Feb 2007
100. Dembinski PH, Lager C, Cornford A, Bonvin J-M, Enron and world finance. A case study in ethics. pp 196–197
101. Sawayda J, Andersen A, An accounting confidence crisis, Daniels fund ethics initiative. University of New Mexico, p 6
102. Sawayda J, Andersen A, An accounting confidence crisis, Daniels fund ethics initiative. University of New Mexico, p 8
103. Dembinski PH, Lager C, Cornford A, Bonvin J-M, Enron and world finance. A case study in ethics, pp 31–23
104. Dharan BG, Bufkins WR (2008) Red flags in enron's reporting of revenues & key financial measures. Soc Sci Res Netw 2, 4
105. Richard A, Oppel Jr, Merrill replaced research analyst who upset Enron. New York Times, 30 July 2002
106. Text of Sherron Watkins' testimony at house hearing on Enron. The New York Times, 14 Feb 2002
107. McLean B, Is Enron overpriced? Fortune, 5 Mar 2001
108. The watchdogs didn't bark: Enron and the wall street analysts, hearing before the committee on governmental affairs United States Senate, One Hundred Seventh Congress, Second Session, U.S. Government Printing Office, Washington, 27 Feb 2002
109. Muolo P, Padilla M, Chain of blame. How wall street caused the mortgage and credit crisis. Wiley, pp 185, 219
110. The financial crisis inquiry report, the financial crisis inquiry commission. Washington, D.C., Jan 2011, pp 117, 119, 170, 278, 339, 393
111. The financial crisis inquiry report, the financial crisis inquiry commission. Washington, D.C., Jan 2011, p 194

112. Richardson M (2012) Why the Volcker rule is a useful tool for managing systemic risk. NYU Stern School of Business, p 8
113. The financial crisis inquiry report, the financial crisis inquiry commission. Washington, D.C., Jan 2011, p 44
114. Zamansky J, The chickens come home to roost for standard & poor's. Forbes Magazine, 5 Feb 2013
115. The financial crisis inquiry report, the financial crisis inquiry commission. Washington, D.C., Jan 2011, p 119
116. The financial crisis inquiry report, the financial crisis inquiry commission. Washington, D.C., Jan 2011, p xxv
117. The financial crisis inquiry report, the financial crisis inquiry commission. Washington, D.C., Jan 2011, p 8
118. The financial crisis inquiry report, the financial crisis inquiry commission. Washington, D.C., Jan 2011, pp xxii, 20
119. The financial crisis inquiry report, the financial crisis inquiry commission. Washington, D.C., Jan 2011, p 20
120. The financial crisis inquiry report, the financial crisis inquiry commission. Washington, D.C., Jan 2011, p 19
121. The financial crisis inquiry report, the financial crisis inquiry commission. Washington, D.C., Jan 2011, p 63
122. The financial crisis inquiry report, the financial crisis inquiry commission. Washington, D.C., Jan 2011, p 18
123. The financial crisis inquiry report, the financial crisis inquiry commission. Washington, D.C., Jan 2011, p 235
124. MacDonald E (2011) Goldman Sachs accused of misleading congress, clients. FOX Business, 14 Apr 2011
125. Press release: Goldman sachs to pay record $550 million to settle SEC charges related to subprime mortgage CDO, firm acknowledges CDO marketing materials were incomplete and should have revealed Paulson's role. U.S. Securities and Exchange Commission. Washington, D.C., 15 July 2010
126. Wilchins D, Brettell K, Chang R (2010) Factbox: how Goldman's ABACUS deal worked. Reuters, 16 Apr 2010
127. The Chernobyl accident: updating of INSAG-1. IAEA Publications, Vienna, 1992, p 87
128. The Chernobyl accident: updating of INSAG-1. IAEA Publications, Vienna, 1992, p 13
129. Karpan N (2006) Vengeance of peaceful atom. Dnepropetrovsk, pp 294–296
130. The Chernobyl Accident (1992) Updating of INSAG-1. IAEA Publications, Vienna, p 43
131. Karpan N (2006) Vengeance of peaceful atom. Dnepropetrovsk, p 290
132. Dyatlov A (2003) Chernobyl. How it was. Nauchtekhlitizdat, Moscow, p 136
133. Karpan N (2006) Vengeance of peaceful atom. Dnepropetrovsk, pp 262, 398
134. The Chernobyl accident: updating of INSAG-1. IAEA Publications, Vienna, 1992, p 31
135. Excepts from transcripts of the meetings of politburo concerning Chernobyl disaster (April 1986–November 1989). The transcripts was published in book "Within the Soviet Politburo… Records of Anatoly Chernyaev Vadim Medvedev, Georgy Shakhnazarov (1985–1991), The Gorbachev Foundation, Moscow, 2008, ww.gorby.ru/userfiles/protokoly_politbyuro.pdf
136. Karpan N (2006) Vengeance of peaceful atom. Dnepropetrovsk, pp 400–404
137. The Chernobyl accident: updating of INSAG-1. IAEA Publications, Vienna, p 88
138. Medvedev G (1989) Chernobyl notebook. New World Magazine, no 6
139. Karpan N (2006) Vengeance of peaceful atom. Dnepropetrovsk, pp 446, 451
140. The Chernobyl accident: updating of INSAG-1. IAEA Publications, Vienna, 1992, p 52
141. Leskov S (2011) Smart guys. Vremya publisher, Moscow, p 65
142. Dyatlov A, Chernobyl. How it was, Nauchtekhlitizdat. Moscow, p 102
143. Unapprehended atom. Interview with Victor Bryukhanov, "Profile" Magazine, Moscow, no 29(477), 24 Apr 2006

144. Shunevich V, Bryuhanov V (2012) I was expelled from the party directly at a meeting of the Politburo. Fakty newspaper, Kiev, 7 July 2012
145. National commission on the BP deepwater horizon oil spill and offshore drilling. Report to the President, Jan 2011, p 2
146. Deepwater Horizon's Blowout, Part 1, CBS, 60 Minutes, 22 Aug 2010, http://www.cbsnews.com/video/watch/?id=6795538n
147. National commission on the BP deepwater horizon oil spill and offshore drilling. Report to the President, Jan 2011, p 102
148. National commission on the BP deepwater horizon oil spill and offshore drilling. Report to the President, Jan 2011, p 4
149. Ingersoll C, Locke RM, Reavis C (2012) BP and the deepwater horizon disaster of 2010. MIT Sloan School of Management, 3 Apr 2012, p 15
150. National commission on the BP deepwater horizon oil spill and offshore drilling. Report to the President, Jan 2011, pp 117, 123, 224
151. National commission on the BP deepwater horizon oil spill and offshore drilling. Report to the President, Jan 2011, p 123
152. Mr Katsumata speech BGM 2003, Reconstruction after misconduct—the pursuit of excellence, TEPCO, http://www.tepco.co.jp/en/news/presen/pdf-1/0310-e.pdf
153. Stecklow S (2007) Nuclear safety reports called into question. Wall Street J
154. Acton JM, Hibbs M (2012) Why Fukushima was preventable, Carnegie endowment for international peace, Mar 2012, p 24
155. 2003 Annual Report, TEPCO, pp 2,19, http://www.tepco.co.jp/en/corpinfo/ir/tool/annual/pdf/ar2003-e.pdf
156. The official report of The Fukushima nuclear accident independent investigation commission, the national diet of Japan, executive summary, 5 July 2012, p 16
157. The official report of The Fukushima nuclear accident independent investigation commission, the national diet of Japan, chapter 1. Was the accident preventable? 5 July 2012, p 23
158. Acton JM, Hibbs M, Why Fukushima was preventable, Carnegie endowment for international peace, Mar 2012, p 12
159. Yoshida R (2011) GE plan followed with inflexibility. The Japan Times, 14 Jul 2011
160. Acton JM, Hibbs M (2012) Why Fukushima was preventable, Carnegie endowment for international peace, Mar 2012, p 29
161. The official report of The Fukushima nuclear accident independent investigation commission, the national diet of Japan, executive summary, 5 July 2012, p 80
162. Hughes JD, Deeper D (2014) A reality check on U.S. Government Forecasts for a Lasting Shale Boom. Part 1: executive summary. Post Carbon Institute, 26 Oct 2014
163. Rogers D, USGS releases troubling EURs for shale, energy policy forum, 7 Sept 2012
164. Quinn J (2014) Fracked up, burning platform blog, 23 Sept 2014, http://www.theburningplatform.com/2014/09/23/fracked-up
165. U.S. EIA cuts recoverable Monterey shale oil estimate by 96 pct, Reuters, Wed 21 May 2014
166. Urbina I, Behind the Veneer, doubt on the future of natural gas. The New York Times, 26 June 2011
167. Oil and gas: information on shale resources, development, and environmental and public health risks, U.S. Government Accountability Office, Sept 2012, p 24
168. Mad and messy regulation. The Economist, 10 July 2013
169. Reedjan S, Eni is said to abandon polish shale aspirations. The New York Times, 14 Jan 2014
170. Czapigo-Czapla M, Natural gas. Polish Geological Institute—National Research Institute, http://geoportal.pgi.gov.pl/surowce/energetyczne/gaz_ziemny
171. Strzelecki M, Almeida I, Fracking setback in poland dims hope for less Russian gas. Bloomberg, 10 Oct 2014
172. Ilie L, Romania does not have shale gas, PM Ponta says. Reuters, 9 Nov 2014
173. Rogers D (2013) Shale and wall street: was the decline in natural gas prices orchestrated? Energy Policy Forum, Feb 2013, p 13

174. Sandrea I, US shale gas and tight oil industry performance: challenges and opportunities. The Oxford Institute for Energy Studies, 21 Mar 2014, p 2
175. Rogers D, Shale and wall street: was the decline in natural gas prices orchestrated? Energy Policy Forum, Feb 2013, p 7
176. Documentary "25 million pounds", Director Adam Curtis, 1999
177. The financial crisis inquiry report, the financial crisis inquiry commission. Washington, D.C., Jan 2011, p 134
178. The financial crisis inquiry report, the financial crisis inquiry commission. Washington, D.C., Jan 2011, p xix
179. Woellert L, Onaran Y, Fuld targeted by lawmakers as surrogate for wall street excess. Bloomberg, 6 Oct 2008
180. McDonald L, Robinson P, A colossal failure of common sense: the incredible inside story of the collapse of Lehman. Random House, 24 Nov 2009, pp 91, 226, 234–236
181. Alman R, Cudmore R, McVeigh N, Lehman brothers: An exercise in risk mismanagement. New England College of Business, 2009
182. The financial crisis inquiry report, the financial crisis inquiry commission. Washington, D.C., Jan 2011, p 243
183. The financial crisis inquiry report, the financial crisis inquiry commission. Washington, D.C., Jan 2011, pp 19, 170, 171
184. The financial crisis inquiry report, the financial crisis inquiry commission. Washington, D.C., Jan 2011, p 27
185. The financial crisis inquiry report, the financial crisis inquiry commission. Washington, D.C., Jan 2011, p 288
186. Andrews EL, Housing slump 'unfolding,' Treasury Chief Says. The New York Times, 16 Oct 2007
187. Gudikunst A, Enron: a study of failures, who, how and why, bryant college working paper series. Faculty Newsletter, Sept 2002, p 3
188. Ryneck D, 10 stocks to last the decade. Fortune, 14 Aug 2000
189. Karpan N (2006) Vengeance of peaceful atom. Dnepropetrovsk, pp 339–342
190. Borovoy A, Velihov E (2012) Experience of Chernobyl, national research center "Kurchatovsky Institute", Moscow, p 11; Medvedev G, Chernobyl Notebook, New World Magazine, no 6, 1989
191. Vasil M (2000) Victor Bruchanov, former director of Chernobyl NPP: "If they could find to me appropriate criminal article, I think, they will execute me", Fakty newspaper, Kiev, 18 Oct 2000
192. Documentary "Chernobyl. Chronicle of silence", Director: Irina Larina, 2006
193. Documentary "The Battle of Chernobyl", Director: Thomas Johnson, 2006
194. Gorbachev M (2006) Turning point at Chernobyl, project syndicate, 14 Apr 2006, http://www.project-syndicate.org/commentary/turning-point-at-chernobyl
195. America speaks to BP, Full transcript: bob dudley interview, public broadcasting service (PBS), 1 July 2010, http://www.pbs.org/newshour/bb/environment-july-dec10-dudleyfull_07-01
196. Milov V, Oversleeping of the accident, Gazeta.RU, 16 June 2010, http://www.gazeta.ru/column/milov/3385462.shtml
197. National commission on the BP deepwater horizon oil spill and offshore drilling. Report to the President, January 2011, p. ix
198. Acton JM, Hibbs M (2012) Why Fukushima was preventable, Carnegie endowment for international peace, Mar 2012, p 28
199. Acton JM, Hibbs M (2012) Why Fukushima was preventable, Carnegie endowment for international peace, Mar 2012, p 27
200. The official report of the fukushima nuclear accident independent investigation commission, the national diet of Japan, executive summary, 5 July 2012, p 44
201. DiColo JA, Fowler T (2012) Exxon: 'losing our shirts' on natural gas. Wall Street J
202. Engdahl FW (2013) The Fracked-up Usa shale gas bubble. Global research, 14 Mar 2013

203. Daugherty D (2013) As shell sells off even more major shale assets, what could be next? Houston Bus J
204. Litvak A, Shell to restructure shale assets in U.S., Pittsburgh Post-Gazette, 13 March 2014
205. Schaps K, Zhdannikov D (2014) Shell cuts spending in U.S. to lower shale exposure. Reuters, 13 Mar 2014
206. Gilbert D, Scheck J, Fowle T (2013) Shale-boom profits bypass big oil. Wall Street J
207. Antonioli S (2014) BHP Billiton eyes sale of U.S. shale gas assets in profit drive. Reuters, 27 Oct 2014
208. Loder A (2014) Shale drillers feast on junk debt to stay on treadmill. Bloomberg News, 30 Apr 2014
209. Loder A (2014) Shakeout threatens shale patch as frackers go for broke. Bloomberg News, 27 May 2014
210. Haldane AG, May RM (2011) Systemic risk in banking ecosystems. Nature 469:351–355
211. Johnson N (2011) Proposing policy by analogy is risky. Nature 469:302
212. Lux T (2011) Network theory is sorely required. Nature 469:303
213. Sornette D, von der Becke S (2011) Complexity clouds finance-risk models. Nature 471:166 (March 10, 2011) (long version at http://www.er.ethz.ch/publications/finance/Systemic_risk_in_banking_It_is_complex_but_not_that_complicated_2.2.2011.pdf)

Climate Change and Disaster Forensics

Ross Prizzia

Abstract "Climate Change and Disaster Forensics" relates and applies forensic theory, insight and analysis to disaster related research and practice. It explores, describes and explains human causality of climate change related disasters and their impact on human and environmental losses. The chapter also identifies and describes new and innovative methodologies and strategies to analyze climate related disasters, reduce disaster risk, and improve disaster mitigation, adaption, and management. Emphasis is given to vigilant monitoring and assessment of Intended Nationally Determined Contributions (INDC) to reduce greenhouse gas (GHG) emissions to limit global warming to 2 °C by 2030, a critical target set to prevent some of the worst impacts of climate change.

Keywords Climate change · Disaster forensics · Disaster risk reduction · Forensic disaster analysis (FDA) · Climate smart disaster risk management (CSDRM) · Intended nationally determined contributions (INDC) · Conference of the parties (COP 21) · Paris 2015

1 Climate Change Related Disasters

Climate change related disasters include *hydrological* events such as floods, storm surges, and coastal flooding, and *meteorological* events such as storms, tropical cyclones, heat/cold waves, drought, and wildfires. Geophysical disasters include earthquakes, volcanoes, dry rock falls, landslides, and avalanches. While the number of geophysical disasters has remained fairly stable since the 1970s, the number of climate change related (hydro-meteorological) disasters has greatly increased. A World Meteorological Organization (WMO) report found that weather and climate change related disasters caused nearly 2 million deaths globally since 1971. The report focused on six types of hazards: floods, droughts, extreme

R. Prizzia (✉)
University of Hawaii-West Oahu, Kapolei, USA
e-mail: rprizzia@hawaii.edu

A.J. Masys (ed.), *Disaster Forensics*, Advanced Sciences and Technologies for Security Applications, DOI 10.1007/978-3-319-41849-0_9

temperatures, storms, wildfires, and landslides from 1971 to 2010. Temperature anomalies during the 2003 European heat wave alone caused more than 72,200 deaths as observed and recorded by the National Aeronautics and Space Administration (NASA).

The Centre for Research on the Epidemiology of Disasters (CRED) has been maintaining an Emergency Events Database EM-DAT since 1988. CRED EM-DAT provides an evidence base to the international community to assist humanitarian action at national and international levels, improve decision-making for disaster preparedness, and provide an objective basis for vulnerability assessment and priority setting. EM-DAT's 2014 annual review of disaster figures outlines information about the EM-DAT International Disaster Database, its definitions, criteria and content and provides regional analysis on Africa, the Americas, Asia, Europe, and Oceania. Findings in the report include that 324 natural hazard related disasters, many of which were climate change related, killed more than 7823 people worldwide, affected over 140.8 million others and caused US$99.2 billion of damages. In 2014, Asia accounted for almost 70 % of worldwide reported disaster victims as well as the most damages (64.6 %) [18].

The scientific opinion on climate change is that the Earth's climate system is unequivocally warming, and it is likely that humans are causing most of it through activities such as deforestation and burning fossil fuels that increase concentrations of greenhouse gases (GHG) in the atmosphere. This scientific consensus is expressed in synthesis reports by scientific bodies of national and international standing and by opinion surveys among climate scientists. Individual scientists, universities, and laboratories contribute to this overall scientific opinion via peer-reviewed publications [44].

The rate of global warming is accelerating. The 20th century's last two decades were the hottest in 400 years and possibly the warmest for several millennia, according to a number of climate studies. The United Nations' Intergovernmental Panel on Climate Change (IPCC) reports that 11 of the past 12 years are among the dozen warmest since 1850. The Arctic is feeling the effects the most. Average temperatures in Alaska, western Canada, and eastern Russia have risen at twice the global average according to the multinational Arctic Climate Impact Assessment report compiled between 2000 and 2004. After studying these climate data collected over many years, most of the leading scientific organizations worldwide have issued public statements acknowledging certain undisputed facts about the Earth's climate. The Earth's climate is warming rapidly and much of it in the past century is likely due to human activities. Also, as noted by Ian Burton, Emeritus Professor at the University of Toronto, Scientist Emeritus with the Meteorological Service of Canada, and coordinating lead author of Chapter "Disasters and Mishaps: The Merits of Taking a Global View" for the IPPC 2011 report, "Linking disasters to climate change adaptation has helped push forward thinking about disasters so that the event itself cannot be viewed in isolation…The size of the losses and the cause of the disaster are dependent on human decisions" [37].

2 Climate Change Disaster Losses

Disaster forensics studies rely primarily on loss data and available tools such as the Damage and Loss Assessment (DaLA) methodology developed by the Economic Commission for Latin America and the Caribbean Global Facility for Disaster Reduction and Recovery (GFDRR) [16]. The DaLA builds on loss data collection, recording, and analysis with the purpose of identifying root causes of disasters and determining recovery and reconstruction needs. The DaLA identifies the causes of the disaster through measuring relative contribution of exposure, vulnerability, coping capacity, mitigation and response to the disaster, with the goal to improve disaster management from lessons learned. Documenting the details and trends of hazard exposure and loss data enhances the ability to measure progress towards resilience, risk management, and disaster risk reduction. Disaster forensics data and findings can inform and improve climate change disaster management as it engages in risk assessment and reduction that require an accurate recording of previous disasters, especially associated losses such as human casualties and property, environmental, and economic damage and loss.

Disaster loss data recording links the science of risk assessment to policy making for reducing disaster risks [8]. Loss data collections are useful for identifying trends and patterns in the data over time and for tracking relationships between development and disaster risk [21]. For example, the Global Assessment Report (GAR) which recorded loss data in national and global disaster databases is increasingly used within risk modeling platforms to guide policy and decision-making processes for Disaster Risk Reduction (DRR) [29]. When combined with other data such as disaster risk management expenditures or demographic information, disaster loss data indicate the relevance of DRR policies in the broader context of development and climate change [13].

According to the 2013 United Nations International Strategy for Disaster Reduction (UNISDR) Global Risk Assessment, the worst disasters are yet to occur. Generally, future disaster losses are estimated through probabilistic risk models which infer vulnerabilities, loss exceedance curves, and fragility or damage curves to calculate different risk metrics, such as annual expected loss (AEL) and probable maximum loss (PML). The AEL is an estimation of how much an insurer can expect to pay in the coming 12 months. The PML is an estimation of the most likely worst case event for any given year, which differs from the Maximum Possible Loss, a figure indicating the entire portfolio being destroyed, which is very unlikely. The AEL and PML are used to complement historical analysis and are especially useful for decision makers in assessing the probability of annual and maximum loss that can result from major future disasters. These assessments can also incorporate climate change scenarios to help governments in developing new and advanced adaptation strategies. For example, property insurers rely on many types of risk analysis tools including the complex CAT (catastrophe) model. CAT modeling uses computer-assisted calculations to estimate the potential losses due to a catastrophic natural event such as flood, hurricane (wind damage and storm surge), earthquake, tornado, hail, wild fire and winter storm. CAT models function to help insurers prove their financial solvency [17]. Insurers' obligations are the claims they need to pay over the next year as the properties they are insuring are damaged. The problem is that nobody knows how many claims they will need to pay. CAT models address this problem by calculating several statistical figures, including the AEL and PML, based on a carrier's portfolio of insured properties.

CAT models calculate different expected and probable risk and loss figures by utilizing three interconnected modules: The *hazard module* estimates the risk of natural catastrophes affecting each of the locations within an insurer's portfolio. Which hazards are evaluated depends on the location of the properties, with flood, earthquake, and wind/hurricane being the most common. The *exposure module* takes the hazard estimates and translates the chances of damage or loss into monetary values based on the physical attributes of the properties. Construction material, building use, contents inside the building, and other features are used in calculating the exposure. The *financial module* converts the exposure calculations into the output specifically needed to convey the solvency of the company; i.e. the balance between money available to pay claims and the number of claims they might need to pay. These calculations are based on policy information (premiums, deductibles, and limits).

As sophisticated as they are, CAT models are not always accurate because it is impossible to estimate how an actual natural catastrophe will impact property. Every event tends to expose a gap in the models. For example: For super storm Sandy, the CAT models expected the ports of New York to suffer flood losses only on the containers on the bottoms of the stacks on their docks. However, all the containers were positioned on the docks to avoid the wind blowing the stacks over. The additional loss incurred was unaccounted for. During the same storm, the CAT models underestimated the potential losses by overlooking that many buildings in Manhattan have their IT infrastructures located in flood-prone basements.

3 Climate Change Adaptation and Disaster Risk Reduction

The Intergovernmental Panel on Climate Change (IPCC) defined adaptation as adjustment in natural or human systems in response to actual or expected climatic stimuli or their effects that moderates harm or exploits beneficial opportunities [27]. From its beginning, the international climate effort has focused primarily on mitigation—reducing or preventing GHG emissions. However, in recent years, more attention is being given to adaptation—adjusting to and dealing with the impacts of climate change. A successful adaptation can reduce vulnerability by building on and strengthening existing coping strategies. In general, the more mitigation there is, the less will be the impacts of climate change that we must adapt to and the less the risks we must prepare for. Further, the greater the degree of preparatory adaptation, the less may be the impacts of any given degree of climate change.

This potential and capacity to adapt is referred to as "adaptive capacity." Human societies have varying adaptive capacity to respond to or anticipate climate changes. However, there are climate changes in temperature, storm frequency, flooding, and other weather events that previous experience has not prepared them for. Adaptation measures can help to reduce vulnerability to climate change by building adaptive capacity as well as allowing populations to benefit from opportunities resulting from climatic change, such as growing new crops in areas that were previously unsuitable [22].

Adaptive strategies encourage actions to decrease, spread, or transfer the risk of damages. Although it may be preferable to take a proactive approach to eliminate the cause of the risk, reactive post-harm compensation may be a last resort effort to rebuild communities after adverse impacts from extreme weather events. Insurance for flooding and other catastrophes allows for post disaster resilience of individuals and communities. Where there are failures in the private insurance market, the public sector creates residual market mechanisms (RMM) to encourage individual risk reduction by subsidizing premiums. A study in 2013 identified key equity issues for policy considerations [36]:

- Transferring risk to the public purse does not reduce overall risk,
- Governments can spread the cost of losses across time rather than space,
- Governments can force home-owners in low risk areas to cross-subsidize the insurance premiums of those in high risk areas,
- Cross-subsidization is increasingly difficult for private sector insurers operating in a competitive market, and
- Governments can tax people to pay for tomorrow's disaster.

Government-subsidized insurance such as the U.S. National Flood Insurance Program is criticized for providing a perverse incentive to develop properties in hazardous areas, thereby increasing overall risk. This possibility may be countered with appropriate land use policies that limit new construction in areas that have potential climate risks and/or encourage the adoption of climate resilient building codes to mitigate potential damages.

Adaptive capacity and sound development policy and strategies are strongly intertwined. Climate change is now central to national planning processes and to development assistance. It is a major consideration in Climate Risk Management (CRM), a generic term referring to an approach to climate-sensitive decision making that is informed by a large and growing body of work bridging the climate change adaptation, disaster management, and development sectors. CRM seeks to promote sustainable development by reducing the vulnerability associated with climate risk. It covers a broad range of potential management tools including early response systems, strategic diversification, dynamic resource allocation rule, financial instruments, infrastructure design, and capacity building. CRM employs strategies that maximize positive and minimize negative outcomes for communities in fields such as agriculture, food security, water resources, and health. In addition to avoiding adverse outcomes, a CRM strategy may maximize opportunities, such as the use of scientific technologies that allow farmers in climate-sensitive economic sectors to use favorable seasonal forecasts to maximize their crop productivity [20].

4 Disaster Management and Risk Reduction Technology

The Karlsruhe Institute of Technology/Center for Disaster Management and Risk Reduction Technology (CEDIM) located at the University of Karlsruhe, and the Helmholtz-Centre Potsdam—German Research Centre for Geosciences (CEDIM), are interdisciplinary research centers dedicated to fundamental and applied research in the field of disaster management and risk reduction. These centers work to advance the science of natural and man-made hazards assessment and develop disaster management solutions for the early detection and reduction of risk. In 2011, CEDIM began a new research program, CEDIM Forensic Disaster Analysis (FDA), to examine disasters in near real time using an interdisciplinary approach with a focus on complex interactions among (1) the natural hazard, (2) the technical installations, facilities, and infrastructures, and (3) the societal structures, institutions and capacities. Previous CEDIM research projects which included Flood Hazard in a Changing Climate and Natural Risk in a Changing Climate, launched in 2007, continued in the context of CEDIM Forensic Disaster Analysis (FDA) [23].

Recent technological advances have created unprecedented opportunities for earthquake loss assessment. Realistic assessment of potential damages is possible once the hazard and impact parameters have been roughly estimated. The internet usually provides information from various sources less than an hour after a major sudden onset earthquake somewhere on the globe; modern crowd sourcing allows users to quickly assess initial information about the disaster; remote sensing data of high quality are commonly available 1–3 days after the disaster. Methods to acquire and assess disaster loss data are available not only for earthquakes but also for other natural disasters. For example, the Global Disaster Alert and Coordination System (GDACS)—a joint initiative among the United Nations, the European Commission, disaster managers worldwide (www.gdacs.org/), and CoreLogic Catastrophe Modeling (EQECAT), a private sector company (www.eqecat.com/)—is readily accessible and it provides highly relevant risk and disaster related information on a timely basis.

Forensic disaster analysis (FDA) has been identified as a research target by the Integrated Research on Disaster Risk (IRDR) (www.icsu.org) initiative located in Beijing. IRDR's forensic investigation of disasters (FORIN) program studies natural hazards to uncover the root causes of disasters through in depth analysis that go beyond the typical reports after disasters: "Thoroughly analyzing cases, including both success stories and failures, will help build an understanding of how natural hazards do—or do not—become disasters" (www.irdrinternational.org) [1].

By generating a comprehensive portrait of disasters, researchers and practitioners learn not only about specific aspects of a disaster, but fundamental lessons that can inform and improve disaster management. The FORIN program investigates the circumstances, causes and consequences of losses in disasters as well as the conditions that have limited or prevented loss. The research is conducted in the context of the latest scientific developments and approaches to disaster risk reduction. Also, FORIN program case studies and updates are published on the IRDR website (http://www.irdrinternational.org/projects/forin/) to increase understanding of disaster causes, communication, advocacy, and public awareness in IRDR's FORIN investigations [24].

The Center for Disaster Management and Risk Reduction Technology (CEDIM) adopts this comprehensive understanding of disasters and applies its methodology of near real-time forensic disaster analysis (FDA) as a complementing component of the FORIN program of IRDR [50].

5 CEDIM Forensic Methodology

CEDIM's Forensic Disaster Analysis (FDA) uses the term "forensic" in the sense of scrutinizing disasters closely by using the most advanced modern empirical and analytical methodologies available in science, engineering, remote sensing, and crowd sourcing to estimate potential disaster risks and impacts. The data derived by different modeling techniques are combined, updated, and compared against information coming from the ground. The interpretation of this information is the starting point for comprehensive science-based assessments and judgments of disasters in near real-time. The science-based assessments are compiled based on multi- and interdisciplinary expertise and include the critical evaluation, assessment, validation, appraisal, and quantification of an event. The near real-time FDA approach is justified based on the following considerations [50]:

- Time criticality is important as many pieces of information emerge within the first days of a disaster
- Interaction with the many actors is most intense and open during these days.
- Potential user interest (emergency services, tourism industry, insurance industry, economic cooperation agencies, relief agencies, etc.) is also at peak at this initial stage.
- Initial hypotheses (with little information after the first days) on loss evolution and implications can be tested in the following days.
- Understanding of natural disasters within their respective socio-economic contexts is significantly increased.

The analysis of data available from various sources is complemented by CEDIM's models for near-real time loss estimates. CEDIM's near real time FDA covers methods, models and tools for rapid information extraction from social media (twitter), rapid impact assessments of atmospheric events and floods, and rapid assessments of socio-economic impacts, direct and indirect economic losses, status of transportation, and interruption to supply chains. Currently CEDIM is developing near-real time forensic disaster models for three types of hazards: geophysical, meteorological, and hydrological. The forensic approach focuses on strengthening the capability to rapidly generate a portrait of a disaster to reveal its main characteristics, causes of loss and potential loss estimates, the short and long-term impacts on regional and national scales; to track the evolution of the disasters; and to develop a framework for future loss and risk reduction.

One critical condition for near-real time forensic analysis is the availability of data bases that can be consulted in case of a disaster. For earthquakes, CATDAT the worldwide damaging earthquakes and secondary effects (tsunami, fire, landslides, liquefaction and fault-rupture) database created in 2003, is the most comprehensive data base that is globally available [4, 6]. It includes data on 12,400 damaging earthquakes of which 7100 occurred after the year 1900. It refers overall to 21,000 different sources in 60 languages, all of them translated to English. It also includes socio-economic analysis data and tools that use more than 100 socioeconomic parameters. Each disaster in the data base is validated to the extent possible. Other data bases in use are the Weather hazard—early warning web service of CEDIM, and for regional studies the CEDIM risk explorer. Another tool called the EQUATOR allows for the automatic internet search for relevant information in the event of a disaster. For earthquakes, the real-time multi-source reporting web site on all worldwide earthquakes, the Earth Quake-Report (http://earthquake-report.com/), can be readily accessed.

6 Examples of CEDIM Forensic Disaster Analysis (FDA)

The first forensic disaster analysis (FDA) activities are documented on the CEDIM webpage. They include a number of geological and hydro-meteorological disasters. The October 23, 2011 Van earthquake which occurred in eastern Turkey in 2011 was the first test case using the near-real time FDA approach. So far, CEDIM covered the following disasters to further refine the near-real time FDA approach:

- The March 11, 2011 Tohoku earthquake in Japan was analyzed for weeks in terms of the evolving direct and indirect economic losses [4, 6, 32], and the direction and velocity of a possible radioactive cloud was also meteorologically monitored.
- The November 2011 Thailand flood was monitored with a view on the supply chain crisis that emerged for computer parts (hard discs), seriously affecting the global market of those goods and several major electronic and car manufacturing companies.

- Between May 20 and 29, 2012 the Ferrara (Northern Italy) earthquake sequence, caused 'only' 24 fatalities but 15,000 homeless people, more than a billion € direct economic loss (0.6 % of Emilia-Romagna's GDP), significant heritage losses as well as large industry and residential losses. Around 500 million € damage to the cheese, ham and other agricultural industries in the region are estimated.
- From October 22 to 29, 2012, Hurricane Sandy made its way from the Caribbean Sea into the Atlantic Ocean and finally entered the U.S. on the morning of October 30, not far from New York. According to the Saffir-Simpson Hurricane Wind Scale with a 1–5 rating, Sandy was a category 2 Hurricane (154–177 km/h). Along its path Sandy caused many fatalities in Jamaica, Haiti and Cuba and left many people homeless. The interaction between Sandy and an extra-tropical weather system created a huge storm in the U.S., affecting large areas; it was associated with high impact weather as far as the Great Lakes and even beyond in southern and southeastern Canada. Due to the huge spatial extension and intensity, Sandy caused massive damage and losses in the densely populated East Coast states.

FDA focused on fatalities, casualties, shelter needs, power loss, business interruption, and social vulnerability of the disasters. Two FDA reports were issued on the 2012 Western U.S. summer drought that occurred after a record-breaking hot spring. The heat persisted in the U.S. and North America through summer 2012. July 2012 was the hottest July on record in the U.S. with the summer (June–August) ranking 3rd hottest since 1895. Analysis focused on a social vulnerability index for droughts to assess the aggravated impact by social and economic conditions based on an Adaptive Capacity index (AC index) with three major components that characterize the economic capacity, human and civic resources, and agricultural innovation [22]. The index is flexible and can be applied to managed and natural ecosystems as well as to socio-economic systems. The overall vulnerability is determined by combining vulnerability caused by the direct exposure to drought, and vulnerability to drought caused by social and economic conditions. The disaster data generated by CEDIM FDA are invaluable for disaster risk management.

7 Climate Smart Disaster Risk Management (CSDRM)

Climate smart disaster risk management (CSDRM) is an inclusive concept that considers immediate and long term aspects of managing disaster risk related to climate change. CSDRM focuses on fluctuating disaster risks and uncertainties, increasing adaptive capacity, and addressing poverty and vulnerability and their structural causes. CSDRM recognizes that risk associated with climate change is a result of human vulnerability as well as local and global environmental change.

In a disaster context, vulnerability refers to the exposure or susceptibility of a community to a particular hazard [10]. It refers to the extent to which a natural or social system is susceptible to sustaining damage from climate change and considered to be the outcome of the effect that an event may have on humans, referred to as capacity or social vulnerability, *and* the risk that such an event may occur, often referred to as exposure [42]. A growing number of climate scholars conclude that human vulnerability is related to the increase in natural disasters and that severe weather, flooding, wildfire, earthquakes, tsunami, and volcanic eruptions are hazardous natural events that can seriously affect vulnerable populations [13, 39]. Risk and loss associated with any natural hazard depend on the scale, severity and timing of the event, *and* the affected communities' vulnerability to the hazard. This relationship is sometimes expressed mathematically as some variant of: "Disaster Risk = Hazard x Vulnerability" [10]. The net impact of an external risk such as a disaster and an internal risk such as the lack of means to cope may be positive (resilience) or negative (vulnerability). According to the IPCC, vulnerability is a degree to which a system will respond to a given change in climate including beneficial and harmful effects. It is also a degree to which a system is susceptible to or unable to cope with adverse effects of climate change including climate variability and extremes [28]. Among the human activities, agricultural activities are considered to be the most sensitive to climatic conditions and to climatic variability [30].

As a component of adaptation, CSDRM addresses disaster risk and loss using the latest technologies and methodologies to reduce human vulnerability. Further, it assumes that adapting to climate change requires that risk management be integrated into development planning and practices. The CSDRM approach documents and applies the information from lessons and limitations learned from disaster research and practice to inform more effective climate change adaptation of disaster risk reduction. Pelling and Schipper [43] describe the following:

- governance—social capacity can extend coping capacity beyond financial limits and large scale risk reduction is best handled by government in partnership with civil society. However, the potential of civil society to support DRR is limited.
- insurance and risk transfer—enhance risk reducing behavior when supported with information and regulation and spread risk over time, place and between risk categories. The limitation is whether commercial insurance can be extended to the poor without increasing their poverty.
- perceptions—risk and the propensity for risk reducing behavior are distorted by social media. Organizations and individuals tend to be slow in acknowledging new risks, but can change behavior quickly after a disaster. The limitation is whether there are forms of education and learning arrangements that can stimulate risk reducing behavior and values.
- rights and ethics—those more at risk tend to spend on immediate or more tangible needs rather than invest scarce resources to reduce potential future risk. The planning tools to open dialogue between those at risk and the risk managers are limited.

Pelling and Schipper [43] advise that living with climate change requires two modes of adaptation and that both modes of adaptation require greenhouse gas (GHG) emission mitigation and the consideration of existing challenges of social justice and ecological sustainability. One mode involves the internalization of risk management into the development agenda which responds to increased levels of uncertainty associated with greater levels of climate variation and larger extremes in weather events. The second mode requires climate proofing of development decisions for long-term investments to take gradual changes in climatic regimes into account alongside other future components such as demographic change affecting investments from agricultural science to infrastructure projects.

A 2012 report on the advantages of and recommendations for use of the CSDRM approach noted that CSDRM is unique because it provides policy makers and practitioners with a means to identify the processes needed to build resilience to climate and disaster risks, taking into account the pressing challenges of sustainable development, social justice, environmental sustainability, and other competing community needs. CSDRM recognizes that the three pillars of sustainability—economic, social, and environmental—are collectively a powerful tool for defining the complete sustainability problem. If any pillar is weak, then the system as a whole is unsustainable. CSDRM recognizes that dynamic sets of risks emerge from physical, environmental, economic, political and social sources, and that multiple and often simultaneous shocks and stressors are part of the lived reality for many communities and households [19]. CSDRM helps to:

- Evaluate which existing tools and frameworks in Disaster Risk Management (DRM), Climate Change Adaptation (CCA) and development are right for particular programs, policies and projects.
- Develop the ability to identify and form strategic partnerships in a multi-sectoral working environment.
- Reflect, review and evaluate progress through concrete indicators.
- Establish that the program/policy/project is supporting the realization of climate smart and disaster-proof sustainable development.
- Think through the implication of climate and disaster risks at each step of a project management cycle.

The following is the recommended process for applying the Climate Smart Disaster Risk Management (CSDRM) approach [19]:

- Steps 1 and 2: "Where are we now?" These steps are taken before starting. It involves using action points and guiding questions to assess and reflect on an organization's capacities. Indicators are then used to review existing programs or policies or to plan for new ones.
- Steps 3 and 4: "Where do we want to be?" "What do we need to do differently?" These steps involves identifying potential entry points to apply the CSDRM approach, map out integration pathways, develop action plans and, select indicators to measure progress.

- Step 5: The CSDRM Journey—"Are we moving towards integration?"
  This step involves monitoring and reviewing the progress and understanding the internal and external factors that enable or constrain integration efforts. Doing so helps to identify new opportunities and/or corrective actions.
- Step 6: Looking Back—"What has changed, why and how?" This is an important focus of the approach and involves looking at the progress made, evaluating it, and reflecting on what has worked (or not) and what needs to be changed.

Challenges of the CSDRM approach include maintaining a balanced consideration of the three pillars of sustainability—economic, social, and environmental—in each of the steps of the recommended process. The CSDRM approach enables organizations to learn about, reflect upon, and integrate the challenges of climate and disaster risks as they impinge upon organizational mission and goals. CSDRM is flexible and adaptable to unique and complex environments. It can help to determine which of the many existing tools and frameworks in DRM, CCA, and development are appropriate for any given situation; identify and form strategic partnerships; and provide concrete indicators to measure progress. CSDRM assists and supports the realization of sustainable development that is climate smart and disaster proof [46]. The importance of integrating the three pillars of sustainability—economic, social and environmental—in resilience management in complex environments is gaining attention and serious consideration of many climate scholars. It is acknowledged that, "Resilience management goes beyond risk management to address the complexities of large integrated systems and the uncertainty of future threats, especially those associated with climate change" [34]. The urgency of dealing with the threats of climate change is global. According to Tom Mitchell, this may require "elected representatives and international organizations to put disaster risk management at the top of their policy agenda and to take these messages into international negotiations" [38].

8 Advances in Natural Disaster Forensics

Several advances relevant to natural disaster forensics include "causal loss analysis", "information gap analysis" and "near real time assessment of economic impact." The "causal loss analysis" project focuses on creating methodologies to estimate and identify the key indicators and root causes after the immediate impact of a major event in terms of structural damages and shelter needs based on widely available data. This allows for the quantification of the scale of the disaster immediately after a hazard impacts a population, providing data about the level of impact and the emergency response needs. Often, immediate response is required before complete data are available. The causal loss analysis project analyzes historic catastrophic events (geophysical and hydro-meteorological) of the past 40–50 years to understand the aggravating factors (socio-economic, regional building practices, weather, etc.) that affect the impact of a hazard. Using the parsimonious modeling approach, socio-economic fragility functions are calculated and calibrated with a selected database of historic events from CATDAT to develop a standard relationship between the hazard intensity, loss and other regional data that are widely available (e.g. human development index (HDI), population density) and the total impact of the event [5]. This project also has synergies with the CEDIM "Earthquake Loss Analysis" project. Similar methodologies were used in quantification of post-disaster losses in the Bohol earthquake and Haiyan typhoon in 2013. The causal loss analysis project analyzes individual and groups of natural disaster events where data are available at a local scale in order to determine the influence of individual factors on disaster impacts beginning with shelter needs (building damage, homelessness, utilities, etc.). The key indicators then serve as a proxy for the potential scale and impact of a disaster and the time aspect leading to a potential catastrophe. The purpose and intent of the "causal loss analysis" project is to identify some of the key indicators required for study in FDA in the "near real-time" of a disaster. Identifying such indicators provides a focus for a holistic view of the post-disaster shelter needs as well as other potential insights on aid, recovery and reconstruction needs [5].

"Information gap analysis" is designed to improve the quality of a disaster response which can affect how severely a society is harmed by a disastrous event. While an efficient response can help to reduce casualties and suffering, a less efficient response may aggravate the situation. In order to include the disaster response as a potential contributing factor to the overall disaster impact, information gap analysis uses a methodology that analyzes disaster response in near-real-time. A disaster response data classification scheme has been developed to facilitate analysis of the publicly communicated disaster response information issued within the first 0–5 days of a disaster. Potential deficiencies in communicating critical disaster response information to the public are identified, thus facilitating disaster response. Near-real-time analysis of numerous disasters has established best practice standards for information content and timing. Comparison to these standards results in the identification of information gaps; this process is referred to as "information gap analysis" [15]. Information gap analysis was used following Cyclone Phailin in India and the Bohol Earthquake and Typhoon Haiyan in the Philippines as part of CEDIM FDA activities.

The "near-real time assessment of economic impact" project utilizes a method that enables the near-real time assessment of direct and indirect economic impact for all phases of a disaster. Natural disasters can result in considerable economic costs, which can have important consequences for the affected regions and their recovery [31]. Often more important than the direct damages (e.g., destruction of buildings and physical infrastructures) are the indirect damages that occur due to business and supply chain interruptions. As the interdependencies within and across modern supply networks are complex and information about them is typically limited, an assessment of indirect disaster impact is difficult. In addition, early available disaster information may be incomplete, uncertain, lacking or conflicting. The "near-real time assessment of economic impact," however, identifies and focuses on the most relevant information or most vulnerable areas affecting the afflicted community. It is easily adaptable, so that new information can be integrated rapidly. The analysis can then be enriched successively providing near real time (as the situation unfolds) more detailed levels of analysis and more precise results. As a result, disaster managers can make more appropriate and timely decisions. It must be possible to quickly adapt the method to different information systems and infrastructures to ensure that first results can be determined shortly after any potential disaster, wherever it occurs. Given these requirements, the near-real time assessment of economic impact project chose an indicator-based method to assess sector specific vulnerability against business interruptions. To rapidly adapt the method when new data become available, the project uses *key indicators*. These key indicators are calculated on the basis of national input-output tables, which are available in any country, so that a first rough estimation of indirect costs is available near real time. The method can adapt to integrate additional indicators to enable more detailed or precise assessment by considering other information such as the production strategies or adaptive capacity that successively refine the first rough indirect cost estimate [3, 9]. The fastest tools are used to communicate the uncertainties present and to explain the possible consequences of the lack of information in a transparent manner. The project also links with other related projects and use "Case-Based-Reasoning" to identify the most relevant indicators, to determine cause-effect chains and interdependencies between indicators as well as to calibrate the model [31].

9 Agenda 21—Kyoto to Lima to Paris

Agenda 21 was made public at the UN Conference on Environment and Development (Earth Summit), held in Rio de Janeiro on June 13, 1992, where 178 governments voted to adopt this action plan. The "21" in Agenda 21 refers to the 21st Century. It is a nonbinding voluntarily implemented action agenda/plan of the United Nations with regard to sustainable development that has been affirmed and modified at subsequent UN conferences. The Kyoto Protocol is an international treaty, which extends the 1992 United Nations Framework Convention on Climate Change (UNFCCC) that commits State Parties to reduce greenhouse gases (GHG) emissions, based on the premise that (a) global warming exists, and (b) human-made CO_2 emissions have caused it. The Kyoto Protocol was adopted in Kyoto, Japan, on December 11, 1997 and entered into force on February 16, 2005 [48]. The Kyoto Protocol set binding emission reduction targets for industrialized countries for the first commitment period 2008–2012. There are currently 192 Parties to the Protocol. Subsequent Agenda 21 climate change environmental and development activities, assessments, and initiatives followed the first commitment period. Agenda 21s climate action agenda/plan for the UN, other multilateral organizations, and individual governments around the world can be executed at local, national, and global levels.

The Intergovernmental Panel on Climate Change (IPCC) analyzed written comments and observations submitted by participating governments and organizations for a Fourth Assessment Report [24]. The report was delivered in stages, beginning with Working Group I's report on the physical science basis, based on 9200 peer-reviewed studies [41, 45]. The IPCC Fifth Assessment Report was finalized in 2014. It was developed through a scoping process which involved climate change experts from all relevant disciplines, users of IPCC reports, and in particular, representatives from governments. The report, Climate Change, The Physical Basis, involved worldwide collaboration including 259 authors and 39 countries [26]. It is anticipated that the Fifth Assessment Report will pave the way for a global, legally binding treaty on reducing carbon emissions at the UN Climate Change Conference in Paris during December 2015.

On December 14, 2014, at a meeting in Lima, Peru, after more than 36 straight hours of negotiations, top officials from nearly 200 nations came to an agreement committing every country in the world to reducing the fossil fuel emissions that cause global warming. The resulting "Lima Accord" represents a breakthrough in the two-decade effort to forge a significant global pact to counter climate change. It is the first time that all nations, rich and poor, have agreed to cut back on the burning of oil, coal, and gas. The driving force behind the agreement was not the threat of legal sanctions or other punitive consequences. It was global peer pressure. The strength of the Lima Accord is that every country pledged to put forward a plan to reduce emissions at home; the weakness is that there are no legally binding requirements that countries cut their emissions by any particular amount. This omission was to get every country to agree to the deal, including the U.S., historically the world's largest carbon polluter.

With no language requiring the significant cuts scientists say are needed to stave off the costly effects of global warming, countries can put forth weak plans that amount to little more than business as usual. Countries can even choose to ignore the deal and submit no plan at all [7]. However, experts say by asking countries to put forward plans dictated by their own economies and domestic politics rather than a top-down mandate, the Lima Accord secured the agreement of every nation to some kind of carbon-cutting action. Each nation agreed to enact domestic laws to reduce carbon emissions and put forth a plan by March 31, 2015 laying out how much each one will cut emissions after 2020, and what domestic policies it will pass to achieve the cuts. Countries that miss the March 2015 deadline are expected to report their plans by June. The plans from every country, known within the UN as "Intended Nationally Determined Contributions," (INDCs) form the basis of a sweeping new deal to be signed in Paris in December 2015.

The United Nations Climate Change Conference will be held in Paris, France from November 30 to December 11, 2015. The international conference will include the 21st yearly session of the Conference of the Parties (COP 21) that dates back to the 1992 United Nations Framework Convention on Climate Change (UNFCCC) and the 11th session of the Meeting of the Parties (CMP 11) that dates back to the 1997 Kyoto Protocol. COP 21 and CMP 11 meetings determine leaders of negotiations and work on finalizing details of the negotiations. The objective of the 2015 United Nations Climate Change Conference is to achieve, for the first time in over 20 years of UN negotiations, a legally binding and universal agreement on climate from all the nations of the world including the major emitters of GHGs. In order for the agreement to come into effect in 2020, at the end of the second commitment period of the Kyoto Protocol, COP 21 and CMP 11 meetings will be held to move the agreement forward [14].

In July 2015, Pope Francis published an encyclical titled *Laudato si'* intended in part to influence the approaching United Nations Climate Change Conference. The encyclical calls for action against human-caused climate change. *Laudato si*' (Medieval Central Italian for "Praise Be to You") has the subtitle "*On care for our common home*." In it, the pope critiques consumerism and irresponsible development, laments environmental degradation and global warming, and calls all people of the world to take "swift and unified global action" [53].

10 Intended Nationally Determined Contributions (INDCs) to Emissions Reduction

The process for Intended Nationally Determined Contributions (INDCs) pairs national policy-setting—in which countries determine their contributions to GHG emissions reduction in the context of their national priorities, circumstances, and capabilities—with a global framework that drives collective action towards a low-carbon, climate-resilient future. The INDCs can create a constructive feedback loop between national and international decision-making on climate change. They are the primary means for governments to communicate internationally the steps they will take to reduce GHG emissions and address climate change in their own countries. The largest GHG emitters contribute a majority of global emissions with the top 10 emitters contributing 72 % of the total while the lowest 100 emitters contribute less than 3 %. Significant mitigation actions are needed by the largest emitters within their capacity to do so. The energy sector contributes to more than 75 % of global GHG emissions which necessitates a rapid transformation of the energy sector by 2050 to avoid the worst impacts of climate change. Records show that six of the top 10 top emitters are developing countries. China contributes approximately 25 % of global emissions, making it the top emitter. India, Indonesia, Brazil, Mexico and Iran contribute relatively large shares of global emissions as their economies grow. Per capita emissions are still distributed unequally as are per person emissions which still vary among the top 10 emitters, with the United States' per capita emissions eight times those of India [12]. INDCs will reflect each country's intent for reducing GHG emissions, taking into account its domestic circumstances and capabilities. Countries may address how they will adapt to climate change impacts and what support they need from, or will provide to, other countries to achieve a low-carbon, climate-resilient future.

In 2014–2015, the United Nations Development Program (UNDP) and the United Nations Framework Convention on Climate Change (UNFCCC) convened a series of Regional Technical Dialogues to support countries in the process of preparing and putting forward their INDCs. The countries that participated in those dialogues requested additional detailed guidance on INDC preparation. In partnership with the UNDP, the World Resources Institute (WRI) prepared a guidance document to assist in the design and preparation of INDCs including mitigation and adaptation components and explanations of the criteria for fairness. The guide captures ideas shared during the dialogues, reflects the current state of negotiations, and puts forward options for the preparation of INDCs based on research from recent literature and relevant UNFCCC documentation. A 2015 report by the UNDP and WRI, "Designing and Preparing Intended Nationally Determined Contributions (INDCs)" provides detailed directions for technical and process-related considerations. It walks Parties through the choices they will face in preparing and designing their INDCs, laid out in five general steps: identifying the benefits of an INDC, organizing the INDC process, identifying data and analysis to inform the INDC, designing the INDC, and communicating the INDC [33].

Well-designed INDCs will signal to the world that the country is doing its part to combat climate change and limit future climate risks. Countries should engage in a transparent process when preparing their INDC in order to build trust and accountability with domestic and international stakeholders. A strong INDC is ambitious, leading to transformation in carbon-intensive sectors and industry; transparent, so that stakeholders can track progress and ensure countries meet their stated goals; and equitable, so that each country does its fair share to address climate change. It is important that INDCs be clearly communicated so domestic and international stakeholders can anticipate how these actions will contribute to global emissions reductions and climate resilience in the future. An INDC should also articulate how the country is integrating climate change into other national priorities such as sustainable development and poverty reduction, and send signals to the private sector to join in these efforts.

WRI is working on a variety of projects that assist governments in preparing INDCs and help stakeholders understand and evaluate INDCs [52]. The WRI Open Climate Network (OCN) is working with partners in eight focus countries to evaluate current emissions trends and abatement potential out to 2030, with a view to informing initial INDCs. Following the release of the INDCs, OCN and its partners will evaluate mitigation pledges based on insights from the GHG Protocol and other tools. These assessments will provide decision-makers in the world's largest economies with data that will enable them to develop ambitious GHG reduction targets in their INDCs (WRI 2015). WRI also works on INDCs through the WRI Open Book initiative that enhances INDC transparency. The Agreement on Climate Transformation 2015 (ACT 2015) project works to catalyze agreement at COP21, including the ways in which countries' commitments and other actions from their INDCs are finalized and linked to the 2015 agreement.

On July 1, 2015, after a number of countries submitted their INDCs, U.S. Secretary of State John Kerry commended the progress in a White House Press release stating, "This week, the U.S. and Brazil—the two countries with the most absolute emissions reductions in the world since 2005—released a joint statement establishing a joint climate change working group and outlining key areas of cooperation on climate change, with Brazil committing to new renewable energy and sustainable land use goals…. The targets announced this week—in addition to those from the U.S., EU, Mexico, Canada, Switzerland, Norway, Russia, and Japan—represent more than two-thirds of global greenhouse gas emissions…. While much work remains to be done to secure a durable climate agreement in Paris, I commend these leaders for helping to build momentum towards this goal" [51].

Many other countries are expected to submit their INDCs before October 2015. The WRI Climate Analysis Indicators Tool (CAIT) Paris Contributions Map tracks and analyzes INDCs as countries submit them (WRI 2015). Data from the CAIT Paris Contributions Map enable users to explore, compare, and assess GHG mitigation plans in each country's INDC. The data are structured according to a framework based on the Lima Decision, the GHG Protocol Mitigation Goal Standard, and GHG Protocol Policy and Action Standard (WRI 2015). The CAIT Paris Contributions Map will be updated as new INDCs are released throughout 2015. A map view in the tool allows you to see all countries at a glance, and examine what types of contributions they have put forth, and what sectors and GHGs are covered, among other options. All data on the site are available for direct download. The data are derived directly from documents submitted to the UNFCCC by national governments. All INDCs submitted to the UNFCCC Secretariat by October 1, 2015 will be included in a synthesis report that will be released by November 1, 2015. The report will reflect the aggregate emissions reduction from all the participating countries.

11 The Sendai Framework for Disaster Risk Reduction 2015–2030

Related to the INDCs, the Sendai Framework for Disaster Risk Reduction 2015–2030 was adopted at the Third UN World Conference in Sendai, Japan, on March 18, 2015. The Sendai Framework is a 15-year, voluntary, non-binding agreement which charts a global course over the next 15 years to ensure engagement and ownership of emissions reduction action by all stakeholders, and to strengthen accountability in disaster risk reduction. It emphasizes that the State has the primary role to reduce disaster risk but that responsibility should be shared with local government, the private sector, and other stakeholders.

Paragraph 48 (c) of the Sendai Framework calls upon "the United Nations Office for Disaster Risk Reduction (UNISDR), in particular, to support the implementation, follow-up and review of this framework through […] generating evidence-based and practical guidance for implementation in close collaboration with States, and through mobilization of experts reinforcing a culture of prevention in relevant stakeholders [...]" [11]. In order to support the process, a number of Sendai Framework implementation guides were developed based on the experience and use of the similar "Words into Action" guide that was created during the Hyogo Framework for Action (HFA) decade from 2005 to 2015. The guides focus on a specific topic, sector, or stakeholder group, such as guidance on aligning and measuring targets and indicators, meeting the priorities for action, development of risk-informed plans and strategies, investment opportunities, innovative action and implementation since HFA, and working with relevant partners and stakeholders [11].

Climate experts warn that limiting global warming to 2 °C by 2030 is mandatory to prevent some of the worst impacts of climate change. However, according to a new report from the MILES Project Consortium, the national climate plans (INDCs) that countries have submitted so far in preparation for COP 21 in Paris in December 2015 would put the world on an emissions pathway that would lead to about 54 $GtCO_2e$ being emitted in 2030 [35]. This falls short of the emissions reductions necessary to reach the stabilization target of 44 GtCO2e by 2020 recommended by the United Nations Environment Program (UNEP) [49]. If no action is taken the world is currently on a path that will lead to around 60 GtCO2e in 2020. There must be far more ambitious emissions reduction action proposed in the INDCs to limit global warming to 2 °C by 2030 [12].

The Columbia Climate Center (CCC) developed an Energy and Emissions Model (EEM) to estimate the effect of climate, energy, and land-use policies on GHG emissions in countries around the world. The CCC uses EEM to quantify whether the existing suite of policies can achieve targeted emissions reductions. In collaboration with Deutsche Bank Climate Change Advisors (DBCCA), CCC reported on the expected gap in estimates in GHG emissions [2]. The United Nations Environment Program (UNEP) also published The Emissions Gap Report 2013: A UNEP Synthesis Report. The series of international agreements since Agenda 21 was made public at the UN Conference on Environment and Development (Earth Summit) held in Rio de Janeiro on June 13, 1992 have yet to guarantee the minimum acceptable level of global warming limited to 2 °C by 2030. The gap between intended and necessary emissions reduction looms large. The hope is that the international commitment to INDCs to effectively reduce the impact of GHG on climate change will culminate in a strong, equitable, legally binding, final universal agreement at the December Climate Conference (Summit), Paris 2015, 21st Conference of the Parties on Climate (COP) [40].

12 Conclusion

Stabilizing the global climate is the greatest environmental challenge of the 21st century. Temperatures have exceeded global annual averages for 38 consecutive years with devastating impacts. Extreme weather events are more frequent and severe, including heat waves and drought that destroy agriculture, increase the risk of wildfires and endanger lives. Rising sea level resulting in flooding and storm surge threatens coastal communities and infrastructure. Nations worldwide suffer untold destruction and loss from the onslaught of natural disasters that have called attention to the climate crisis.

To slow or reverse the trend of increasing natural disasters that are exacerbated by human caused GHG emissions and other polluters, common misconceptions about climate change related disasters must be dispelled. These misconceptions include that natural disasters are caused by unpredictable forces of nature and not humans, and there is little that can be done about it. The accumulated data by climate scientists over decades provide undeniable evidence that burning fossil fuels, tropical deforestation, and other human activities continue to discharge greenhouse gases (GHGs) such as carbon dioxide (CO_2), methane (CH_4), nitrous oxide (N_2O), ozone (O_3) and chlorofluorocarbons (CFC) into the atmosphere and these gases trap energy that would normally be radiated into space, thus perilously raising the Earth's surface temperatures. Moreover, the rapidly rising GHG concentrations in the atmosphere are long-lasting in terms of hundreds of years [47].

As ominous as the prevailing climate conditions are, the national/international global community possesses the resources (e.g. three pillars of sustainability, human, technological, etc.) that can stem climate change threats and destruction. These include but are not limited to:

- The new approaches and technologies advanced by disaster forensics that not only work to reduce GHG emissions but also mitigate the impact of their known causes. As covered above, these include:
  - The Integrated Research on Disaster Risk (IRDR) forensic investigation of disasters (FORIN) program which studies natural hazards to uncover the root causes of disasters through in depth analysis that go beyond the typical reports after disasters.
  - CEDIM Forensic Disaster Analysis (CEDIM FDA) research that analyzes disasters and their impacts in near real-time to identify major risk drivers and the root causes. Findings are applied in disaster mitigation with a focus on disasters triggered by geophysical (earthquakes, volcanoes) and hydro-meteorological (floods, storms) extreme events.
  - The "causal loss analysis" project which focuses on creating methodologies to estimate and identify the key indicators and root causes after the immediate impact of a major event based on widely available data to address structural damages and shelter needs.
  - The "information gap analysis" which uses a methodology to analyze disaster response in near-real-time utilizing a disaster response data classification scheme developed to facilitate analysis of the publicly communicated disaster response information issued within the first 0–5 days of a disaster.
  - The "near-real time assessment of economic impact" project which utilizes a method that enables the near-real time assessment of direct and indirect economic impact for all phases of a disaster.

- The influence and support of the private sector, particularly at this time, in positively influencing climate policy at the December 2015 Paris conference. The private sector is already taking more aggressive climate actions than before, such as the adoption of emissions-reduction targets, programs to improve energy efficiency, and fuel switching to renewable energy sources. However, the pace and scale of these changes are insufficient to avoid injurious consequences in the future. Still, the private sector can play a positive role in negotiations by encouraging national governments to make a firm commitment to the Paris climate agreement by adopting short- and long-term corporate emissions-reduction targets, consistent with climate science, that lead to a phase-out of GHG emissions. The Paris climate agreement can create a policy environment that can provide appropriate incentives to enable businesses, investors, and innovators to benefit significantly through increased cost savings, lower investment risk, more innovative products and markets, greater investment in research and development (RD&D), and use of low-carbon technologies.
- The political will to maximize the possibilities for addressing climate change and protecting the global environment. Governments, policymakers, businesses, and civil society at the local, national and international levels *can* and must advance transformative solutions that mitigate climate change and help communities adapt to its impacts. The global political and environmental leadership and technological means exist to accomplish this.

Every nation in the world is vulnerable to the unabated impacts of climate change and exposed to disaster risks that threaten human and environmental systems. Evidence-based intervention and prevention strategies to slow or even reverse the trend of increasing natural disasters intensified by human caused GHG emissions and other polluters are made available through ongoing basic and applied research. The body of knowledge from forensic theory, methodology, insight, and analysis enriches the possibilities of disaster related research and practice to reduce disaster risk and protect communities across the globe. The challenge is to act boldly and decisively and the time is now.

Postscript:

On December 12, 2015, at the Paris climate summit—21st Conference of Parties (COP21)—the final draft of an international agreement capping the increase in global average temperatures to 2 °C above pre-industrial temperatures was accepted by consensus by delegates from close to 200 countries of the world. International leaders including UN Secretary General, Ban Ki Moon, and French Foreign Minister Laurent Fabius, president of the talks, hailed the agreement as a "historical accord" that would stem global warming and strive for a more ambitious limit of 1.5 °C above pre-industrial temperatures.

Weighing the objectives of the Paris agreement against the reality and resources to achieve them, climate scientists and activists as well as policy experts responded with cautious optimism at best. Unless the current "Intended Nationally Determined Contributions" (INDC) to curb carbon emissions become far more ambitious, the targets set to reverse the tide of global warming remain just that, targets. However, the text of the Paris agreement among other mandates calls for transparency, international reviews, funding to assist developing nations, and the dissolution of distinctions between developing and developed nations.

Unlike earlier accords where developed nations divided responsibility to curb carbon emissions, the Paris agreement calls for all countries to cut carbon emissions as determined by their own domestically developed plans based on national circumstances. The agreement commits developed nations to raise $100 billion annually beginning 2020, to empower developing countries to fight global warming by strengthening their technical capacity to transition to renewable energies and sustainable development.

The Paris agreement set up a framework that mandates transparency and international reviews and established a process whereby every 5 years countries are required to present stronger, more ambitious emissions reduction plans. Although there is nothing in the agreement that compels countries to do this and the means to guarantee compliance are weak to nil, the nationally driven carbon cutting plans and meaningful efforts of some major emitters over the last couple years indicate movement in the right direction. The planned long-term, regular public monitoring of countries' progress in emissions reduction exacts international pressure on countries to act on their INDC and to set increasingly more ambitious carbon cutting targets on a regular basis.

Much of the details of the Paris agreement regarding financing, transparency measures, and public review processes are still in developing stages. The difficult work laid out by the international Paris agreement ultimately requires strong and sustained political leadership and policy action agendas at the national and local levels where plans are mobilized. It is too early to judge the Paris agreement, but at the outset it appears to show more promise than earlier international climate summit agreements such as the Kyoto Protocol and Copenhagen Accord, where much the same issues were negotiated and eventually fizzled.

In the years leading up to COP 21 and the Paris Agreement, advancements in the sciences (natural, physical, biological, social, etc.), business and finance fields, and research and technology in some measure expanded the human capacity to limit global warming. The need to reduce greenhouse gas emissions (GHG) and to counter the destructive impacts of climate change fostered cooperation in local communities among public and private sectors including government, business, education, investors, and others, recognizing the urgency of global warming and the peril of inaction. The particular focus of disaster forensics in uncovering and dealing with the root causes of climate change through innovative methodologies, technologies, and strategies is a significant contribution in the efforts to achieve the goals of the international Paris agreement at the national and community levels where the real work must begin.

References

1. Burton I (2010) Forensic disaster investigations in depth: a new case study model. Environment 52:5
2. Columbia Climate Center (CCC) (2012) Earth Institute, Columbia University
3. Cutter S, Burton C, Emrich C (2010) Disaster resilience indicators for benchmarking baseline conditions. J Homel Secur Emerg Manage 7(1)
4. Daniell JE, Khazai B, Wenzel F, Vervaeck A (2011) The CATDAT damaging earthquakes database. Nat Hazards Earth Syst Sci 11:2235–2251. doi:10.5194/nhess-11-2235-2011
5. Daniell JE (2013) The development of socio-economic fragility functions for use in worldwide rapid earthquake loss estimation procedures, Ph.D. Thesis (unpublished), Karlsruhe Institute of Technology, Germany, 2013
6. Daniell JE, Wenzel F, Vervaeck A (2011) The Socio-economic effects of the 2011 Tohoku earthquake. In: Geophysical research abstracts, vol 13, EGU2011-14270
7. Davenport C (2014) A climate accord based on global peer pressure. The New York Times, 14 Dec 2014
8. De Groeve T, Poljansek K, Ehlrich D, Corbane C (2014) Current status and best practices for disaster loss data recording in EU member states: a comprehensive overview of current practice in the EU member states. Report by joint research centre of the European Commission, JRC92290, doi:10.2788/18330
9. Dottore ML, Zobel CW (2014) Analyzing economic indicators of disaster resilience following hurricane Katrina. Int J Bus Analytics 1(1):17
10. Dykstra J (2011) Climate change, natural disasters and human vulnerability. Science Media Center
11. Fowler J (2015) Sendai framework words turn into action, United nations office for disaster risk reduction (UNISDR), 18 Aug 2015. http://www.unisdr.org/archive/45452. Accessed Sept 2015
12. Friedrich J, Ge M, Damassa D (2015) Infographic: what do your country's emissions look like? World Resources Institute, 23 June 2015, (http://www.wri.org). Accessed Sept 2015
13. Gall M, Emrich T, Cutter S (2015) Who needs loss data? Input paper prepared for the 2015, global assessment report on disaster risk reduction, available at www.preventionweb.net/gar
14. Geneva Environmental Network (GEN) (2014) Préparations and major challenges of the UNFCCC COP 21/CMP 11 (French), H.E. Jacques Lapouge, French Ambassadeur for Climate Negotiations
15. Girard T (2014) Information gap analysis: near real-time evaluation of disaster response (Geophysical Institute, Karlsruhe Institute of Technology, Karlsruhe, Germany, EGU General Assembly 2014, held 27 April–2 May, Vienna, Austria, id.16112, 05/2014)
16. Global Facility for Disaster Reduction and Recovery (GFDRR) (2010) Damage, loss and needs assessment. Guidance notes. vol 1–3. The World Bank, Washington
17. Grossi P, Kunreuther H (eds) (2005) Catastrophe modeling: a new approach to managing risk (Huebner international series on risk, insurance and economic security). Springer Publishing
18. Guha-Sapir D, Hoyois P, Below R (2015) Annual disaster statistical review 2014: the numbers and trends, centre for research on the epidemiology of disasters (CRED), 54 p
19. Harris K, Seballos F, Silva Villanueva P, Curmi P (2012) Changing climate, changing disasters: pathways towards integration strengthening climate resilience. Institute of Development Studies (IDS), Brighton
20. Hellmuth ME, Moorhead A, Thomson MC, Williams J (eds) (2007) Climate risk management in Africa: learning from practice. International Research Institute for Climate and Society (IRI), Columbia University, New York, USA
21. IFRC (2005) World disasters report—data or dialogue? The role of information in disasters
22. Iglesias A, Moneo M, Quiroga S (2007) Methods for evaluating social vulnerability to drought Dpto. de Economía y Ciencias Sociales Agrarias, E.T.S. Ingenieros Agrónomos, Universidad Politécnica de Madrid, Avenida Complutense, s/n, 28040 Madrid, Spain

23. Integrated Research on Disaster Risk (IRDR) (2013) CEDIM's annual research report 2012—focus on forensic disaster analysis in near real time http://www.cedim.de/download/CEDIM_Annual_Research_Report2012_online.pdf. Accessed Sept 2015
24. Integrated Research on Disaster Risk (IRDR) (2011) Forensic investigations of disasters: the FORIN project (IRDR FORIN publication No. 1). Integrated Research on Disaster Risk, Beijing
25. IPCC (2011) Special Report on Renewable Energy Sources and Climate Change Mitigation. Cambridge University Press, Cambridge, United Kingdom and New York, NY, USA, p 17
26. IPCC (2013) Climate change 2013: the physical science basis, working group contribution to the fifth assessment report of the intergovernmental panel on climate change (IPPC)
27. IPPC (2007) Fourth assessment report "synthesis report" topic 1 observed changes in climate and their effects and topic 2 causes of changes: intergovernmental panel on climate change (IPPC)
28. IPPC (2001) Climate change: Impact, adaptation and vulnerability, summary for policy makers and technical summary of working group, Geneva, Switzerland: II. Available from http://www.ipcc.ch/pdf/climatechanges-2001/impact-adaptation-vulnerability/impact-spm-en.pdf. Accessed Sept 2015
29. ISDR (2013) Global assessment report on disaster risk reduction. United Nations, Geneva, Switzerland
30. Javid A (2004) Vulnerability mapping mitigates climate change. Available at http://www.geospatialworld.net/index.php?option=com_contentandview=articleandid=20718andItemid=2004. Accessed Sept 2015
31. Karlsruhe Institute of Technology (KIT) (2014) Methodology for a rapid assessment of the economic impact of natural disasters: business and supply chain interruptions Karlsruhe Institute of Technology, Germany
32. Khazai B, Daniell JE, Wenzel F (2011) The March 2011 japan earthquake—analysis of losses, impacts, and implications for the understanding of risks posed by extreme events. Technikfolgenabschätzung Theorie und Praxis 20(3):22–33
33. Levin K, Rich D, Tirpak D, McGray H, Waskow D, Bonduki, Y, Comstock M, Noble I, Mogelgaard K (2015) Designing and preparing intended nationally determined contributions (INDCs), United Nations Development Program (UNDP) World Resources Institute (WRI), www.wri.org
34. Linkov I, Bridges T, Creutzig F, Decker J, Fox-Lent C, Kroger W, Lambert J, Levermann A, Montreuil B, Nathwani J, Nyer R, Schatre B, Scheffler A, Schreurs M, Theil-Clemen T (2014) Changing the resilience paradigm, commentary: change, vol 4, June, 2014. www.nature.com/natureclimatecahange. Accessed Sept 2015
35. MILES Project Consortium (2015) University Hospital, Cologne Center for Integrated Oncology (CIO), Köln Bonn
36. McAneney J, Crompton R, Musulin R, Walker G, McAneney D, Pielke R (2014) Reflections on disaster loss trends, global climate change and insurance. Applied studies in climate adaptation. Wiley, pp 26–33
37. Mclean D (2011) IPPC authors speak out on special report, United Nations office for disaster risk reduction (UNISDR), 25 Nov 2011, available at http://preventionweb.net/go/23798. Accessed Sept 2015
38. Mitchell T, Wilkinson E (2012) Disaster risk management in post-2015 policy frameworks: forging a more resilient future. Overseas development institute briefing papers 75, June 2012. Available at http://www.odi.org.uk/resources/details.asp?id=6663title=disaster-risk-management-sustainable-development-policy-post2015. Accessed Sept 2015
39. Mitchell T, Guha-Sapir D, Hall J, Lovell E, Muir-Wood R, Norris A, Scott L, Wallemacq P (2014) Setting, measuring and monitoring targets for disaster risk reduction: recommendations for post-2015 international policy frameworks
40. Morgan J, Levin K, Song J (2015) Sending the right signals: how the Paris agreement can close the emissions gap, World Resources Institute (WRI), Oct 2015
41. Nesbit J (2013) Settled science, US News, Dec 5, 2013

42. Pandve HT, Fernandez K, Singru SA (2015) Climate change vulnerability: index and mapping. Indian J Occup Environ Med 15(3):142–143. doi:10.4103/0019-5278.93207, PMCID: PMC3299101
43. Pelling M, Schipper L (2009) Climate adaptation as risk management: limits and lessons from disaster risk reduction. International human dimensions of global environmental change (IHDP) www.uni-bonn.de/ihdp/ (Note: IHDP is an international, non-governmental, interdisciplinary science program dedicated to promoting and coordinating research aimed at describing, analyzing and understanding the human dimensions of global environmental change.)
44. Prizzia R (2015) Climate change and disaster management. Sentia Publishing, 240 p
45. Readfearn G (2013) IPCC climate change report by numbers. The Guardian
46. Strengthening Climate Resilience Institute of Development Studies (SCRIDS) (2012) Changing climate, changing disasters, pathways to integration. Brighton
47. U.S. Global Change Research Program (USGCRP), Our Changing Planet: the U.S. Global Change Research Program for Fiscal Year 2015
48. United Nations Treaty Collection (UNTC) (2005) Kyoto protocol to the united nations framework convention on climate change. Kyoto, Japan 2005
49. United Nations Environment Program (UNEP)—Emissions Gap Report 2013—November 2013, The Emissions Gap Report 2013 A UNEP Synthesis Report, Published by the United Nations Environment Program (UNEP), November 2013 Copyright © UNEP 2013 ISBN: 978-92-807-3353-2 DEW/1742/N
50. Wenzel D, Comes F, Fortier F, Geldermann J, Müller T (eds) (2013) Near real-time forensic disaster analysis. in: Proceedings of the 10th international ISCRAM conference—Baden-Baden, Germany, May 2013, CEDIM, www.cedim.de
51. White House Press Release (2015) Submission of intended nationally determined contributions (China, R.O.K., Iceland and Serbia) & U.S.-Brazil Climate Cooperation, Secretary of State John Kerry, Washington, D.C
52. World Resources Institute (WRI), Agreement on Climate Transformation (2015) (ACT) 2015. wri.org/our-work/project/act-2015. Accessed Sept 2015
53. Yardley J, Goodstein L (2015) Pope Francis, in sweeping encyclical, calls for swift action on climate change. The New York Times

The Complex Dynamic Causality of Violent Extremism: Applications of the VERA-2 Risk Assessment Method to CVE Initiatives

D. Elaine Pressman

Abstract This chapter examines a model of dynamic causality of violent extremism and a concomitant a structured professional judgment (SPJ) individual risk assessment tool for the spectrum of violent extremism. The analytical approach, referred to as the VERA-2, an acronym for Violent Extremist Risk Assessment- Version 2, is a systematic, consistent, and reliable evidence-based tool. The approach assists an analyst to arrive at a reliable adjudicated level of overall risk for an individual through the application of a rigorous, scientific and transparent methodology. A risk level rating is obtained through an evaluation of relevant risk promoting and risk mitigating indicators supported by an evidence-base. Assessments are undertaken at specified points in time employing the comprehensive set of the risk related indicators. Repeated use of this tool over successive time points provides the analyst with a trajectory of risk as well as risk levels judgments at a specified time on each of the included risk indicators as well as for overall risk. With dynamic systems, there is a need to monitor behavioral changes in an individual on discrete risk elements. The approach facilitates this task and identifies the most salient elements of the individualized risk pattern. This facilitates the development of relevant and focused intervention programs for countering violent extremism (CVE) initiatives, supports investigations and provides defensible prioritization of persons of interest. The tool is effective for determining efficacy of intervention programs via the ability to measure change in individuals on risk related indicators from a pre-determined baseline. The tool includes a comprehensive list of 31 indicators. Limitations and benefits of the VERA-2 approach are discussed in the chapter.

Keywords Violent extremism · Risk assessment · VERA 2

D.E. Pressman (✉)
Carleton University, Ottawa, Canada
e-mail: epressman@rogers.com

A.J. Masys (ed.), *Disaster Forensics*, Advanced Sciences and Technologies for Security Applications, DOI 10.1007/978-3-319-41849-0_10

1 Introduction

Attacks by violent extremists have been occurring with increasing frequency over the past years from a spectrum of ideological objectives. Incidents have occurred in the United States, Canada, The United Kingdom, Spain, France, Denmark, Norway, Australia, Pakistan, Iraq, Afghanistan and other countries around the globe. In 2013 more than 9700 terrorist incidents were recorded in 93 countries [1]. These incidents claimed more than 18,000 lives and 33,000 injured. In 2014, a rise in lone offender attacks was observed. Many of these attacks were inspired by ISIL, Al-Qaeda and other extremist groups. Others do not appear to have been specifically directed by a terrorist organization, extremist group or their affiliates. In the future, centralized leadership of terrorist organizations may be less important than the radicalization process itself, the individual identity of the perpetrator, and the narratives believed [2].

It is individuals who plan and conduct these attacks. These individuals have been radicalized at home or abroad as members of a group or by a self-initiated process. The ideological basis for the extremism and radicalization may differ but regardless of ideology, the damage may be equally lethal. Preventing and countering such violent extremism is a high priority for most nations in the world today.

At a "High Level Meeting of the United Nations Security Council" held in May 2015, grave concern was expressed over the flow of foreign terrorist fighters (FTF) to and from conflict zones and the targeting, recruitment and radicalization of youth [3]. In 2014, over 25,000 individuals from more than 100 countries were cited by the UN as foreign terrorist fighters. This represented a significant increase over the previous year. When youth high risk travelers are detected and detained, they require individualized and focused countering violent extremism (CVE) efforts.

The Radicalization Awareness Network (RAN) in Europe released a report in 2015 that recommended four key areas of focus for governments to address radicalization. These included: (1) investing in prevention interventions, (2) involving and training first line practitioners to be alert to signals of risk, (3) ensuring multi-agency co-operation, and (4) using tailor-made interventions that require a case-by-case approach. Implementation of these key recommendations would benefit from the knowledge obtained from a relevant, nuanced and scientific individual risk assessment approach [4].

Individual risk assessment protocols for radicalization have two objectives. The first objective is to determine an overall individual level of risk of radicalization and extremism based on evidence. The second objective is to identify the specific elements of risk and need present that are related to the individual. This detailed understanding of the causal elements specific to an individual and obtained from the risk assessment allows for more informed decisions for a case-by-case specialized intervention. In addition, details obtained from the assessment allow for the use of optimal intervention strategies and realistic goals.

The Violent Extremist Risk Assessment Approach, known by the acronym VERA, was developed in 2009 to meet the needs for a standardized and scientific approach to risk assessment for violent extremism [5]. The VERA uses the Structured Professional Judgment (SPJ) methodology. This methodology was deemed to be the most appropriate methodology available to assess the dynamic nature of indicators relevant to radicalization to violence. The SPJ methodology was subsequently supported as "best practice" by risk assessment expert opinion [6]. The VERA includes an empirically justifiable set of causal dynamic elements of radicalization to be evaluated on an individual basis with a final estimation of overall risk determined through structured judgment. The VERA-2, a revised version [7], was informed by the empirical experience obtained from applications of the VERA tool and feedback from experts. This included use with convicted terrorists in the high security correctional setting and from feedback from experts in the field of violent extremism and terrorism. Feedback relating to the tool was received from psychologists and psychiatrists in the justice system, from intelligence and security analysts using the tool operationally, from professionals working with radicalized individuals in the prison setting, from users in local and national police agencies and from risk assessment and terrorism experts. The VERA-2 risk assessment tool has been used in countries on four continents including North America, Europe, Asia and Australia.

The results obtained by the VERA-2 will apply to the key recommendations identified above by RAN. The VERA-2 will identify the individualized causal risk elements in the radicalization process that can support prevention initiatives by identifying case-specific focus. Exposure to the VERA-2 risk indicators and the methodology can assist first line contacts to increase awareness of early warning signals. A common risk analysis approach used by professionals in multiple agencies can foster improved communication which can serve to facilitate collaboration. When professionals from different sectors use the same dynamic indicator language, improved understanding and co-operation is fostered. Finally, it is the risk assessment results that produce the in-depth understanding of the vulnerable individual. This understanding is essential to the development of appropriate specialized and individualized programs that are recommended to be fine-tuned on a case by case basis.

2 Countering Violent Extremism and the VERA-2 Approach

Countering violent extremism (CVE) programs in the international arena have suffered from a lack of program efficacy information and outcome data [8]. Where data has been available, it is most often related to the "output" of programs rather than the "outcome" of programs and initiatives. Output is determined by activity such as the number of interventions provided or the numbers of persons engaged. Such data has been criticized for not reflecting the effectiveness of the program. Outcome is related to the efficacy element of the initiatives provided. It is outcome that is the fundamental question to be answered in terms of CVE programs, that is, did the program achieve the desired results.

Outcome data can only be produced with standardized and reliable assessment protocols. The VERA-2 is one such reliable approach. Outcome data requires advanced planning because both pre-program and post-program data is required on each of the participants. Baseline data must be obtained on individuals at the beginning of the program or at a designated Time 1 (T1) and at subsequent times, which may be at interim points or at the end of the program (T2). Assessments undertaken during the course of the program are useful for obtaining feedback on objective progress. Interim assessments provide information pertinent to the need to modify programs or refocus them in line with the individual's needs and context.

At a "High Level Meeting of EU Ministers of Justice on the "Criminal Justice Response to Radicalisation" (CJRR) in Brussels in October 2015, Christiane Taubira, The French Minister of Justice underscored this need for individual understanding and information. She commented that the more we understand individuals and this new phenomenon of radicalization, the more effective we will be [9].

The French Minister distinguished among different types of violent extremism and terrorism including nationalists, separatists, anarchists, extreme right-wing, religious terrorism and what she called the "new terrorism". She reported that there were 300 persons in France being monitored, 1800 persons identified in France as part of networks, some 500 persons from France known to be in the Middle East notably in Syria. There were 133 reported as having already died in Syria and this included 9 persons in suicide bombings. She further reported on 3000 reports to a prevention centre from families worried about radicalization of their children and over 200 judicial proceedings underway [10]. Prevention and CVE initiatives are relevant to all types of violent extremism. The VERA tools have been designed to apply to this broad spectrum of ideologies.

There was general interest expressed in programs to address prevention and countering violent extremism and general agreement that no one single profile could be identified as "typical" for radicalized individuals. This supported the need for a case-by-case approach. Increasing numbers of de-radicalization programs in EU countries were planned or being initiated such as those announced by the Belgium Minister of Justice. The need to share best practices was underscored [11]. Such best practice includes the application of relevant and reliable risk assessments.

Risk assessments such as the VERA-2 provide a useful vehicle for discovering the complex causality elements driving an individual's radicalization and will assist professionals to obtain a nuanced understanding of the individual with whom they are to be engaged [12]. Risk assessments identify the most important risk factors for program focus which can inform individualized CVE program content. This personalized understanding will enhance effectiveness, as noted by the French Minister of Justice [13].

Risk pertinent indicators in the VERA-2 tool are provided at the meta-level within a conceptual framework. The basic tenet of a meta-analysis is that there is a common truth. This common truth relates to the generalized indicators that are causal elements in violent extremism. The complex dynamic causality model includes a comprehensive set of these meta-constituent elements associated with the radicalization process. During the actual individual risk assessment, the detailed micro-level information for each indicator is delineated in terms of the individual being assessed. Direct engagement with the individual is used to obtain additional information on the indicators when feasible. The meta-analysis in the VERA-2 conceptual framework has application to youth, adults, men and women equally.

With youth, additional assessment tools such as youth violence risk assessment approaches [14] can be used. Micro-level details such as the specific ideologies and beliefs, specific grievances, networks, narratives, attitudes, motivations, personal needs, susceptibility, past violence, individual skills and protective or risk mitigating factors are extracted from available information, reports, observations and other sources. The 31 indicators specified in the VERA-2 are divided into sectors and appear in Table 1 in this chapter.

3 Structured Professional Judgment and Other Potential Approaches to the Risk Assessment of Violent Extremists

The structured professional judgment analytical approach (SPJ) has been described as a composite of empirical knowledge and clinical judgment. The approach uses inductive reasoning to reach probabilistic conclusions. The method guides the assessor to arrive at a risk level judgment through the consideration of all of the structured indicators in the tool. This collection of indicators comprises the set of necessary and sufficient causal elements. In the case of the VERA-2 this comprehensive set of indicators is specifically pertinent to the causal elements of radicalization and violent extremism. The SPJ method is regarded as the most appropriate methodology to use with individuals who are radicalized, violent extremists or terrorists [15].

The structured professional judgment approach does not use additive scores. This is because human inductive reasoning provides the most accurate probabilistic estimate of risk in a complex inter-active causality model. Additive results do not provide for a differentiation of the importance (weighting) of the indicators in terms of their contribution to overall risk. Additive models do not account for differences in the quality and credibility of the evidence available for each unique indicator. The use of sums do not allow for discriminating judgments pertaining to potential deception by the subject of the assessment. It is trained judgment that is more accurate and defensible for probabilistic risk decisions due to the nature of radicalization to violence.

New risk analysis techniques are being explored for CVE initiatives. Some of these are being pushed by the demand for simplicity, quick fixes and ease of administration. Some approaches being developed are analogous to score cards or color signals such as red light-green light approaches. Although some approaches also claim to be SPJ, they do not represent the standard accepted structured professional judgment method. Scores that are used must be meaningful and pertinent to the complexity of the process under evaluation.

Risk assessment results with methods that violate the inherent complexity of radicalization will mask critical distinctive features and significant contextual elements. This is projected to reduce the effectiveness of CVE initiatives. Ultimately, it is a comprehensive analysis that is required for effective CVE initiatives with youth and adults. It is a deep-level understanding of the individual that will provide the

Table 1 VERA-2 indicators (N-31)

	Pressman and Flockton [7]	Low	Moderate	High
BA.	*Beliefs and Attitudes*			
BA.1	Commitment to ideology justifying violence			
BA.2	Perceived victim of injustice and grievances			
BA.3	Dehumanization/demonization of identified targets of injustice			
BA.4	Rejection of democratic society and values			
BA.5	Feelings of hate, frustration, persecution, alienation			
BA.6	Hostility to national collective identity			
BA.7	Lack of empathy, understanding outside own group			
CI.	*Context and Intent*			
CI.1	Seeker, consumer, developer of violent extremist materials			
CI.2	Identification of target (person, place, group) for attack			
CI.3	Personal contact with violent extremists			
CI.4	Anger and the Expressed intent to act violently			
CI.5	Willingness to die for cause			
CI.6	Expressed intent to plan, prepare violent action			
CI.7	Susceptible to influence, authority, indoctrination (If a leader, specify and rate as high risk)			
HC.	*History and Capability*			
HC.1	Early exposure to pro-violence militant ideology			
HC.2	Network (family, friends) involved in violent action			
HC.3	Prior criminal history of violence			
HC.4	Tactical, paramilitary, explosives training			
HC.5	Extremist ideological training			
HC.6	Access to funds, resources, organizational skills			
CM	*Commitment and Motivation*			
CM.1	Glorification of violent action			
CM.2	Driven by criminal opportunism			
CM.3	Commitment to group, group ideology			
CM.4	Driven by moral imperative, moral superiority			
CM.5	Driven by excitement, adventure			
	Additional considerations such as being coerced to act, search for significance or status			
P.	*Protective Items* *Note reverse rating for protective items*			
P.1	Re-interpretation of ideology less rigid, absolute			
P.2	Rejection of violence to obtain goals			
P.3	Change of vision of enemy			
P.4	Involvement with non-violent, de-radicalization, offence related programs			
P.5	Community support for non-violence			
P.6	Family support for non-violence			
SPJ	*VERA Final Judgment*	*Low*	*Moderate*	*High*

information necessary to support and encourage disengagement, de-radicalization and assist the individual to seek "life significance" replacements.

The VERA-2 administration requires a commitment of time and resources but provides the opportunity for optimal understanding of the radicalized individual. Resources committed to surface analyses have not provided the expected or intended gains for investment. They may have contributed to government sponsored countering violent extremism program disappointments.

In summary, deception potential looms large with cognitively intact persons who have been radicalized to the point of engaging in purposeful acts of extremist violence. The possibility of deception in communication exchanges should be anticipated and the impact of provided information by the subject and its credibility should be considered in the analytical process leading to the overall risk judgment. Not all indicators bear the same level of importance in terms of their impact to overall risk. The weightings of the individual indicators in the context of the assessment should not be ignored. Indicators for which evidence is available as well as for indicators where evidence is not available needs consideration when adjudicating the overall probability of risk. When information thought to be highly relevant is absent, efforts to obtain the missing information should be undertaken. Missing information can be critical to risk decisions.

The indicators in this VERA-2 complex dynamic model of causality are divided into 3 sectors: (a) risk elements, (b) risk descriptive elements (personal needs) and risk mitigating elements. Risk elements are further divided into three categories: (1) beliefs, attitudes and ideology; (2) contextual features and intention to act; (3) historical background and capacity. The protocol includes the use of specific rating criteria available with user training. The risk indicators appear in Table 1.

Formal structured professional judgment approaches are more defensible scientifically than unsupportable scores. Unless additive scores enhance rather than detract from analytical precision, quick fixes should be avoided in lieu of full scale risk assessments for CVE interventions. This is because the blurring of important distinctions should be avoided whenever possible. Basic tools can be considered for triage or screening applications rather than for comprehensive risk formulations. CVE initiatives tailored to the individual, as recommended best practice, require in-depth risk analyses for effectiveness.

Best practice SPJ methodologies include the following features: (1) risk promoting and risk mitigating factors; (2) the assessment of severity levels per indicator using a standardized and consistent rating system rather than subjective impressions; (3) use of analytical professional judgment with inductive logic principles to generate final risk decisions.

Other models of risk assessment such as actuarial or unstructured analysis (clinical unstructured analysis) have been identified but are less relevant for violent extremism applications. Unstructured assessments are known to be less accurate than structured assessments [16] and actuarial models typically use a binary rating system that is not sufficiently nuanced for violent extremism antecedents. Actuarial approaches also typically include static rather the dynamic factors more relevant to the process of radicalization to violence.

Actuarial risk assessment tools have been criticized due to the use of group base rates for recidivism to make determinations about an individual's propensity for re-offending. These group rates may not be relevant to the case in question [17]. As a result, predictive validity of actuarial approaches, where it is claimed, has been considered dubious by some experts [18]. Due to the volitional nature of radicalization and the individualized nature of the process, statistically predictive validity is not considered the "holy grail" of risk assessment protocols for violent extremism. Other types of validity that support the VERA-2 model and protocol are discussed later in this chapter.

The VERA-2 tool is a bias free protocol being relevant to the spectrum of ideological violence, males and females, youths and adults. This neutrality is important as bias is a criticism often directed against tools that are developed for a particular population, race or religious group. The relevance of the VERA tool to the spectrum of violent extremism has been supported by independent researchers in the United Kingdom [19].

The VERA-2 is not a silver bullet. There is no certainty in risk assessment. A structured analysis, nonetheless, provides a systematic analysis of a comprehensive list of indicators in order to be able to make the best possible judgments in a reliable manner. This approach may permit better monitoring of "persons of interest" in addition to a consistent and informed approach to CVE initiatives and prevention of violent extremism. The use of a common tool can support information sharing even across borders.

No risk assessment approach is infallible but more comprehensive approaches that are consistent and account for the complexity of the radicalization process are best practice today. Training is required in the use of the protocol to ensure consistency in users, proper understanding of the strengths and limitations of the approach, and correct interpretation of the results. Training informs users about the dangers of assumptions and the importance of using evidence. The requirement of sufficient information for the risk assessment is also underscored in training.

The information obtained from the use of the VERA-2 assists community and other officials in the construction and delivery of individualized countering violent extremism programs. The structure and definition of the tool permits a generation of an individual's trajectory of risk over time. This is useful for security analysts responsible for monitoring individuals as well as professionals in agencies directing and implementing CVE programs. The use of the VERA-2 can provide risk classifications of cohorts to support prioritization. The systematic and transparent method used to identify risk elements that are increasing or decreasing provides important feedback to agents responsible for CVE activities.

4 Dynamic Complex Causality and Violent Extremism

David Hume, the 18th century philosopher argued that causation added up to nothing more than the psychological anticipation of a previously experienced conjunction and that sense impressions do not account for any forceful necessitating connections between cause and effect [20]. He argued that although there are situations where some things do regularly follow other things there are cases where they do not. Cause and effect, he argued, occur when a cause [X] and the effect [Y] are contiguous in time and space; and [X] precedes [Y]. Causality can be inferred from the constant conjunction between [X] and [Y].

Human behaviour is not subject to a rigid behaviorist explanation or regularity between cause and effect. When behaviour is volitional and complex as in the antecedents of violent extremism, Hume's caution is especially relevant. Volitional human behaviour does not have constant or statistical regularity. Causality in such situations is imperfect. There is no strong causality in regularity theories.

Regulatory theories apply when there is a constant or statistical regularity between some antecedent event(s) [A] and some later event(s) [Y]. In such cases, there are attempts to analyze causation in terms of invariable patterns of succession. The difficulty identified with regularity theories motivated more probabilistic approaches to causation. The SPJ approach used in the VERA risk assessment of violent extremism is based on such probabilistic notions using a form of logical induction [21].

Human behavioral causality is characterized by imperfect regularities for two different reasons. The first reason is that there may be heterogeneity of circumstances in which the cause arises. An example of this heterogeneity can be illustrated by two individuals with the same grievances and beliefs in a common narrative. One of the individuals is susceptible to influence and is able to be encouraged to engage in acts of terrorism while the second person who is not susceptible to influence is not able to be persuaded to act despite his beliefs and grievances. Second, there may be imperfect regularities due to a failure of physical determinism. In this case, an individual may decide volitionally at a given point in time not to engage in a planned event. This may be due to a volitional cognitive decision or be the result of a strong risk mitigating factor. In such a case, no other conditions will exist for that event for that person at that time.

This suggests the need for a theory of causation that does not presuppose "causal determinism". Probabilistic theories are more appropriate for complex human behaviour because causes or antecedents can change the probability of their effects. An effect may occur in the absence of a cause or fail to occur in its presence. A Bayesian framework is being explored to further the opportunities for using mathematical models and probabilities for the VERA-2 approach [22] but the importance of the role of human inductive reasoning noted earlier may impede progress with Bayesian models. Logical induction offers more flexibility and consideration of the complexities consistent with the argument used above. Finally, the underpinning of a probabilistic complex dynamic theory of causation (CDC) for violent extremism comes from philosophical "Action Theory". This is a construct that is consistent with the complex nature of human behavior motivated by political, religious or other ideological elements. Action theory is concerned specifically with marking the boundary between action and the rest of behaviour which is not intentional [23]. This includes a role for attitudes, ideas and beliefs as reasons or explanations of action. The concept of the free will and individual responsibility for the action is also an aspect of Action Theory.

Only voluntary behaviour is considered "action" by this theory and action must be purposeful. The distinction between behaviour and action is explained by the difference between the "wink and the blink" where the wink is intentional action and the blink is involuntary behavior. One can convincingly argue that if ideological purpose is not involved in an action, it is not violent extremism or terrorism.

Violent extremism and terrorism are defined as purposeful actions. They are planned or undertaken to further a political, religious or other ideology. They have the intention to cause harm to individuals or property and coerce change. Terrorist acts are planned to cause fear and psychological distress in a civilian population, to intimidate the public with regard to security, including economic security, and to compel a person, government or organization (whether in or outside a country) to do or refrain from doing any act [24].

A Complex Dynamic Casualty model can be built based on a quasi- Venn concept where there are logical relations between the risk indicators. These indicators can be classified into three different sets for a representational model: Set A (representing attitudes, beliefs and ideology); Set B (representing bonds, context, intention, friendships, networks, groups/personal associations) and Set C (representing capacities, historical experiences, training). All the indicators in the VERA-2 tool can be restructured into these three sets for the model. It is possible to restructure all the VERA-2 indicators into these three sets. Motivation, for example, based on moral values would be restructured into Set A, whereas criminal opportunism as a driver would be reallocated to Set C (capacity and historical elements).

These sets can and do intersect but the intersections are individual in nature rather than universal. Friendships can influence beliefs and attitudes; skills and capacities can influence personal associations; historical experiences can influence intention and personal needs. Other relationships indicators can similarly be described. A large number of permutations and combinations between indicators in the sets exist. The causal elements in Set A, Set B and Set C and the interactions of the constituent elements of the sets impact the complex causality of radicalization to violence. These interactions are not subject to regularity and rigid behavioral explanation as noted earlier.

The indicators included on the VERA-2 are a comprehensive set where each indicator is not necessarily a condition of causality but a potential condition. The risk of radicalization results from the idiosyncratic and complex interactions of risk promoting and risk mitigating indicators active in an individual. The active indicators and their interactions are delineated by the micro-level risk analysis and assessment. This risk assessment provides a distinctive features analysis of the complex interactions and presence of the risk elements contributing to violent extremism. This entails evaluation of the severity of each present indicator, the establishment of an importance level of each indicator to overall risk, and the determination of a final probabilistic risk judgment, based on the evidence and the structured analysis.

5 Risk and Threat

Although risk is commonly present in our lives, it is not usually well understood. It is also not able to be predicted with certainty. This is despite other violence risk assessment tools which claim the ability to be able to predict the risk of recidivism of violence.

A commonly held definition of risk is the exposure to danger due to injury, damage or loss, or other adverse or unwelcome circumstance [25]. This definition is often applied to organizations, agencies, locations, identifiable groups or individuals exposed to danger of injury or damage or adverse circumstance rather than to the individual who is the initiator of the risk. Risk is related to the probability of an event (attack or injury) combined with the harm or severity of the loss expected to occur. In this case, *Risk = probability of injury or damage × severity of the loss (harm)* or $R = p \times h$ [25]. Risk assessments for individuals are intended to estimate the likelihood that an identifiable person or actor will generate the harm. The individual is the focus of the assessment such that *R = probability of the individual acting to cause the injury × the severity of the harm or loss*. The individual under investigation or suspicion or identified for CVE programs is the focus of the risk assessment for violent extremism.

Risk assessments for violent extremism should be undertaken by trained persons lawfully charged with making such risk judgments. The outcome of assessments for radicalization to violence is not prediction of recidivism but an analysis of radicalization and extremism and a logical inductive estimate of the risk of violent extremism with an analysis that presents risk and risk mitigating indicators. Individuals who present a risk for violent extremist action may also represent a threat. This threat may be limited due to the objectives of the individual or the threat may be substantial such as when the intention of the individual is to cause catastrophic loss of life. The term “threat” is often confused with “risk” but is distinct from it.

Threat assessments are defined as the function of “capability” and “intent”. Threat is used in the context of attacks but individuals may represent a threat. Risk and threat often co-exist in the case of individuals suspected of being violent extremists and terrorists. There are indicators pertinent to capacity and intention in the VERA-2 risk assessment tool. The results of the risk assessment provide information on both risk and threat of an individual.

6 Validity

Over the past five years, research has been undertaken on the VERA risk assessment tools, using individuals convicted of terrorist offences, in non-institutional research studies, by user questionnaires and by single case study analysis. Such studies have investigated different aspects of the validity and reliability of the

VERA risk assessment protocols. Predictive validity, although desirable, was not considered an appropriate or realistic goal based on the dynamic character of the process of radicalization to violence.

There is now evidence from prison populations that the risk characteristics for violent extremists and terrorist differ significantly from those of ordinary violent offenders [26]. As a result, tools developed specifically to assess the predictive validity for the population of general violent offenders should not be assumed to apply to violent extremists. Most experts in terrorism also agree that it is not feasible to predict with precision who will become a terrorist because terrorists often are essentially normal functioning individuals with volitional control in contrast to many ordinary violent criminals [27]. There are, however, many types of validity that can and do support risk assessment approaches, in particular the VERA and VERA-2.

6.1 Content and User Validity

An independent study on the VERA risk assessment approach was undertaken by researchers in Europe in 2013. This work supported the use and relevance of the VERA risk assessment tool for the spectrum of violent extremism and also supported the content relevance (validity) and user ease of the VERA [19]. A high presence of most of the attitude items included in the VERA protocol were found to be present in all of the individual terrorists who were studied. These terrorists represented different ideological motivations of extremist behaviour.

High scores on indicators were obtained on the subjects assessed for alienation, perceived injustice, lack of empathy and dehumanisation. Glorification of violent action was found to be a particularly important indicator. All the terrorists studied showed some evidence of this factor which incorporates beliefs that violence is noble, legitimate and justified.

These results support the hypothesis that attitudes, ideology and belief indicators are applicable broadly to terrorists and violent extremists and can exist in a bias-free risk assessment tool. Results revealed that the risk factors identified in the VERA could be applied with equal efficacy to terrorists of different ages, lone-actors, individuals who were members of groups and those who supported the spectrum of ideological causes and objectives.

The contextual items in the VERA were observed to be relevant to terrorists. High ratings were noted for the indicators "direct contact with other violent extremists" and "community support for violent action". High scores were also noted in items related to "anger at the political decisions and actions of the country". Although further research is recommended, the majority of factors in the VERA were found to be "relevant and important to risk assessment" [19]. The indicators applied to lone-actors and members of extremist militant groups equally. The tool was found easy to apply across the variety of terrorists in the sample. The protective items were found to be important for the accurate identification of extremists and to identify which individuals were less likely to commit future terrorism. The VERA was identified as having important applications for prisons where violent extremists are incarcerated and where program impacts are necessary to assess, where early release decisions are required and where counter-violent extremism programs are needed. Overall, the majority of the VERA factors were appropriate, easy to apply and the VERA was found to be a useful risk assessment guide with content validity and documented user ease [28].

The content validity of the VERA 2 was also evaluated on the basis of information obtained from subject matter experts in terrorism. Content validity refers to the extent to which a measure represents all facets of a given construct. Specifically, the content validity refers to the set of elements in a measure and whether this set of elements is a comprehensive and representative set, that is, does the instrument contain the set of "necessary and sufficient" indicators for the construct, in this case, that of violent extremism. If there is content validity, then the instrument is expected to be able to measure what it is supposed to measure. Importance ratings on the risk promoting and risk mitigating indicators were obtained from over 60 professional security and intelligence analysts working in the area of counterterrorism and familiar with the population and to identify the comprehensiveness of the set of indicators. Professionals were asked to rate each of the risk indicators for their perceived importance in terrorism analysis and to identify missing indicators. The items were rated as being highly relevant and important, moderately important or not important. Raters reported that the majority of the "risk indicators" used in the VERA-2 and the VERA were highly important or very important. These subject matter experts were solicited for risk or mitigating risk items that were missing based on their empirical experience. Recommendations were received related to expanding cyber-behaviour indicators and increasing the motivational indicators to identify instances where coercion is present, or the search for life significance or status were motivating elements. These elements will be addressed in subsequent revisions.

6.2 Deductive Validity of the VERA

Deductive validity is a form of mathematical induction that links premises with conclusions and is referred to as propositional calculus. If premises are true, and the rules of deductive logic are followed, then the conclusion must follow by the rules of mathematical logic. In this case, the conclusion must be both true and valid [29]. Deductive reasoning is used in the legal system where lawyers present evidence and illustrate, using a form of deductive logic, that a given individual is culpable. When a specific behavior is defined as illegal on the basis of applicable statues, and if a person is shown to have engaged in such specific and defined behaviour, then the person has engaged in illegal behaviour and is culpable. In deductive validity, the conclusion is included in the premises used.

Deductive validity in the case of the VERA-2 works as follows. If a violent extremist (x) is defined as an individual who exhibits (a) + (b) + (c) where (a) refers to the commitment and intention to use violence to further an ideology, and (b) refers to the observed behavior of engaging in an illegal action such as planning an attack to further an ideological cause; and (c) refers to the action that is intended to coerce a political change and cause fear in the population, then by definition (x) = (a) + (b) + (c). If (y) represents a specific person who then is found to exhibit (a) + (b) + (c) then by logical deduction (y) = (x), that is (y) is a violent extremist. The risk indicators of the VERA tools are consistent with legal definitions typically used to define terrorist actions in selective jurisdictions in North America and the European Union as well as other Western democracies. When the evidence base identifies specific actions of an individual (included in indicators on the VERA-2 tool) such as "planning terrorist actions and these action(s) are defined by statutes in a legal jurisdiction as constituting behavior defined as terrorism, the outcome judgment that the individual is a high risk for terrorist activity or violent extremist action has deductive validity.

If the sufficient information is not available to produce a logical deduction, the information that is available can provide early warning signals based on logical pragmatic inductive reasoning. Inductive logic provides legitimate general probabilities which would correspond to the low, moderate or high risk levels identified in the VERA-2 as based on true evidence on individual indicators.

6.3 Face Validity

Face validity is defined as the extent to which a test or tool is viewed as covering the concept it purports to measure. It refers to the transparency or relevance of a test as it appears to users or participants. A tool can be said to have face validity if it "looks like" it is going to measure what it is supposed to measure [30].

The VERA-2 was developed to measure risk of violent extremism, and if the people you show it to all agree that it looks like a good test of violent extremism, then you have demonstrated face validity of the tool. Generally, face validity is used to refer to the validity of a "test" determined by non-experts or naïve observers. If you ask an expert for their opinion as to whether a tool seems like it is going to measure what it is supposed to measure, it could be argued that you are testing more than basic face validity. Generally, face validity means that the test "looks like" it will work, as opposed to "has been shown to work".

The face validity of the VERA and VERA Version 2 has been demonstrated via empirical evidence obtained from expert operational users over the past 5 years. National intelligence analysts, national police services officers, psychiatrists and psychologists in multiple countries in Europe, North America and Asia who are working in national security positions and in terrorism prevention activities have been using the VERA 2 and have reported that it is relevant to their analytical needs. Although face validity is considered to be the simplest form of validity, when such validity is provided at the subject matter expert level in violent extremism and terrorism, and when this face validity is recognized by national and international counterterrorism agencies, the robustness of the face validity is significantly higher. This can be classified as "Expert Adjudicated" and Operational Validity.

In the official Ontario Supreme Court Record of a case prosecuted under Canada's terrorism laws, the VERA was provided with what could be termed "expert adjudicated face validity" in the court record. The indicted individual, who had entered a plea of guilty to two terrorism offences and was in court for sentence determination, applied through his lawyer to have a psychiatric report provided by a defence psychiatrist. The opinion of the psychiatrist was ruled admissible. The psychiatrist testified on the application for the defence. He was described as a general adult outpatient psychiatrist and claimed to be familiar with extremist ideology. After reportedly consulting with a colleague, he testified that that the ordinary risk assessment tools for violence are not able to capture the clinical criteria in the case given its unique factors and "especially because the accused was not considered to have any anti-social psychopathic personality traits [31].

In section [30] of the court record, the psychiatrist states that after familiarizing himself with the VERA protocol, he concluded that "his interviews had encompassed most of the items identified" thereby offering his judgment of the expert adjudicated face validity of the VERA tool [32].

In section [40] of the official court record, the psychiatrist further stated that he covered the major areas suggested in the VERA in his assessment report. In [47] it was noted that both counsel (defence and prosecution) seemed to agree to the relevance (face validity) of the indicators in the VERA and that there was agreement "that terrorism offences involve different motivations and underlying characteristics of offenders than those that occur in other criminal offences.

This supports the underlying rationale of the VERA risk assessment protocol. In preparing his report, the consulting psychiatrist did not use "any traditional actuarial risk assessment tools" because he testified that they were not considered sufficiently relevant. He did consult the VERA indicators without objection from the opposing counsel supporting the significant level of legal adjudicated expert face validity of the indicators in the VERA. This provides further support for the previously documented content validity of the VERA [33].

6.4 Consumer (User) Validity

Experts originally demonstrated interest in the VERA and VERA-2 tools because of the face validity. When they reported that their use of the tool with suspected or detained individuals assisted them in their analytical tasks and that the results supported their professional judgment of risk and threat, this validation again exceeds face validity. It becomes what can be termed consumer or expert user validity. Over 90 % of the services trained in the VERA-2 have requested additional training for colleagues due to the consumer validity of the approach. This includes police forces in North America and Europe and security agencies in South East Asia. Agencies have requested in-house training of experts to ensure sustainability of the VERA-2 protocol in their organizations due to the positive results they have documented over periods of months or years in CVE applications. Adaptations for community workers have been undertaken based on the perceived benefits of the approach and countries have invited partner countries to training programs to facilitate collaboration and communication of results of assessments. Violent extremists do not respect borders and countries benefit from a common assessment language.

6.5 Construct Validity

Construct validity refers to the degree to which a test measures what it claims to be measuring. The VERA 2 claims that it will assess the risk of violent extremists and further that the risk measured is different from other ordinary tools that assess violence in non-ideological motivated offenders.

The question posed is whether the VERA 2 construction can demonstrate differentiation between a group of violent extremists motivated by the interest in furthering ideological objectives and unlawful violent criminals not motivated by ideological issues. The question centers on whether the set of indicators is sufficiently sensitive to differentiate these groups statistically.

A number of violence risk assessment tools have been developed and are typically used in the prison setting. These include the Historical/Clinical/Risk Management 20 (HCR-20) instrument, The Violence Risk Scale-Screening Version (VRS-SV), The Psychopathy Check List-Screening Version (PCL-SV) and the Level of Service Inventory-Revised (LSI-R). To evaluate construct validity, a group of high security inmates convicted of terrorist and/or violent extremist offences were compared to a group of known ordinary violent and non-ideological motivated offenders on the VERA-2 risk assessment instrument and the tools identified above that evaluate general violence and psychopathy. These groups were matched as closely as possible for sex, age, ethnic background and religion. The results are provided on Tables 2 and 3. The results document the significant differences found between the sample/experimental group consisting of violent extremists/terrorists and the control group consisting of ordinary violent criminals on all the tests used. This supports the construct validity of the VERA 2 and illustrates the ability of the instrument to provide discriminating information on risk for the population of violent extremists. These results provide preliminary empirical evidence that the VERA 2 contains an adequate set of sensitive indicators to differentiate among the two groups and demonstrate that the terrorist group shows higher risk on the tool with indicators that are relevant to violent extremism [34].

A reverse pattern was demonstrated for the tools constructed for ordinary violence and psychopathology. On these tools, the ordinary violent offenders

Table 2 Group differences scale on VERA-2, risk measures and psychopathy measure

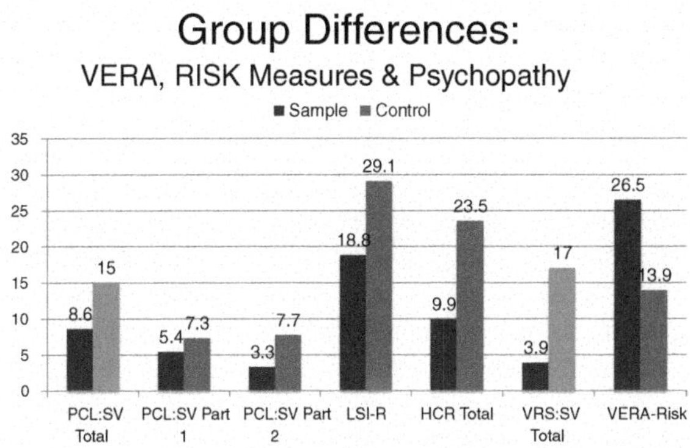

Table 3 Mean scores and significance for the sample and control groups

Mean Group Differences:
Risk Scales

Measure	Sample Mean (SD)	Control Mean (SD)	T-Test
PCL:SV	8.6 (3.6)	15.0 (5.7)	3.69, p < .01
Part 1	5.4 (2.1)	7.3 (3.1)	-1.69, n.s
Part 2	3.3 (3.1)	7.7 (3.9)	-2.95, p < .01
LSI-R	18.8 (8.4)	29.1 (8.7)	-2.83, p < .05
HCR-20	9.9 (4.8)	23.5 (7.9)	-4.85, p < .001
VRS:SV	3.9 (5.6)	17.0 (8.7)	-4.20, p < .001
VERA-Risk	26.4 (7.9)	13.9 (8.0)	3.69, p < .01

demonstrated higher risk levels than the terrorists supporting the premise that terrorists have different risk characteristics than ordinary violent offenders. It supports the premise that there is likely to be flawed results in risk obtained when ordinary violence risk tools are used to assess the risk of violent extremists. The results suggest that specific tools with appropriate indicators for ideologically based violence are required for this population. The sole exception to findings of significance between the experimental and control groups was found on Part 1 of the Psychopathy Check List–SV. On this sub-test both groups were found to share the trait of narcissism. Despite the lack of a significant effect on Part 1 of the PCL-SV, noted above, the overall PCL-SV scores demonstrated significant difference [35].

This result can be explained by the fact that individuals who are ideologically motivated are often known to demonstrate elements of narcissism and self-importance.

Terrorists are known to be able to demonstrate empathy for persons in their in-group, family members and friends in comparison to psychopaths. Individuals who have engaged in violent extremist acts are often described as good friends, active members of their community and they are often engaged in charitable activities in their communities. The kind of empathy they demonstrate may best be described as selective empathy. This is not an absence of empathy. It is not universal and tends to be focused on family and friends and the in-group.

These results support the contention that the VERA 2 assesses a different kind of risk, specifically the risk of violent extremism and terrorism, compared to tools designed to measure ordinary criminal violence heavily dependent on addiction issues (drugs, alcohol) mental instability, psychopathy, and personal gain. Additional testing with a larger group is recommended. This first data provides support for the important aspect of construct validity of the VERA-2.

7 Reliability

One of the fundamental measures of the utility of a risk assessment tool is the extent to which raters can independently assess the same individual(s) and arrive at the same outcome. This level of agreement between assessors/raters is referred to as inter-rater reliability. High levels of inter-rater reliability are an indication that the items that comprise the risk assessment and their coding rules have been clearly articulated.

The inter-rater reliability of the Violent Extremism Risk Assessment protocol was examined by independent researchers. The level of agreement between the two raters that was found was 85.7 %. Inter-rater reliability analyses were performed for each of the terrorists using Cohen's Kappa and the results revealed that every value of Kappa was 0.76 or greater ($p < 0.001$). This demonstrated a high degree of consistency between the two raters across all of the terrorists chosen for analysis in the study and suggests that the VERA tool can produce high inter-rater reliability [19].

8 Summary

The VERA-2 approach is a promising individual risk assessment tool for radicalization to violent extremism. The results obtained from the application of the tool will provide information pertinent to the overall individual risk of radicalization and extremism. The results will also provide a detailed and differential analysis of the risk indicators that are significant and specific to an individual. This information will inform the development of case specific interventions for countering violent extremism, recommended by professional bodies and government officials. In addition to CVE applications, the VERA-2 risk assessment model has relevance to national security analyses of "persons of interest".

The VERA-2 method can be applied to returning and prospective foreign terrorist fighters to classify cohorts according to risk and threat. The tool can also be used to provide early warning signals and risk trajectories for individuals "on the radar" and being monitored. Applications of the VERA-2 in the prison setting are increasing and apply to individuals detained prior to trial, inmates convicted of terrorism and violent extremism offences, program decisions in the prison and for early release decisions. The VERA-2 is a bias-free instrument which has been successfully used with both youth and adults. It is relevant to men and women motivated by the spectrum of ideologies.

Additional studies are recommended to support the established validity and reliability of the VERA 2. The results obtained to date have been promising despite the limited data base. The empirical feedback from expert users in national security, CVE and intelligence services has provided strong consumer and expert adjudicated face validity. The VERA-2 is not a "silver bullet". There is no simple solution to countering violent extremism and radicalization. There is also no substitute for a nuanced understanding of the risk elements that are driving an individual's radicalization. For this, specialized tools are required.

The VERA-2 risk assessment protocol is not intended to replace professional judgment. It is intended to support professional judgment through the application of a structured, systematic, consistent, scientific and transparent approach. The VERA-2 is based on a complex dynamic model of radicalization causality that is broken down into its constituent elements to assist decision making. Inductive logic is used with structured professional judgment method to arrive at an overall probabilistic risk judgment which is more appropriate than predictive validity for this population.

No risk assessment approach is infallible. Although uncertainty will exist in all risk assessment systems of analysis, uncertainty does not necessarily imply a lack of knowledge. The protocol used in the VERA-2 provides knowledge through its consistency and structure. Any risk assessment approach is necessarily dependent upon the quality and availability of a sufficient amount of information pertinent to identified causal factors. There must be adequate and credible evidence on a sufficient number of relevant indicators to permit reasoned assessment.

New and ongoing international collaboration will support the VERA-2 protocol's development in the future and increased data collection is planned. The research studies to date indicate that the VERA tools have acceptable validity and reliability but additional research is needed to support the evidence to date. The results of the VERA-2 risk assessment tool are situation and time dependent. The VERA-2 has benefits but also limitations that require understanding. It is not a tool to be used to screen individuals in a general population. The VERA-2 is to be used with identified individuals due to suspicion or cause. Intelligence Services, law enforcement agencies and CVE programs are charged with safeguarding and supporting national security. Hopefully, the VERA-2 today and in the future, can play a role in supporting these national security objectives to safeguard all citizens.

References

1. United States Department of State Country Reports on Terrorism (2013)
2. Kaidanow TS (2015) At-large and coordinator for counterterrorism, US state department special briefing. Washington, DC. 19 June 2015. http://www.state.gov/r/pa/prs/ps/2015/06/244030.htm
3. United Nations Security Council (2015) Report of the 7453rd meeting: action against threat of foreign terrorist fighters must be ramped up, security council urges in high-level meeting. 29 May 2015. http://www.un.org/press/en/2015/sc11912.doc.htm [Retrieved 2 January 2016]
4. RAN (2015) Preventing radicalization to terrorism and violent extremism. http://ec.europa.eu/dgs/home-affairs/what-we-do/networks/radicalisation_awareness_network/ran-best-practices/docs/ran_collection-approaches_and_practices_en.pdf
5. Pressman DE (2009) Risk assessment decisions for violent political extremism. Public safety Canada, Government of Canada, Ottawa. Cat. No.: PS3-1/2009-2-1E-PDF. https://www.publicsafety.gc.ca/cnt/rsrcs/pblctns/2009-02-rdv/index-en.aspx [Retrieved 3 January 2015]
6. Monahan J (2012) The individual risk assessment of terrorism. Psychol Public Policy Law 18 (2):167–205
7. Pressman DE, Flockton JS (2012) Calibrating risk for violent political extremists: the VERA-2 structural assessment. British J Forensic Pract 14(4):237–251
8. Romaniuk P, Fink NC (2012) Evaluating terrorism prevention programs. Center on global counterterrorism co-operation. http://www.globalcenter.org/wp-content/uploads/2012/10/CGCC_EvaluatingTerrorismPrevention.pdf [Retrived 6 January 2016]
9. High Level EU Ministers of Justice Conference Criminal Response to Radicalization (2015) Brussels. 19 October 2015. http://www.eu2015lu.eu/en/actualites/articles-actualite/2015/10/19-conf-radicalisation-debat/ [Retrieved 3 January 2015]
10. High Level EU Ministers of Justice Conference Criminal Response to Radicalization (2015) Brussels. 19 October 2015. http://www.eu2015lu.eu/en/actualites/articles-actualite/2015/10/19-conf-radicalisation-debat/ [Retrieved 3 January 2015]
11. Ibid
12. Pressman DE (2015) Invited comments on risk assessment and the VERA-2 to the EU high level ministerial conference on the criminal justice response to radicalisation. Presentation at EU convened meeting in Brussels. 19 October 2015. http://ec.europa.eu/justice/events/response-radicalisation-2015/files/official_agenda_high_level_conference_19_october.pdf
13. Op. Cit
14. Borum R, Bartel P, Forth A (2006) SAVRY: professional manual for structured assessment of violence risk in youth. Psychological Assessment Resources Inc., Lutz, Florida
15. Monahan J (2012) The individual risk assessment of terrorism. Psychol Public Policy Law 18 (2):167–205
16. Ogloff JRP, Davis MR (2005) Assessing risk for violence in the Australian context. In: Chappell D, Wilson P (eds) Issues in Australian crime and criminal justice. Chatswood, Lexis Nexis Butterworths
17. James N (2015) Risk and needs assessment in the criminal justice system. Congressional research service, R44087. Available at: https://www.fas.org/sgp/crs/misc/R44087.pdf
18. Ibid
19. Beardsley NL, Beech AR (2013) Applying violent extremist risk assessment (VERA) to a sample of terrorist case studies. J Aggr Conflict Peace Res 5:12
20. Hume D (1969) A treatise of human nature. Penguin Books, London

21. Copi IM, Cohen C, Flage D (2007) Essentials of logic, 2nd edn. Pearson, London
22. Barbieri D, Pressman DE (2015) Violent extremist risk assessment: a Bayesian framework. Conceptual paper presented to the "Intelligence in the knowledge society" conference, Romanian National Intelligence Academy. 17 October 2015
23. Juarrero A (2002) Dynamics in action: intentional behavior as a complex system. The MIT Press, Cambridge, MA
24. Public Safety Canada (2014) Public safety 2014 report on the terrorist threat to Canada. http://www.publicsafety.gc.ca/cnt/rsrcs/pblctns/2014-pblc-rpr-trrrst-thrt/index-eng.a [Retrieved 6 January 2016]
25. USDOD (2009) Dictionary of military and associated terms. First published 12, April 2001 and amended up to 2009. http://jitc.fhu.disa.mil/jitc_dri/pdfs/jp1_02.pdf [Retrieved 3 January 2016]
26. Motiuk L (2015) Presentation to international corrections and prison association. Melbourne Australia. 29 October 2015
27. Post JM (2007) The mind of the terrorist: from the IRA to Al-Qaeda. Palgrave Macmillan, New York
28. Ibid
29. IEP (2016). Deductive and inductive argument. http://www.iep.utm.edu/ded-ind/ [Retrieved 4 January 2016]
30. Holden RB (2010) Face validity. In: Weiner IB, Craighead WE (eds) The Corsini encyclopedia of psychology, 4th edn. Wiley, Hoboken, NJ
31. Amara RV (2010) ONSC 251, Court File no: 2025/07 Date 20100112. http://caselaw.canada.globe24h.com/0/0/ontario/superior-court-of-justice/2010/01/12/r-v-amara-2010-onsc-251.shtml [Retrieved 7 January 2016]
32. Ibid
33. Amara RV (2010) ONSC 251, Court File no: 2025/07 Date 20100112. http://caselaw.canada.globe24h.com/0/0/ontario/superior-court-of-justice/2010/01/12/r-v-amara-2010-onsc-251.shtml [Retrieved 7 January 2016]
34. Pressman DE, Duits N, Rinne T, Flockton J (2015) REM: risicoanalyse extremisme voor management: VERA^{v2}NL met addendu. Nederlands Instituut voor Forensiche Ppsychiatrie en Psychologie/Netherlands Institute for Forensic Psychiatry and Psychology, Ministerie van Veiligheid en Justitie/Ministry of Security and Justice, Government of The Netherlands
35. Ibid
36. A discussion of logical syllogisms and Venn diagrams. http://www.philosophypages.com/lg/e08a.htm

Author Biography

D. Elaine Pressman, PhD is a Distinguished Senior Fellow at the Netherlands Institute of Forensic Psychiatry and Psychology (NIFP), in the Ministry of Security and Justice and an Associate Fellow at the International Centre for Counter-Terrorism (ICCT), in The Hague, in The Netherlands. She is also a Senior Fellow at the Canadian Centre for Intelligence and Security Studies, at the Norman Paterson School of International Affairs at Carleton University, Ottawa.

Staying Alive in the Business of Terror

Gisela Bichler and Stacy Bush

Abstract Staying alive (and at large), is a career advantage when you manage an insurgent group. If instead, your objective was to detonate a suicide bomb, success would be measured differently. These divergent goals must be considered when examining the social network within which individual actors are embedded, as each outcome may require a different supporting structure, warranting the application of different theory and associated metrics. Breaking from the extant literature that is principally concerned with assessing the cellular structure of attack groups and the centrality of actors, this chapter applies a business model of competitive advantage to examine how varied egonet structures correlate with the operational success of command staff—here the objective is to stay alive. Investigating the utility of Burt's (Structural holes: the social structure of competition. Harvard University Press, Cambridge, 1992), Burt (Adm Sci Q 42(2):339–365, 1997). theory of structural holes, we find that the communication patterns of central leaders of Al Qa'ida and the Islamic State of Iraq (ISI), who were active since 2006 and survived at-large until November 2015, involved smaller egonets that had fewer non-redundant ties, lower density, and were significantly less likely to involve reliance on a central actor for information. In short, less social capital *and* lower constraint improved the likelihood of survival.

Keywords Terrorism · Networks · Al Qa'ida · ISIS

G. Bichler (✉)
California State University, 5500 University Parkway, San Bernardino, CA 92407, USA
e-mail: gbichler@csusb.edu

S. Bush
San Bernardino County Probation Department, San Bernardino, USA

A.J. Masys (ed.), *Disaster Forensics*, Advanced Sciences and Technologies for Security Applications, DOI 10.1007/978-3-319-41849-0_11

1 Introduction

Terror attacks cause untold tragedy and no region of the world is immune. In a quest for new insight on how to disrupt terror operations and prevent future attacks, many are turning to social network analysis. By studying the underlying structure of the group and how it morphed into an organization capable of launching an attack, it may be possible to identify critical social structures upon which to focus counter-terrorism efforts. For this reason, much of the previous research in this area examined the set of individuals involved in the terror attacks, most of which are suicide bombings. While the tactical utility of this research is not disputed, we argue that adding another stream of research would benefit the field.

Some individuals within insurgent groups[1] intend to remain active indefinitely (e.g. command staff); whereas, others seek only limited or temporary involvement (e.g. suicide bomber). This means that within the attack network are individuals fulfilling substantively different roles. Individuals fulfilling different roles may exhibit dissimilar patterns of relations (local networks): to be a successful suicide bomber may require an entirely different network configuration than to be an effective financier or emir. Thus, we argue that depending on what their role is within the organization, different local network structures are important for ensuring successful outcomes.

In this chapter we examine the structure of communications at the managerial level of two related terrorist groups, Al Qa'ida and the Islamic State of Iraq (ISI). Redefining success as being able to stay alive and disaster as being killed or captured, we apply Burt's structural holes theory of social capital [1, 2], arguing that terror group "managers" with greater information benefits (able to access more diverse information pools) and less dependency on a single information source will be more successful in avoiding death or capture. Our results show some support for this argument: individuals with greater ability to use their social capital (information benefits accrued through their network) were significantly more likely to remain alive and at-large. However, before we say any more about our findings, there are a few topics that require discussion.

This chapter begins with a brief review of a recurring debate among network criminologists about the structure of covert networks. Then, a quick tutorial is offered on Burt's structural holes theory of social capital [1]. Before discussing our results, we explain our methods and highlight the potential shortcomings of the data used. After presenting and interpreting our findings, the chapter concludes with a few notes about how the field of applied social network analysis is advancing.

[1]While we are aware of the differences between insurgent and terror groups, the terms are used interchangeably in this chapter.

2 Literature Review

2.1 Recurring Debate: Efficiency/Security Trade-off

Investigating several different types of covert networks, Raab and Milward [3] argue that while studies find considerable structural variation, one commonality exists; networks need to remain flexible so as to be able to complete their objectives while under attack and in the face of changing conditions. Operating within a hostile environment, insurgents need to remain hidden from view and resilient to counter-terrorism efforts. In this context, redundancy is both an asset and a liability. Transmitting information through central communication hubs embedded within a dense network (where everyone is connected to everyone else) ensures maximum efficiency. Any orders issued will reach their intended target directly, and indirectly (for emphasis or assurances), with the greatest celerity. However, there is a drawback. If one actor is captured and information seized, the entire operation is compromised. Greater interconnectivity among actors now becomes a liability.

To better secure the operation, communication chains may be elongated. Passing information through several actors reduces the potential network exposure that may arise due to counter-terrorism efforts. Taking down one actor, will not reveal the larger organization, as each person only knows a few other people in the group. The downside to this network structure is the loss of efficiency. Time-to-task ratios are extended, as each message takes more time to travel through the network.

Described as the security/efficiency trade off, Morselli et al. [4] suggest that depending on the status of the group and the events that need to occur, the network may shift to be more efficient or more secure. At times when security is paramount, long communication chains emerge; forcing information to pass through more people to reach their intended recipient (e.g., [5, 4]). When time is critical, communication paths shorten to increase efficiency. The conventional explanation is that insurgent networks accomplish this shift by activating latent ties. Relations acting as "transitory short cuts" lie dormant for months, if not years, until being conjured up for a particular attack; thereby, enabling the group to morph as needed from an extremely secure structure to one that is more efficient (e.g., [5, 4, 3]).

Due to its shifting nature, the true structure of the group is not evident in a single attack or cross section of communication. Rather, insurgent networks need to be considered over time in order to reveal the full company of actors, as well as the dynamic nature of the operation (e.g., [6]). This omnipotent perspective is critical to understanding what structural characteristics are associated with operational success and organizational longevity, and to predict how the organization will respond when under attack. For instance

- Xu et al. [7] found that over the course of 14 years, the Global Salafi Jihad network evolved scale-free network structures as smaller hubs were removed by counter-terrorist operations. Older actors, those present longer within the network, were able to stay secure due to their limited connectivity and greater amount of experience.

- Bush and Bichler [8] showed that during a period of heavy drone strikes key Al Qa'ida leaders and support staff assumed central brokerage positions and transitivity (local density) increased as communications became more direct.
- Studies of the 9/11 attack network [4] and the Bali bombing [9] reveal that as activity increased to launch a pending attack, density increased.
- Magouirk et al. [10] concluded that Jemaah Islamiyah (JI) became fractured and decentralized as key actors were removed; visible leaders of attacks did not have control over general operations, nor were they exerting control over affiliates.

Generating a comprehensive record of network structure and its evolution, that offers tactical benefits, requires pooling and analyzing data extracted from multiple sources. Whether this involves deploying sophisticated data mining software (e.g., [11]) or simply going beyond a case study approach and pooling information from several sources (e.g., [12]), such data enables us to project how an insurgent group may respond to the elimination of key leaders. For example, integrating information from a multitude of sources, Wu et al. [13] determined that network position was not sufficient to predict Bin Ladin's successor; rather, human capital (i.e., social status and ideology) was also of material importance. Moreover, as the methods used to capture data about terrorists develop, so too must we apply a wider range of SNA theory and analytics to better understand how insurgent groups organize for success. One theory gaining traction among social network scholars of terrorism is Burt's theory of structural holes [1, 2].

2.2 *Rethinking Success: Structural Holes Theory*

> The structural holes in which competition develops are invisible relations of nonredundancy, relations visible only in their absence [1: 4]

Burt's [1, 2] structural holes theory about competition and the role of social capital in maintaining an advantage, offers a useful framework for rethinking the concept of success within the context of insurgent activity.[2] Social capital is a property of relationships and is jointly owned by both people involved. Within a competitive environment, like work or a social group, your relationships are the conduits through which you receive information about new opportunities to acquire

[2]The reader should be aware that there are different conceptions of social capital. Our adoption of Burt's [1] perspective (and suggested metrics) does not indicate that it is the only useful conceptualization. Moreover, the theory did not remain in a vacuum: several additions were made (see [2]). Constrained by the limits of what can reasonably be covered in a chapter, we encourage readers to see Borgatti et al. [14] for a brief synopsis of the political and sociological discussion on this topic and the various measures available to study the attendant network structures. Readers are also advised to explore Granovetter's [15, 16] discussion of the information benefits of weak ties, as well as some of Burt's more recent writings.

assets or experience, deploy resources, influence others, and transmit information. By investing time and energy into maintaining certain contacts, you can improve your access to these opportunities, thereby ensuring that you are the first to learn of new developments, and thus, first to act on them. This position of maximum information benefits also better insulates the individual, and ultimately group they work for, from pending disasters [2]. Burt [1] writes

> [...] information benefits of a network define who knows about these opportunities, when they know and who gets to participate in them. Players with a network optimally structured to provide these benefits enjoy higher rates of return on their investments, because such players know about, and have a hand in more rewarding opportunities (13).

While having more direct contacts tends to generate a larger social network, bigger networks do not always increase your exposure to new, and potentially, more *valuable* information. Instead, Burt [1] suggests that greater information benefits are gained when existing or added relations diversify information streams. By maintaining relationships with individuals who are part of different social clusters, you are able to access varied pools of information, thereby boosting your social capital and increasing your "success factor." However, it is best to ensure that these clusters do not interact. If bridging relationships form, joining the different clusters, the network thickens (becomes denser) and your social capital decreases.

In a dense network, many, if not all of your direct contacts, know each other. Either, within group cohesion is high or you may have links with structurally equivalent people—that is people with all of the same friends. Burt's [1, 2] warns that dense networks provide less diverse information: the same data circulates, sometimes repeatedly, through the group. In fact, you may be better off dropping all but one relationship because you would still receive all of the information benefits by maintaining only a single association with someone from the original group. It is a matter of efficiency—maximizing the benefits provided by your connections by maintaining only non-redundant contacts. Why? Because, every relationship is an investment, and if the investment fails to deliver information benefits, it is a drain on your resources.

3 Research Hypothesis

Right about now, our readers may be wondering why this is relevant to the study of insurgency. Burt's [1, 2] theory of structural holes is applicable for two reasons. Our first declaration is that *competition is not restricted to legitimate enterprise.* If as Burt [1, 2] asserts, competition emerges naturally from interactions between people and is a property of the relation, not the individuals involved, then it is reasonable to extend the theory to competitive relations among members of an insurgent group. Members of all groups compete on some level. Whether driven by praise, profit, a desire for promotion, or aspirations of a higher sort, the gaps in social structure—defined as disconnections or non-equivalencies between players—can be thought of as opportunities that garner competitive advantage; advantage that leads to greater success, individually and for the group, to which the individual belongs [1].

Middle managers within insurgent groups can be thought of as illicit entrepreneurs, who are just as reticent to engineer and exploit their social relations to maximize their success margins, as any other manager within legitimate enterprise. In fact, we are not the first to argue that entrepreneurial tendencies within criminal enterprise groups contribute to illicit success. See for example Malm and Bichler's [17] work on money laundering; Morselli's analysis of a broker in the cannabis smuggling trade [18]; or Raab and Milward's [3] investigation of arms trafficking during conflict. Acknowledging that criminal enterprise is a highly competitive arena, from which insurgency is not immune, we join a growing number of researchers advocating that the structural holes theory of social capital has explanatory value in the study of terrorism (see for example [19, 13]). What we add to this emerging discourse is that we are among the first to redefine success. This introduces our second assertion: *the currency of success varies because motivations vary*.

Within a business context, individuals may seek promotion, prestige, or a greater share in profits, illicit or otherwise. However, a political theorist would argue that while some professional insurgents may be motivated by conventional "employment" goals, the objective, and thus the definition of success, is substantively different for theological jihadists involved in suicide bombing, or central command waging a war to gain political control of territory. Hence, instead of considering profit margins, promotions or bonuses, we recast what information benefits bring to the managers of terror groups, arguing that longevity (living to fight another day) is the most successful outcome accrued to positions of network advantage. Moreover, when under targeted attack, the most salient marker of success may simply be to avoid death or capture.

Applying, Burt's [1] theory to the business of insurgency, we ask, "Who stays alive?" To the best of our knowledge, no one to date has thought about whether structural holes in personal networks correlate with insurgent success as measured by evading capture or death. Considering the most central players within Al Qa'ida and Islamic State of Iraq (our proxy for organizational managers), we hypothesize that individuals with a higher level of social capital (more structural holes in their personal network) survive longer, avoiding death or capture, than those who have lower social capital.

4 Current Study

4.1 Data Source

The current study uses two assemblages of documents made available to the public under the Harmony Program, by the Combating Terrorism Center (CTC) at West

Table 1 Communication chains generated by each intelligence source

Source	Documents	Pages	Year					Total chains
			2006	2007	2009	2010	2011	
Al Qa'ida intelligence	13	154	30	99	11	409	101	650
ISI intelligence	17	58	72	92	29	0	0	193
Combined total	30	212	102	191	40	409	101	843

Point. The first set of documents containing intelligence about the Al Qa'ida organization, released to the public on May 3, 2012, were part of an undisclosed number of files seized, when the United States Special Operations Forces raided the residence of Usama Bin Ladin on May 2, 2011 [20]. This cache of intelligence consisted of 17 declassified and translated documents totaling 197 pages of English text. From these items, 13 documents (154 pages of text) were selected, dated between September 2006 and April 2011. These items contained sufficient information to generate a picture of how communications flow among authors and recipients. This set of materials and the resulting communications network is referred to as *Al Qa'ida intelligence*.[3]

The second set of translated and original documents, about 700 in total, released some time later by the CTC, again under the auspices of the Harmony Program, included intra-organizational communications of the Islamic State of Iraq (ISI) and its subsequent and prior incarnations. Similar to the Al Qa'ida intelligence, these documents provide transaction memos, field updates, and operational strategy (for other analyses of these documents see: Al-'Ubaydi et al. [21]; Jung et al. [22]). To be used in this study, the document must have been written within the time period under investigation (2006–2011). With limited information about the context of each item, we soon discovered that only 59 documents were confirmed to have been written within the time period, fewer still included details about the author and intended recipient. Consequently, we found that 17 documents (58 pages of text) were usable. This batch of intelligence, referred to hereafter as *ISI intelligence*, included communications occurring between 2006 and 2009.

Pooling these intercepted documents provides an opportunity to explore the tactical utility of a non-random set of information, for each document contained the details of messages transmitted along communication chains. Generating a network from the messages may provide valuable insight into the structure of the groups. The content and nature of messages ranged from general operational missives (e.g., internal communications, public discourse, and political activity) to specific instructions (e.g., requests for leave, financial transactions, tactical support, media strategy, security protocol, reports from the field). As reported in Table 1, we extracted 843 communication chains from the 30 documents examined. While we

[3]For a more detailed description of this coding process see Bush and Bichler [8].

had a greater number of ISI intelligence documents, the brevity of these items resulted in fewer coded communications.

4.2 Network Construction

Actor-to-actor links were generated by dissecting each message. Each person mentioned in the message was recorded as such: the sender or author of a directive was the ego and was linked in a chain to subsequent intermediaries, who in turn connected to the intended recipient. A passage of text illustrates the coding process. For instance, in document SOCOM-2012-0000003-HT, authored by Usama Bin Ladin, to his associate Atiyya Abdul Rahman; Usama writes

> Regarding what brother Basir mentioned relating to Anwar al-'Awlaqi, it would be excellent if you inform him, on my behalf in a private message to him, to remain in his position where he is qualified and capable of running the matter in Yemen. (SOCOM-2012-0000003-HT; p. 2)

Two communication chains can be extracted from this text: a prior message had been received by Usama, BASIR → ATIYYA → BIN LADIN; and, he responds with a message to be delivered to Basir, BIN LADIN → ATIYYA → BASIR. Converting these information chains into links between pairs of people we obtain the following ties: (1) BASIR and ATIYYA, (2) along with a second link from ATIYYA to BIN LADIN and then, (3) BIN LADIN to ATIYYA and (4) ATIYYA to BASIR.

Table 2 reports the number of ties, actors, and descriptive statistics for each network independently, as well as the combined network. Due to the reciprocal nature of communications flow, messages traveling back and forth between pairs of actors, the combined network includes 1342 ties but this represents only 541 unique communication channels. These networks are characterized as having very low density: less than 2 % of all possible connections among pairs of actors were observed. Moreover, the degree centrality values suggest that on average, actors were connected to less than 2 people. It is notable that the main component of the

Table 2 Description of networks by intelligence source

Network by source	Total ties	Unique ties	Actors	Components	Density (%)	Avg. degree
Al Qa'ida intelligence	1132	362	195	1	1.00	1.86
ISI intelligence	215	180	115	11	1.40	1.57
Combined (full data)	1347	541	302	10	0.60	1.80

Notes One unique tie was found in both intelligence sources and thus, the total number of unique ties is 541 instead of 542. Eight people are common to both networks; consequently, the total number of actors in the combined dataset does not equate to a sum of the two intelligence sources

ISI network connected with the Al Qa'ida network through 8 actors: Abu Abdullah Al-Shafi'i, Abu Ayyub Al-Misri, Abu Layth, Abu Sajjad, Abu Umar Al-Baghdadi, Mukhtar Abu Al-Zubayr, Shaykh Abu Hind, and Shaykh Ayman Al-Zawahiri.

While network construction captured the direction in which the information flowed, we symmetrized and dichotomized the networks when generating the final combined network that was used in the analyses reported shortly. Our justification for removing directionality rests with one of the primary data limitations. No information is available as to why only a small number of files were released to the public and we do not know how these items were selected from the thousands of files captured by allied forces. Thus, we take the existence of a communication chain to represent a potential communication channel, capable of transmitting information in either direction. Moreover, we are uncertain whether the number of messages captured in our coding process reflects the true volume of communications passing among actors. For this reason, we felt that it was more appropriate to dichotomize the network. However, to check the sensitivity of our analysis we report some findings for the valued (symmetrized) network. These additional results appear in the table notes.

Figure 1 depicts the main component of the combined network. Individuals named in Al Qa'ida intelligence materials are illustrated as grey circles, individuals unique to ISI intelligence are shown as black circles, and the 8 overlapping individuals are numbered and shown as white circles. In addition, the individual with

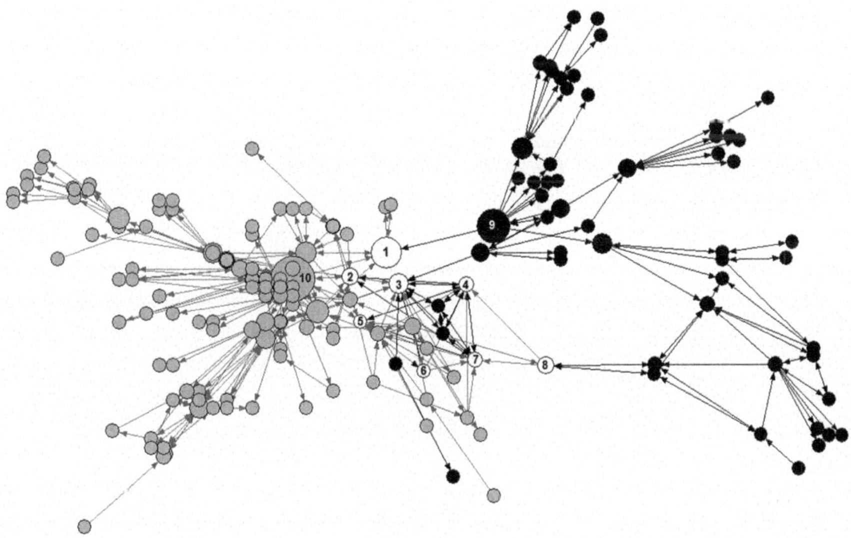

Fig. 1 MDS visualization of main component of the combined network. *1* Mukhtar Abu Al-Zubayr, *2* Shaykh Ayman Al-Zawahiri (at large), *3* Abu Umar Al-Baghdadi, *4* Abu Sajjad (at large), *5* Abu Ayyub Al-Misri, *6* Abu Abdullah Al-Shafi'i, *7* Shaykh Abu Hind (at large), *8* Abu Layth, *9* Abu Walid (at large), *10* Atiyya Abdul Rahman. *Note* Individuals are noted as being at large as of November 2015. The figure was created in NetDraw 2.159 [23]

the highest betweenness centrality (most critical information broker) from each group is also identified. The network is arranged using non-metric multi-dimensional scaling (MDS) which places actors that are more similar closer together. With 274 actors in the main component, similar actors are placed on top of each other in the figure. Similarity represents having comparable shortest paths (geodesic distances) to all others in the network and a comparable set of direct contacts.

Three distinct patterns are visible in this image. First, not only are the 8 individuals named in both intelligence sources situated in between the two communication networks, but most of them are part of a dense, interconnected cluster that contains several others named in only one source. Second, three individuals, Mukhtar Abu Al-Zubayr, Abu Umar Al-Baghdadi, and Abu Layth were critical brokers linking the two networks: they are all deceased. Third, most of the actors in this network are linked through chain-like structures with limited redundancy.

4.3 Sample Selection

The analysis reported in this chapter focused on 41 of the most central actors. In order to ensure equal representation of ISI and Al Qa'ida operatives in selecting the sample, centrality statistics were calculated on each network separately. Actors received a value of 1 for being ranked in the top 15 for each centrality measure: degree, Bonacich's power, eigenvector, and betweenness.[4] Summing across the four measures generated scores ranging from 0 to 4. Twenty-five of the highest scoring actors from each network were selected for the analysis. Then, after resolving duplicates[5] we were left with 41 centrally positioned actors: 15 from Al Qa'ida, 18 from ISI, and 8 actors named in both intelligence sources.

Egonets were produced for each of the 41 centrally position actors (using the combined intelligence network). An ego network is an extraction from the larger network. It includes the actor of interest, called the ego, and all others the ego is directly connected to (called alters), along with any ties that may exist among alters. On average, central egos were directly connected to about 9 others (alters) and 20.64 % of the possible ties among actors in the egonet were observed (see Table 3). Individually, this group of actors is embedded within denser networks than the overall density we found for the network taken as a whole (recall from Table 2 that the combined network with all actors had a density of 0.6 %).

[4]Degree centrality measures the number of direct connections an actor has; Bonacich's power and eigenvector centrality identifies individuals who may have only a few connections, but those associates are highly connected; and, betweenness centrality can identify which individuals are situated along the shortest paths among all other pairs in the network. For more information about these centrality measures see Borgatti et al. [24], Hanneman and Riddle [25].

[5]One journalist also appeared in the top 25 listing for Al Qa'ida. Since this person did not have a known role within an insurgent group they were excluded from this analysis.

Table 3 Description of the indicators of egonet structure

Measures	Average	Median	SD	Minimum	Maximum
Degree centrality	9.34	7.00	10.23	2.00	66.00
Egonet density (%)	20.64	10.91	25.63	0.00	100.00
Effect size (WNET)	8.35	5.67	0.18	0.34	1.00
Hierarchy (WNET)	0.03	0.01	0.08	0.00	0.40

Note Structural holes measures are calculated on the whole network

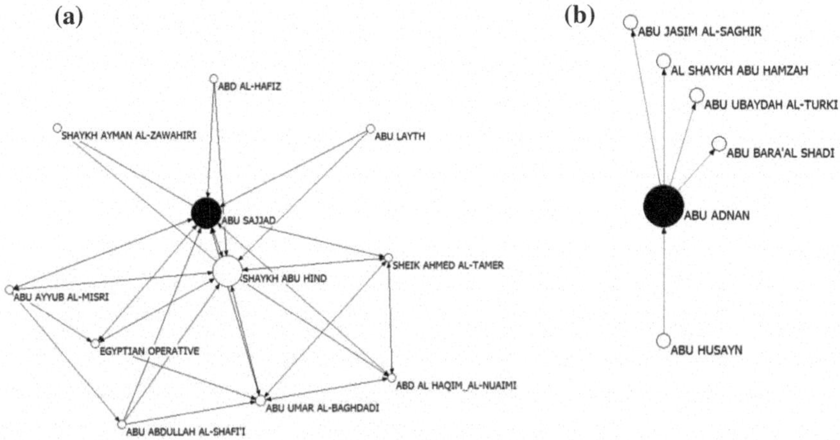

Fig. 2 Comparison of Egonets for Abu Sajjad and Abu Adnan. **a** Dense egonet of Abu Sajjad, **b** sparse egonet of Abu Adnan. *Note* The directionality of the information flow captured in these egonets was displayed for reader interest

As you can see by the standard deviations, and the range between minimum and maximum scores, these averaged descriptive statistics obscure important structural variation within the sample. For instance, as shown in Fig. 2, Abu Sajjad is not the most instrumental person in his own egonet. In fact, Abu Sajjad could be removed from this group and communications would continue because Shaykh Abu Hind has all of the same direct ties. Abu Sajjad is redundant in his own network. He is also socially constrained; he is hemmed in by having connections that are themselves linked, and thus, he does not appear to have access to unique information. Abu Adnan enjoys a better position: Adnan appears to function as a bridge in the flow of information. He is positioned to receive information from Abu Husayn, and then, he *controls* the dispersal of information to others.

According to Burt [1, 2], the structure of these personal networks tell us something about the social capital of each person: people surrounded by gaps, referred to as structural holes, have a greater influence on those around them because they can

Table 4 Description of the dependent variable, success

Insurgent group	Actors	Number successful	Percent successful (%)	Avg. months alive (for actors who died or were captured)
Al Qa'ida	15	4	26.67	82.00
ISI	18	17	94.44	106.00
Overlaps (both)	8	3	37.50	56.40
Total	41	24	58.54	75.88

act on and provide unique information.[6] Adnan is more socially competitive than Sajjad, as he is vital to the flow of information in his network because structural holes separate his direct contacts from each other. Thus, Abu Adnan has more social capital, and as such, should be more successful, than Abu Sajjad.

4.4 Variables

Success. The dependent variable, success, was measured two ways. First, a dichotomous variable indicated whether the central actor survived (valued at 1) or had been arrested or killed (valued at 0) during the research period.[7] The research period begins in January 2006 and extends to November 2015.[8] The second variable reflects the number of months at large and active during the study period. A person who survived the entire study period received a value of 119 months. For all others, we counted the number of months they remained alive or at large before the month they were reported to have died or been captured. These values range from 24 to 113 months; averages are reported in Table 4.

Indicators of Structural Holes. Generally, we can decipher whether a person's communication network includes many structural holes by examining degree centrality (the number of direct contacts) and density [24]. Density is a simple

[6]Recall from our previous discussion that social capital is a dynamic byproduct of competition borne from relational structures that affords some individuals significantly more access and control of resources than others [1].

[7]Admittedly, searching the status of each person using publicly available sources is a limitation for two reasons. First, some individuals go by several names and it is possible that a notice of death or arrest was missed. Second, many individuals assume names of others who have fought before them. To determine which person was killed or arrested, we considered their known locations, role within their respective organizations, and age. Again, we may have missed something. However, we compared our coding with published findings (see Wu et al. [13]) which confirmed our classification of Al Qa'ida operatives.

[8]While the documents used to code communication chains are dated between September 2006 and April 2011, we decided that coding death or capture required a longer time frame. All actors active in September 2006 were alive in January 2006, so extending back from the documents was reasonable. Moreover, we wanted to code the ability of individuals to remain alive, and thus, expanding the window forward to when this chapter was written seemed appropriate.

measure of cohesion that tells us the proportion of ties within the egonet that are observed compared to what could exist. While relevant to the characterization of personal networks, these measures do not tell us anything about the existence of redundant contacts. Fortunately, Burt [1] offers a couple of additional indicators that can be used to assess the structural differences among egonets that accrue from redundancy—effect size and hierarchy.

Effect size is taken to reflect the non-redundant nature of ego's contacts.[9] The information that flows through your contacts is redundant if the people you are tied to are also connected. This is calculated by summing the number of alters, minus their average degree centrality scores. Ties to the ego are not included. This measure is used to show the ego's unique influence. Higher scores suggest that the direct contacts of the ego have less redundancy and thus, the ego has more influence in controlling the flow of unique information in their personal network. Essentially, higher effect sizes mean more structural holes. What follows is the theoretical explanation for this measure, notably it assumes that a valued network is used.[10]

$$ES_i = \sum_j \left[1 - \sum_q p_{iq} m_{jq} \right], \quad q \neq i, j \tag{1}$$

where

p = ego's contacts linking to q divided by the sum of ego's contacts (proportion)
m = strength of j's connection with q divided by the strongest other association (proportion)
q = every other actor in the egonet.

The second indicator of structural holes selected for this analysis is associated with constraint. Constraint measures the extent to which the actor of interest, the ego, can avoid or negotiate a demand placed on them by someone in their network. For instance, if your mother has many indirect (through several 3rd parties) relationships linking to you, it would be very difficult to deny knowing that she wants you to do something for her on this coming Saturday, even when you let her phone call go to voicemail. Burt [1] suggests that in assessing the proportion of someone's relational investment that leads to reaching another person, we must factor in the absence (lack) of structural holes around them. If this person has a lot of other associations with people in your network, they would be easier to reach. The equation for constraint follows.

[9]Efficiency is a standardized version of effect size that controls for the size of an egonet. Generally, it is better to use this measure. However, we found that efficiency was highly correlated with density (Pearson Correlation of −0.992), and thus, it was not used here.

[10]Borgatti [26] demonstrates that when applied to a dichotomous network (as we have done here) the effect size is simply the egonet density scaled by a factor of n − 1, such that effect size can be calculated as: $n - \frac{2t}{n}$; where, t equals the number of ties in the egonet, excluding all ties to the ego.

$$c_{ij} = \left(p_{ij} + \sum_{q} p_{iq} p_{qj} \right)^2, \quad q \neq i,j \tag{2}$$

where, i is the ego, j represents an alter, q is a third party, and p is the proportion of links.

Rather than use this raw form of constraint, we opted to use hierarchy. *Hierarchy* is an adjustment of constraint that tells us the extent to which the aggregated constraint that an ego faces is concentrated in a single alter; Burt [1] adopts the Coleman-Theil disorder index for this purpose. Considering the ratio of the constraint for one relation to the average of all relations provides a way of capturing how much more one relationship is a source of constraint than others. Burt [1] explains that by summing this ratio multiplied by its natural logarithm and controlling for the maximum value possible, we obtain a measure ranging from 0 to 1, where 1 indicates that constraint is concentrated in a single tie. Essentially, hierarchy would have a high score if the ego is dependent on one primary contact for their information, and that contact is also central to many other people that are in the ego's network.

$$CT\ disorder\ index = \frac{\sum_j \left(\frac{c_{ij}}{C/N} \right) \ln \left(\frac{c_{ij}}{C/N} \right)}{N \ln(N)} \tag{3}$$

where

$$c_{ij} = \left(p_{ij} + \sum_{q} p_{iq} p_{qj} \right)^2 O_j, \quad i \neq q \neq j, \tag{4}$$

C = the sum of constraint across all N relationships
C/N = the mean level of constraint per contact

Simply stated, this statistics tells us something about dependency. Insurgents with communication networks anchored on a single relationship should be at a distinct competitive disadvantage. A larger hierarchy score would suggest their access to information benefits is heavily constrained by their relation with one person and thus, they have fewer opportunities for action. Essentially, the ego is dependent on one person.

Descriptive statistics for these variables are reported in Table 3. Please note that these values are whole network statistics. Structural holes indicators can be computed with consideration of actors and ties within the egonet, or, scores for alters can be modified to consider their position within the entire network. Whole network statistics are used here because the relative importance of alters is important when considering the social capital of the ego. Comparing the average degree centrality with effect size suggests that egos have relatively few redundant relations; and, the low hierarchy scores indicate that most egonets are not centered on a single alter.

5 Results

5.1 ANOVA

Examining the within and between group differences, we find that central individuals who successfully avoided capture or death during this study period had smaller communication networks (lower degree centrality), significantly lower hierarchy scores (less dependency), and substantively fewer information benefits (fewer non-redundant contacts as indicated by the smaller effect size). These bivariate associations are of low to moderate strength (the results are reported in Table 5). Most noteworthy is the significant association between success and hierarchy. These results suggest that individuals who are less dependent on a single person for their information are significantly more likely to survive and remain at large during the study period.

Given the concerted US-led effort to dismantle Al Qa'ida over the last decade, we ran an analysis of variance comparing egonet descriptors for individuals named in Al Qa'ida sources only, ISI sources only, or both. Individuals named in both sources survived for the shortest period of time, averaging 79.9 months; their communication networks had the greatest density; and, all other values were middle range or comparable to Al Qa'ida (see Table 6). Of particular interest, individuals named in Al Qa'ida documents were embedded within larger communication networks with somewhat less redundancy (more structural holes or information

Table 5 ANOVA comparing egonet structure of successful insurgents to those who were captured or killed

Variables	Captured/Killed (N = 17)	Successful (N = 24)	Eta	F	Sig.
General measures					
Density (%)	24.36	18.00	0.123	0.604	0.442
Degree Centrality	12.59	7.04	0.270	3.075	0.087
Structural holes					
Effect size (WNET)	11.23	6.31	0.240	2.377	0.131
Hierarchy (WNET)	0.06	0.02	0.392	7.096	0.011

Note Repeating this analysis using structural hole metrics calculated strictly on the egonet (meaning we exclude the influence of the rest of the network) confirm these findings and indicate greater differences between groups for the association between hierarchy and success (Eta = 0.496; F = 12.76; $p < 0.001$). Whole network hierarchy values calculated on a valued, symmetrized network reveal that among those captured or killed, 43.2 % of their egonets were constrained to one central individual, compared to 26.7 % for actors who remained at large (Eta = 0.258; F = 2.78; $p < 0.10$)

Table 6 ANOVA comparing egonet structures by intelligence source

Variables	Al Qa'ida (N = 15)	Both (N = 8)	ISI (N = 18)	Eta	F	Sig.
Success (%)	26.70	37.50	94.40	0.650	13.866	0.000
Time alive	91.87	79.88	118.28	0.579	9.561	0.000
General measures						
Density (%)	23.35	24.63	16.60	0.142	0.392	0.678
Degree centrality	13.27	8.50	6.44	0.304	1.939	0.158
Structural holes						
Effect size (WNET)	12.06	6.87	5.91	0.281	1.628	0.210
Hierarchy (WNET)	0.05	0.05	0.01	0.377	3.147	0.054

Note Again, running an ANOVA with egocentric measures that do not adjust for the effects of the rest of the network magnifies the association between hierarchy and success (Eta = 0.540; F = 7.82; $p < 0.001$). Using a valued, symmetrized network, we find that among AQ actors, 56.8 % of their networks are constrained to one alter, whereas actors named in both sources exhibit 27.4 % dependency, and ISI actors had 17.1 % dependency (Eta = 0.567; F. 8.99; $p < 0.001$)

benefits), but they also faced significantly greater constraints (higher hierarchy).[11] This suggests that Al Qa'ida operatives have less social capital because on average, they were more dependent on a single person for information. This constraint may prohibit them from capitalizing on information benefits. A caution is warranted; without other collaborative information it is not possible to determine if this result is an artifact of the intelligence documents used or a material difference in organizational structure.

5.2 *Logistic Regression*

Table 7 reports on the four logistic regression models used to determine whether characteristics of egonet structure are correlated with successful outcomes.[12] Variables were selected for this model based on the ANOVA results reported previously. In the first model, we see that there are substantial decreases in the odds of surviving at-large during the study period as one's communication network increases in size (degree centrality) and density. With the small sample size and

[11]The difference in effect size reported in Table 6 would seem to indicate that actors associated with ISI had fewer structural holes, and thus, less redundancy in the network; however the effect size reported does not control for the size of the network. Once network size is controlled for, individuals named in ISI documents exhibited more structural holes in their egonets (AQ = 0.82; both = 0.80; ISI = 0.87).

[12]For brevity, only the logistic regression is reported.

Table 7 Logistic regression for insurgents successfully avoiding capture or death during the study period

Variables	General Model			Structural holes model			Parsimonious full model			Control model		
	B (S.E.)	Sig. (2-tailed)	Exp (B)	B (S.E.)	Sig. (2-tailed)	Exp (B)	B (S.E.)	Sig. (2-tailed)	Exp (B)	B (S.E.)	Sig. (2-tailed)	Exp (B)
Constant	2.17 (0.97)	0.025	8.79	1.34 (0.63)	0.033	3.81	1.44 (0.71)	0.043	4.22	−2.06 (1.29)	0.110	0.13
General measures												
Density (%)	−2.33 (1.49)	0.119	0.10									
Degree centrality	−0.16 (0.09)	0.071	0.86				−0.04 (0.08)	0.610	0.96	−0.01 (0.08)	0.946	1.00
Structural holes												
Effect size (WNET)				−0.03 (0.07)	0.699	0.98						
Hierarchy (WNET)				−26.77 (12.61)	0.034	0.00	−25.61 (12.91)	0.047	0.00	−20.75 (14.63)	0.156	0.00
Intelligence source										1.58 (0.53)	0.003	4.85
Nagelkerke R^2	0.198			0.253			0.257			0.529		
Predicted correct (%)	58.50			70.70			70.70			78.0		

large standard errors, the significance of these patterns is not conclusive; however, using a threshold of $p < 0.10$, degree centrality is arguable significant and was selected to be included in the final model. Only 19.8 % of the variance in success is accounted for by this model.

The second model investigates Burt's [1, 2] indicators of structural holes. Here we see that hierarchy is a significant predictor of success. Actors with high hierarchy scores are significantly less likely to survive the study period. This dramatic result suggests that with every unit increase in hierarchy (meaning more of the actor's information benefit is dependent on one person) there is a 100 % *decrease* in the odds of surviving the study period. The model itself is better at accounting for the variation in success as indicated by the improvement in Nagelkerke R^2 and the percent of cases predicted correctly; the classification table results show that 70.7 % of predicted outcomes (successfully avoiding capture or death) are correct.

In the final two models we compare degree centrality and hierarchy (parsimonious model) and control for the source of intelligence used to generate the communication networks (control model). While there is no substantive improvement in model fit, the importance of hierarchy is confirmed in the parsimonious model. The control model offers a dramatic improvement in accounting for the observed success of central actors. It is notable that while the significance of hierarchy declined, the effect remained strong, suggesting that individuals that are more dependent on a single source of information are at greater risk of dying or being captured. Notably, distance from being named in Al Qa'ida intelligence sources is critical to remaining at large. (Intelligence source was coded as an ordinal variable ranking distance from Al Qa'ida: being named in AQ intelligence = 1, being named in both sources = 2; and named in ISI documents only = 3.)

6 Discussion

6.1 *Who Stays Alive and At-large?*

Success. This study found that relative to those who were captured or killed, the central actors of Al Qa'ida and ISI who remained alive and at-large during the study period were embedded within smaller, less dense egonets characterized as being less likely to include dependencies on a person central to others. Notably, the egonets of successful insurgents scored lower on effect size, however, controlling for network size, this difference became immaterial. This means that more control over the flow of information within the egonet did not equate to significantly greater success as we defined it. However, those individuals able to remain at large exhibited significantly lower constraint. While this finding only partially supports what we expected based on Burt's [1] theory, it is consistent with more recent findings that suggest that an important contingency is at play.

Social capital exists when someone is positioned to use the information to their advantage that they received through their contacts. If the information cannot be acted upon, it offers no material advantage. As Burt [2] argues, having access to information benefits, while being surrounded by many other "middle managers," diminishes your ability to stand unique among your peers. The existence of peers makes it difficult for you to work the structural holes in your network effectively. Peers may move to undermine your efforts. Moreover, since social capital is merely an opportunity to access information benefits, to make use of these advantages, a person must have the human capital (i.e., acumen, social status, accumulated resources, and authority) needed to assess the utility of information flowing through the network and capitalize on the foreseen advantage. A point well illustrated by Wu et al. [13] in their evaluation of potential successors to Bin Ladin.

Distance from Al Qa'ida. In this study, being named in Al Qa'ida intelligence sources significantly increased the odds of death or capture. With larger networks of fewer redundant contacts, actors named in Al Qa'ida intelligence clearly had more structural holes to work with, and should have been able to avoid failing. Several different explanations may account for this finding.

First, Al Qa'ida was a primary target of counter-insurgency measures during the study period and being named in materials produced by this group made it more difficult to remain hidden. This exposure may also have diminished the actor's ability to capitalize on information benefits. Unable to take advantage of opportunities, central figures may have passed such benefits on to other more peripheral actors. Whereas, central individuals named in ISI intelligence documents enjoyed somewhat less notoriety and this may have allowed them to personally exploit opportunities flowing through their sparser networks.

An alternative explanation is that communication flows differently in each group because one is organized for efficiency and the other for security. Al Qa'ida matured under an onslaught of concentrated counter-terrorism measures, and as such, the necessities of maintaining the jihad may have forced the group to increase efficiency, overlapping communications to ensure that when one operative was detained or killed, the mission continued. This may be why central individuals named in Al Qa'ida intelligence sources have more direct contacts and are embedded within denser networks with a tendency to include greater dependency on one individual for information.

Finally, this result may reflect a material difference in communication style. In coding the materials it was readily apparent that Al Qa'ida documents were substantially longer, including more instructions and reporting on prior communications; contained more narrative through which the identity of named individuals could be reconciled (formal greetings along with reference to nicknames); and, covered a wider range of topics (ISI documents tended to be short reports from the field rather than lengthy discussions of operational strategy or organizational management). Being spread throughout a larger geographic region, Al Qa'ida documents may be more akin to the minutes of meetings, that ISI command staff held in person. Thus, these results may reflect how geography affects terror group communication patterns.

6.2 Limitations

As with all efforts to understand the social structures of covert networks, we must issue a word of caution about the generalizability of these results. First, we intentionally restricted our data collection to a set of intelligence documents captured during military operations. Our intent was to simulate real world data constraints in order to explore the tactical utility of SNA. What we discovered was interesting. Using only 30 documents, we identified 843 communication chains revealing 302 actors and 541 unique ties (1347 if reciprocal communications were counted). More importantly, about 91 % (274/302) actors were contained within the principal component. While this is clearly not the universe of operatives, these strategic communications uncovered a relatively well connected set of associations. As a point of comparison, Sageman's [12] Global Salafi Jihad terrorist network, built from an array of materials consisting of only 366 members.

From this point forward, we suggest that since our data is primarily concerned with command staff communications, merging our information with communications networks constructed from studies of specific attacks would materially extend the set of operators included in the network (ranging from central leaders to foot soldiers). Repeating a structural holes analysis on such an integrated database would provide an opportunity to test the robustness of our findings. It is possible that integrating attack cells (and the links certain command staff have to them) may significantly enhance the social capital of some individuals and better explain why some terrorists are able to remain at large despite targeted counter-terrorist efforts.

Second, the results presented here are based on a symmetrized, dichotomized network. Burt's [1, 2] theory of structural holes considers relationships as investments and that maintaining ties requires regular attention and resources. This implies that relationships are valued and in examining the indicators of structural holes and constraint offered by Burt [1], Borgatti [26] and Borgatti et al. [14] show that modifications are warranted when analyzing dichotomous networks. To address this issue, we included analysis of egonets extracted from a valued network and these results confirm our principal findings. However, the number of messages flowing through a communications network does not necessarily reveal whether the information is novel, useful, or pertains to a unique opportunity that may help an insurgent avoid death or capture. Continued testing of social capital and its association with insurgent success should involve looking more closely at the nature of the information flowing through a network. Unfortunately, this raises an issue affecting all criminological applications of social network theory and analysis in pursuit of advancing the field of network criminology—assembling a complete covert network.

7 Advancing the Field

A widely acknowledged caveat to all research into covert networks is the lack of complete information about actors involved and their varied associations. Complicating an already challenging data collection enterprise is the use of aliases, nicknames and partial identifiers, and multiple identities, i.e., maintaining several digital identities, multiple phones, or different incorporations. For this reason, the extant literature examining the social structure of terrorist operations contains more examples of research aimed at understanding the cell structure of attack units, than investigations of the overarching organizational structure of insurgent operations. However, three recent developments stand to advance the field considerably—data warehousing, advances in data mining, and expanded analytic frameworks.

Data Warehousing. A significant step forward in the study of insurgency accrues from several parallel efforts to promote data integration and sharing. Three notable repositories exist.

1. The ***Harmony Program*** launched in 2005 by the Combating Terrorism Center (CTC) at West Point continues to provide new information that will help "contextualize the inner-functioning of al-Qa'ida, its associated movement, and other security threats" through the analysis and release of primary source documents (https://www.ctc.usma.edu/programs-resources/harmony-program). Not only does this program promote access to hundreds of primary source documents, but the accompanying reports and briefs provide rich detail that can be used to augment database construction.
2. The open-source ***Global Terrorism Database*** (GTD) provides information on terrorist events occurring in 1970 through 2014. Compiled and maintained by the National Consortium for the Study of Terrorism and Responses to Terrorism (START), located at the University of Maryland, this database currently offers details about 140,000 attacks (http://www.start.umd.edu/gtd/).
3. The START program at the University of Maryland also captures information about crimes committed by violent extremists and terrorists in the United States. This data is contained in three related databases: the ***US Extremist Crime Database*** (ECDB) records information about violent crimes committed by extremists 1990–2010 (N = 650+ cases); ***Profiles of perpetrators of Terrorism in the United States*** (PPT-US) includes profiles on 140 organizations involved in terrorist attacks from 1970 to 2012; and, the ***American Terrorism Study*** (ATS) offers detailed information on incidents and pre-incident activities associated with federal terrorism cases between 1980 and 2013.

Combined, this consolidated information enables a wide set of scholarship that will advance our understanding of modern terrorist activity.

Data Mining. Emerging from interdisciplinary partnerships are improved data mining algorithms and more user-friendly dynamic mega-data analytic systems. For instance, Carley et al. [11] showcase an interoperable set of scalable software tools designed to comb through raw text and extract actor-based information about covert networks. Demonstrated on LexisNexis, the CASOS toolkit can then be used to generate and analyze multi-mode, multi-link and multi-time period relational data. More recently, Bouchard et al. [27] demonstrate the utility of deploying a web-crawler called the Terrorism and Extremism Network Extractor (TENE) to gather and generate relational information about extremist activities from the internet. Searching the text of electronic media, i.e., chat rooms, webpages, and social media, is particularly important given the rising importance of these media in radicalizing and facilitating terror attacks. Aptly argued by Xu and Chen [28], technological innovations such as these, stand to address methodological challenges affecting the quality of covert network data (e.g., incompleteness, incorrectness, and inconsistency of data) that exacerbate the difficulties associated with network generation (e.g., entity resolution and data transformation, fuzzy group boundaries).

Expanded Analytic Frameworks. Developments in social network theory and associated analytic tools offer the potential to reorient how we think about and respond to criminal networks. One of the most promising streams of development is dynamic modeling. For instance, Xu et al. [7] illustrate the advantages of applying a dynamic approach to studying group structure. Using Sageman's [12] data about the Global Salafi Jihad (GSJ) terrorist network, these authors show how the relational structure shifted significantly across a 14 year period: social structures changed as the organization evolved through the combined effect of recruiting efforts based on operational needs and the simultaneous loss of key personnel. By the end of the study period, a scale-free structure gradually emerged that was resilient to counter-attacks. Other examples of dynamic analysis of terrorist networks include: Bush and Bichler's [8] use of stochastic actor-based modeling to test the effect of drone strikes on the ability of Al Qa'ida to maintain communications between command staff and other levels of operation; and, Everton and Cunningham's [29] reassessment of the security-efficiency debate by investigating how exogenous and endogenous shocks over time affect the structural integrity and central positioning of actors within the Noordin Top terrorist network. These examples of dynamic modeling highlight how SNA can help us to understand, and ultimately disrupt, terrorist networks.

8 Conclusion

Using social network theory advances the study of terrorism because it allows us to rethink our explanations about how insurgents interact, and in so doing, generates ideas about how being embedded in particular network structures offers operational and strategic advantages. And, because network techniques are scalable, they can be applied to an array of actors, i.e., individuals, subgroups, groups, and organizations. Adding to prior efforts to understand the structure of terrorist groups, this chapter applied Burt's [1, 2] theory of structural holes. Using this theory of competitive advantage, we found that the structural properties of egonets correlated with the ability of terrorists to stay alive and at-large. As revealed by the communication patterns of central leaders of Al Qa'ida and the Islamic State of Iraq (ISI) who were active since 2006 and survived at-large until November 2015, being less constrained by contacts, having fewer information benefits, and remaining unnamed in Al Qa'ida intelligence improved the likelihood of survival.

These results did not conform entirely to the theory because as noted by Burt [2], being situated amongst structural holes may not be sufficient to increase your social capital because being surrounded by many peers with comparable information benefits diminishes your position, and ultimately, your likelihood of success. Moreover, in order to better assess one's ability to capitalize on information benefits we need to include measurement of their human capital. We advocate that future research continue to explore the utility of structural holes theory by integrating a broader set of information, which enables assessment of actor attributes.

One final statement is warranted. As analytic capabilities develop and data repositories expand, social network analysis will continue to improve our understanding of insurgent groups. But, as we extend our investigation of covert networks it is important to keep in mind the importance of testing the tactical utility of SNA. There is no benefit from using these analyses if they cannot in some way generate actionable intelligence upon which counter-terrorism can be based that will thwart future attacks.

References

1. Burt RS (1992) Structural holes: the social structure of competition. Harvard University Press, Cambridge
2. Burt RS (1997) The contingent value of social capital. Adm Sci Q 42(2):339–365
3. Raab J, Milward HB (2003) Dark networks as problems. J Public Adm Res Theor 13(4): 413–439
4. Morselli C, Giguère C, Petit K (2007) The efficiency/security trade-off in criminal networks. Soc Netw 29(1):143–153
5. Krebs VE (2002) Mapping networks of terrorist cells. Connections 24:43–52
6. Everton SF, Cunningham D (2014) Terrorist network adaptation to a changing environment. In: Morselli C (ed) Crime and networks. Routledge, New York, pp 287–308
7. Xu J, Hu D, Chen H (2009) The dynamics of terrorist networks: understanding the survival mechanisms of global Salafi Jihad. JHSEM 6(1):27
8. Bush S, Bichler G (2015) Measuring disruption in terrorist communications. In: Bichler G, Malm A (eds) Disrupting criminal networks: network analysis in crime prevention. Crime prevention studies, vol 28. First Forum Press, Boulder, CO, pp 177–208
9. Koschade S (2006) A social network analysis of Jemaah Islamiyah: the applications to counterterrorism and intelligence. Stud Confl Terrorism 29(6):559–575
10. Magouirk J, Atran S, Sageman M (2008) Connecting terrorist networks. Stud Confl Terrorism 31(1):1–16
11. Carley KM, Diesner J, Reminga J, Tsvetovat M (2007) Toward an interoperable dynamic network analysis toolkit. Decis Support Syst 43:1324–1347
12. Sageman M (2004) Understanding terror networks. University of Pennsylvania Press, Philadelphia, PA
13. Wu E, Carleton R, Davies G (2014) Discovering bin-Laden's replacement in al-Qaeda, using social network analysis: a methodological investigation. Perspect Terrorism 8(1):57–73

14. Borgatti SP, Jones C, Everett MG (1998) Network measures of social capital. Connections 21 (2):27–36
15. Granovetter M (1973) The strength of weak ties. Am J Sociol 78(6):1360–1380
16. Granovetter M (1978) Threshold models of collective behavior. Am J Sociol 83(6):1420–1443
17. Malm A, Bichler G (2013) Using friends for money: the positional importance of money-launderers in organized crime. Trends Organ Crime 16(4):365–381
18. Morselli C (2001) Structuring Mr. Nice: entrepreneurial opportunities and brokerage positioning in the cannabis trade. Crime Law Soc Chang 35(3):203–244
19. Morselli C (2005) Contacts, opportunity, and criminal enterprise. University of Toronto Press, Toronto, ON
20. Rassler D, Koehler-Derrick G, Collins L, al-Obaidi M, Lahoud N (2012) Letters from Abbottabad: Bin Ladin sidelined? Combating Terrorism Center
21. Al-'Ubaydi M, Lahoud N, Milton D, Price B (2014) The group that calls itself a state: understanding the evolution and challenges of the Islamic State. Combating Terrorism Center, West Point
22. Jung DF, Ryan P, Shapiro JN, Wallace J (2014) Managing a transnational insurgency: the Islamic State of Iraq's "Paper Trail," 2005–2010. Occasional Paper Series, Harmony Program. Combating Terrorism Center, West Point
23. Borgatti SP (2002) NetDraw: graph visualization software. Analytic Technologies, Harvard
24. Borgatti SP, Everett MG, Johnson JC (2013) Analyzing social networks. SAGE Publications, Los Angeles
25. Hanneman RA, Riddle M (2011) Concepts and measures for basic network analysis. In Scott J, Carrington PJ (eds) The SAGE handbook of social network analysis. SAGE, Los Angeles, pp 340–369
26. Borgatti SP (1997) Structural holes: unpacking Burt's redundancy measures. Connections 20(1):35–38
27. Bouchard M, Joffres K, Frank R (2014) Preliminary analytic considerations in designing a terrorism and extremism online network extractor. In: Mago VK, Dabbaghian V (eds) Computational models of complex systems. Springer International Publishing, Switzerland, pp 171–184
28. Xu J, Chen H (2005) Criminal network analysis and visualization. Commun ACM 48(6): 100–107
29. Everton SF, Cunningham D (2015) Dark network resilience in a hostile environment: optimizing centralization and density. Criminol Crim Justice Law Soc 16(1):1–20

Counter-Terrorism and Design Thinking: Supporting Strategic Insights and Influencing Operations

Anthony J. Masys

Abstract The recent terrorist attacks in Paris and Jakarta, Ankara, Ivory Coast and Brussels in 2015 and 2016 respectively, highlight the complexity and challenges associated with counter-terrorist operations. The words of Rosenhead and Mingers (Rational analysis for a problematic world revisited. John Wiley and Sons Ltd., West, Sussex, England, 2001 [48: 4–5]) resonate with the complex space of counter-terrorism and these recent incidents. They argue that we '…are not confronted with problems that are independent of each other, but with dynamic situations that consist of complex systems of changing problems that interact with each other'. Such a complex problem space can be viewed as "wicked problems" or "messes" (Rittel and Weber in Policy Sci 4:155–169, 1973 [46]). Rosenhead and Mingers (Rational analysis for a problematic world revisited. John Wiley and Sons Ltd., West, Sussex, England, 2001 [48: 4–5]) describe 'messy problems' as that which have inherent complex interdependencies and dynamic complexity. They argue that 'Individual problems may be solved. But if they are components of a mess, the solutions to individual problems cannot be added, since those solutions will interact'. Within the context of counter-terrorism, deep uncertainty is the source of surprises and shocks in a system and the main cause of discontinuity in the strategic space of a system. It highlights the complex social factors that require an empathic approach to uncover the connectivity and processes (Masys Exploring the security landscape—non-traditional security challenges. Springer Publishing, 2016 [36]) supporting this emergence of violent extremism. The problem space transcends domain specific analysis to require a more inclusive approach that draws upon insights from sociology, economics, political science, humanities in the problem framing (Masys in Applications of systems thinking and soft operations research in managing complexity. Springer Publishing, 2015 [35]). New methods and methodologies have evolved to address such inherent complexity in problem

A.J. Masys (✉)
University of Leicester, Leicester, UK
e-mail: Anthony.masys@gmail.com

A.J. Masys (ed.), *Disaster Forensics*, Advanced Sciences and Technologies for Security Applications, DOI 10.1007/978-3-319-41849-0_12

spaces. Design thinking, system and complexity approaches emerge as a response to the inability of traditional approaches to handle human and social aspects of problem situations. The epidemiological model (disease model) of terrorism is a metaphor that captures the inherent complexity of terrorism (Stares and Yacoubian in Terrorism as a disease: an epidemiological model for countering Islamist extremism, 2007 [53]). References to terrorism being a "virus" or to al Qaeda "mutating" or "metastasizing" are common. Holistic approaches to medicine and disease have leveraged design thinking approaches to tackle the volatility, uncertainty, complexity and ambiguity. Like the epidemiological model, it is argued in this chapter that design thinking can be instrumental in foresight and strategic intervention to support counter-terrorism. Tetlock and Gardner (Superforecasting: the art and science of prediction. Penguin, Random House, 2015 [59: 123]) ask the question 'following the Charlie Hebdo attack in Paris, will there be another attack carried out by Islamist militants'? To examine such a question Tetlock and Gardner (Superforecasting: the art and science of prediction. Penguin, Random House, 2015 [59: 123]) argue that there is a requirement to generate different perspectives and synthesize them. They liken this to a dragonfly. 'Information from these thousands of unique perspectives flows into the dragonfly's brain where it is synthesized into vision so superb that the dragonfly can see in almost every direction simultaneously' (Superforecasting: the art and science of prediction. Penguin, Random House, 2015 [59: 77]). The elements of perspective taking, dialogue, synthesis, empathy, framing, ideation, prototyping, testing and learning makes design thinking a powerful approach in the counter-terrorism problem space. This chapter examines the counter-terrorism problem space leveraging the epidemiological approach to illustrate how Design thinking can be applied to develop analysis methodologies and intervention strategies to support counter-terrorism and resilience. It is about not only understanding the future but also influencing it.

Keywords Design thinking · Counterterrorism · Epidemiological model

1 Introduction

Events like 9/11, Boston Marathon bombing, 7/7, Charlie Hebdo attacks, attacks in Paris in 2015, Jakarta in 2016, Ivory Coast in 2016, Ankara and Istanbul in 2016 and Brussels in 2016 help us to re-frame how we view risks and how we can examine counter-terrorism strategies. What we learn from these attacks and subsequent counter-terrorism strategies, is that scenarios must look beyond threat, vulnerability and impact and include a more 'systemic view of the terrorist space' with the intent of 'designing' intervention strategies that are holistically applied. It is about not only understanding the future but also influencing it. To create a picture of radicalization and violent extremism and terrorism requires employing the 'dragonfly eye' to leverage multiple perspectives, dialogue and reflection as part of a design thinking approach in order to unearth weak signals and design disruption strategies. These three qualities emerge from analysis of disaster and crisis events and lessons learned to support disaster risk reduction and resilience [31–33, 38].

The inherent uncertainty associated with the counter-terrorism problem space challenges traditional linear 'waterfall' problem solving schemas and now requires a more reflective systems and design thinking approach that recognizes the requirement for problem framing, 'ideation' and synthesis. This is about exploring the possibility and plausibility space [43] and in so doing applying a 'systemic lens' to capture what big data can bring to the table.

The inherent interactive dynamic complexity associated with wicked problems illuminates the shortcomings of a siloed, linear, reductionist approach. A more inclusive and holistic understanding of the problem necessitates multiple perspectives from partners and stakeholders. This requires a type of thinking capable of grasping the big picture, including the interrelationships between the full range of causal factors and strategic objectives. In terms of intervention strategies, you cannot just do one thing; every action or intervention causes changes to the overall system that must be recognized and managed.

If we look at the counter-terrorism problem space as a wicked problem, we see that the inherent issues are imperfectly understood resulting in problem space boundaries being drawn too narrowly. How can we better conceptualize and design counter-terrorism intervention strategies?

2 Characterizing the Complex Problem Space

The concept of wicked problems stems from the work of Rittel and Weber [46] and their paper "Dilemmas in a General Theory of Planning". The wicked problems we face in society (healthcare, business, security, humanitarian crisis) are so challenging because they are both unsolved and unstructured [35]. As cited in Lee [25: 22], Rittel outlined the ten characteristics of wicked problems in 1972, which sets out the basis for collective participation and argumentation:

1. *Wicked problems* have no definitive formulation, but every formulation of a *wicked problem* corresponds to the formulation of a solution.
2. *Wicked problems* have no stopping rules.
3. Solutions to *wicked problems* cannot be true or false, only good or bad.
4. In solving *wicked problems* there is no exhaustive list of admissible operations.
5. For every *wicked problem* there is always more than one possible explanation, with explanations depending on the *Weltanschauung* of the designer.
6. Every *wicked problem* is a symptom of another, "higher level", problem.
7. No formulation and solution of a *wicked problem* has a definitive test.

8. Solving a *wicked problem* is a "one shot" operation, with no room for trial and error.
9. Every *wicked problem* is unique.
10. The *wicked problem* solver has no right to be wrong—they are fully responsible for their actions.

Framing the problem space in terms of 'wicked problems' [35], acknowledges the inherent interconnectedness and complexity and thereby calls upon novel approaches that challenge traditional linear thinking mindsets. What emerges from the recognition of wicked problems is the necessity to actively engage with it by building up an understanding and knowledge of interdependent dynamic systems, feedback loops, emergent features and surprises assessing the impacts and employing reflective practices [38].

The financial crisis of 2008, the spread of infectious diseases, the disruptions to critical infrastructure, the global reach of terrorism highlight the systemic risks that characterize the complex threat and risk landscape [6, 33, 34, 36]. Senge [49] argues that since the world exhibits qualities of wholeness, the relevance of systemic thinking is captured within its paradigm of interdependency, complexity and wholeness. Although events can be considered to be discrete occurrences in time and space '… they are all interconnected. Events can be understood only by contemplating the whole' [14].

Ackoff [1: 21] cited in Head and Alford [16: 3] argues that:

> Every problem interacts with other problems and is therefore part of a system of interrelated problems, a *system of problems* … I choose to call such a system a *mess* … The solution to a mess can seldom be obtained by independently solving each of the problems of which it is composed … Efforts to deal separately with such aspects of urban life as transportation, health, crime, and education seem to aggravate the total situation. (Ackoff [1, p. 21], italics in original)

In framing the problem space, systems thinking emerges as a key approach. Jackson [18: 65] defines systems thinking as "… a discipline for seeing the structures that underlie complex situations … it helps us to see the deeper patterns lying beneath the events and the details". Tackling such complexity requires innovative approaches such as design thinking.

Designers are able to synthesize novel and tangible solutions out of the messiness of daily reality. Design thinking affords a methodology and worldview that supports forging connections between seemingly unrelated issues. As such, design is increasingly seen as a way of thinking.

What makes designers key players in problem framing and solution navigation is their ability to engage in reflective practices and a process of continuous learning. It is a constant, tightly meshed cycle of observation, ideation, prototyping and testing. In that process, designers not only create things but they also create new knowledge. These very qualities of design thinking make it a powerful approach supporting counter-terrorism strategies.

3 Metaphorical Thinking—The Epidemiological Model and Counter-Terrorism

The epidemiological model of terrorism (disease model) is a metaphor that captures the inherent complexity of terrorism [53]. References to terrorism being a "virus" or to al Qaeda "mutating" or "metastasizing" are common. As described in Pourbohloul and Kieny [40: 242] there exists a siloed approach to studying diseases and health resulting in policy formulation within these "disease silo". Boundary setting in addressing health matters often negates the recognition of multiple interacting factors. In this sense Pourbohloul and Kieny [40: 242] argue that '… Health systems defy simple representation. They call for novel ways of thinking to improve our ability to predict and control individual and population-based health outcomes. A holistic framework is needed to capture disparate diseases and health conditions and their intricate relationships into a unified platform'. Holistic approaches to medicine and disease have leveraged design thinking approaches to tackle the volatility, uncertainty, complexity and ambiguity.

Viewing the medical model through the lens of complexity, Urry [62] describes how complexity recognizes the emergent properties that result from the dynamic interaction within a system, thereby developing collective properties that are not reflected in the individual components. As such, complexity argues against reductionism.

As noted in Styhre [56], the complexity perspective recognizes that changes result from a multiplicity of interconnected causes and effects. Within the context of understanding accident and disaster aetiology, and in particular counter-terrorism, the systems approach as a guiding methodology informed by complexity theory facilitates a break from '…mechanistic, linear, and causal methods of analysis towards viewing interdependence and interrelation rather than linearity and exclusion' [11: 140]. Dekker [9: 103] argues, 'Were we to really trace the cause of failure, the causal network would fan out immediately, like cracks in a window, with only our own judgment to help us determine when and where to stop looking, because the evidence would not do it for us'. This causal network suggests that future policies and regulations applied to the health domain or for that matter to counter-terrorism strategies may result in unanticipated consequences. This implies that a systems view of the problem is necessitated.

The description of the health domain and the application of epidemiological methods influence how we can support counter-terrorism strategies. Contextualized from Stares and Yacoubian [53] for the counter-terrorism problem space:

- '***Epidemiologists***' (for counter-terrorism) observe rigorous standards of inquiry and analysis to understand the derivation, dynamics, and propagation of radicalization and terrorist events. In particular, they seek clarity on the origins and geographical and social contours of radicalization: where the violent extremism is concentrated, how it is transmitted, who is most at risk or "susceptible" to radicalization, and why some portions of society may be less susceptible, or, for all intents and purposes, immune.

- '***Epidemiologists***' recognize that radicalization, violent extremism and terrorism neither arise nor spread in a vacuum. They emerge and evolve as a result of a complex dynamic interactive process between people, pathogens, and the environment in which they live. Indeed, the epidemiologic concept of "cause" is rarely if ever singular or linear but is more akin to a "web" of direct and indirect factors that play a lesser or greater role in differing circumstances. To make sense of this complexity, epidemiologists typically employ a standard analytical device that "deconstructs" the key constituent elements of a disease.
- Just as '***epidemiologists***' view radicalization and violent extremism as a complex, multifaceted phenomenon, so counter-terrorism leaders have come to recognize that success in controlling and rolling back an epidemic typically results from a carefully orchestrated, systematic, prioritized, multipronged effort to address each of its constituent elements.

The epidemiological model as a metaphor for counter-terrorism strategy development, lends itself to the application of design thinking to support it.

4 Design Thinking

> Design is an activity, which aims at the production of a plan, which plan—if implemented—is intended to bring about a situation with specific desired characteristics without creating unforeseen and undesired side and after effects. Hors Rittel, 1968 (cited in Lee [25: 19])

Design Thinking is a solution oriented methodology used by designers to solve complex problems. It draws upon logic, imagination, intuition, and systemic reasoning, to explore possibilities of what could be, and to create desired outcomes. A design mindset is not problem-focused, it's solution focused, and action oriented. It involves both analysis and imagination. Design represents a process that embraces innovation, creativity, opportunity analysis and problem framing and solving.

Through the phases of Inspiration, Ideation and Implementation, Design Thinking is operationalized through an iterative (not linear) 5 step process (Fig. 1). This process is described in detail (http://dschool.stanford.edu/redesigningtheater/the-design-thinking-process/).

EMPATHIZE	Work to fully understand the experience of radicalization and terrorism. Do this through observation, interaction, and immersing yourself in their experiences.
DEFINE	Process and synthesize the findings from your empathy work in order to form a point of view that you will address with your design.
IDEATE	Explore a wide variety of possible solutions through generating a large quantity of diverse possible solutions, allowing you to step beyond the obvious and explore a range of ideas.

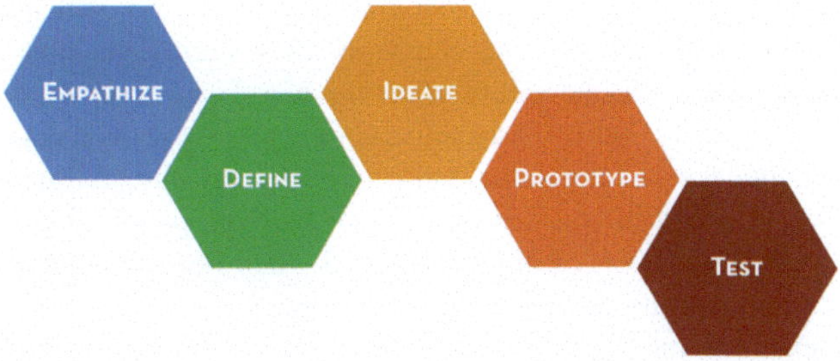

Fig. 1 Design thinking process

PROTOTYPE Transform your ideas into an operational form so that you can experience and interact with them and, in the process, learn and develop more empathy.

TEST Try out high-resolution concepts and use observations and feedback to refine prototypes, learn more about the terrorist problem space, and refine your original point of view.

The application of design thinking in the context of counter-terrorism strategies is apropos. As described in Pourdehnad et al. [41] '…designers have specific abilities to produce novel unexpected solutions, tolerate uncertainty, work with incomplete information, apply imagination and forethought to practical problems and use drawings and other modeling media as means to problem solving. He further argues that designers must be able to resolve ill-defined problems, adopt solution-focusing strategies, employ abductive/productive/appositional thinking and use non-verbal, graphic and spatial modeling media'. These qualities resonate with the counter-terrorism problem space.

5 Discussion

A recent article Berthiaume [3] quotes Canadian Defence Minister:

> 'Sajjan said a lack of understanding, or situational awareness, had undermined allied efforts in Afghanistan, Libya and Iraq. On Afghanistan, in particular, he said early efforts by some countries had helped create corruption, which fuelled the insurgency. A failure to clamp down on the corruption made matters worse. "People have left, patted themselves on the back, didn't even realize the great work that they thought they did had actually created a negative ripple," he said…. "I want to make sure we get this right," he said. "That's why we're making sure we take a bit of time to get this right and create this ripple that's going to lead to some positive outcomes".'

Through this, Defence Minister Sajjan captures the complexity associated with counter-terrorism intervention strategies and the requirement for a 'reflective' approach. Understanding this interconnectedness and complexity is the essence of systems thinking [35] that views the system as a whole rather than its individual component parts, taking into account behaviour of systems over time rather than static 'snapshots' [49], with 'the ability to see the world as a complex system' [54]. As articulated in Atun [2: iv5], '...dynamic complexity arises when the short and long term consequences of the same action are dramatically different, when the consequence of an action in one part of the system is completely different from its consequences on another part of the system, and when obviously well-intentioned actions lead to nonobvious counter-intuitive results'.

When we examine intelligence analysis to support counter-terrorism, it is recognized that reducing the amount of information used, applying linear cause-effect mental models and limiting strategies to creating static options is insufficient. Failing to provide an accurate representation of the real world by ignoring possible wider impacts of policies and decisions can result to unintended consequences. With reference to the epidemiological model, this resonates with the healthcare and counter-terrorism domains.

Systems thinking as part of a design thinking approach can help address the linear and reductionist approaches which prevail in health systems and similarly counter-terrorism, by testing new ideas.

5.1 Counter-Terrorism Design Thinking (Perspective, Reflection, Dialogue)

In shifting from "mechanistic worldview" to a "systemic worldview", we need to change our method of inquiry (and thought process). Systems thinking employs design as a way to better understand the inherent interdependencies and dynamic nonlinear behaviour of system (problem) such as in healthcare or terrorism. The systems approach is a method of inquiry and a way of thinking that emphasizes the whole system instead of component systems and strives to optimize the whole system's effectiveness [41].

To support the systems worldview, three key elements (Perspective, Reflection, Dialogue) comprise a design thinking approach to counter-terrorism that resonates with the 3 epidemiological lessons learned described earlier. It is through these key elements that insights regarding intervention strategies can be formulated with a more holistic perspective.

5.2 Perspective

How we view the world shapes what we see and affects how we manage risk, crisis and disasters. Ramirez and Selin [43: 55] argue that: 'The past is a reasonable guide when growth is stable and past and present trends can be extrapolated reliably to account for emergence. However when change is novel, unexpected, unforeseen, unique, and/or radical, perhaps involving also manifold unintended consequences; the future and the uncertainty it holds means organizations can no longer rely on past or unfolding trends'. Challenging the linear mindset and dominance of perspective that reside in insular organizational cultures, Levine et al. [26: 7] argue that '…a system perspective can often reveal how behaviour that is competent from the standpoint of each individual actor does not contribute to achieving the overall goals which collectively all the actors in the 'system' say they are working towards, in different ways. System problems often result when different actors do not really share the objectives, or when they do not agree on which elements contribute to a single system'.

In order to appreciate the complexity and heterogeneity of problem situations, it is essential for the systems practitioner, in pursuing creativity, to bring to bear radically different views derived from alternative paradigms [19: 136]. Platts and Tan [39: 668] in their work on strategy visualization describe how '…the importance of visual representation to support decision making has been emphasized by many researchers'. Strang and Masys [55] argue that 'Addressing the unique challenges associated with transnational threats as terrorism and organized crime requires creative and collaborative efforts among key intelligence and security stakeholders that facilitate questioning judgments and underlying assumptions, and employing critical and creative thinking in order to explore the possibility space'. Further, Strang and Masys [55] present an argument in support of visual thinking as a design thinking method. They argue that 'Visual thinking in this sense is not necessarily about thinking in images rather than words, but describes "the cognitive processes involved in orchestrating visual information in the mind" [52: 158] and the process of analysis through visually arranging and rearranging information to explore and challenge relationships including cause and effect. This is essentially the same as the concept of visual analytics, defined as "analytical reasoning facilitated by interactive visual interfaces" [27: 2]. The purpose of visual thinking is "to make the complex understandable by making it visible—not by making it simple" [47: 105].

Visual thinking resonates with the notion of perspective which has been an important innovative mindset in art. The art world was dominated by a single perspective until the development of Cubism during the period 1907–1911. Picasso and Braque believed a new visual language was required that would have influence and challenge the status quo. This 'revolution' in the art world challenged notions of perspective that had emerged from the time of the Renaissance. Through their efforts, they introduced a new way of seeing that was both radical and insightful. Leveraging the work of Cezanne, the notion of a single viewpoint picture was overturned thereby reaching beyond the rigid geometry of perspective. Changing ones perspective when viewing an object (like walking around a cube) allows the observer to view the object from above, below and from the sides. In so doing this relativistic perspective proposed by the cubist revolution constructs and presents the object holistically (spatially and temporally) (Fig. 2).

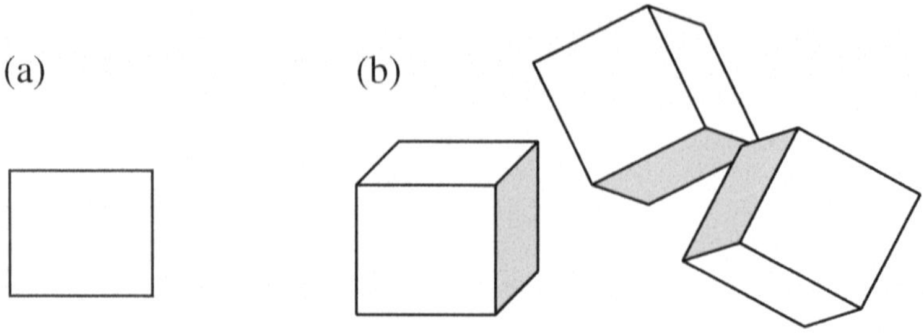

Fig. 2 **a** Single perspective. **b** Multiple perspectives

Shlain [51: 199] describes how 'Cubism breaks with Renaissance perspective. It views objects relatively: that is, from several points of view, no one of which has exclusive authority'. An excellent example of the original cubism is Picasso's work dated 1912 'Violin and Grapes'. Picasso presents the object explored from several different viewpoints thereby creating an image that was multidimensional. The role of multiple perspectives in establishing meaning in complex problems is considered to be of extreme importance. The failure of imagination is a stark reminder of the findings of The 9/11 Commission Report [60: 336] that articulated '…a failure of imagination and a mindset that dismissed possibilities' as a key underlying factor. The Dragonfly eye as a metaphor (like '***Picasso thinking***') resonates with the notion of perspective. Multiple perspectives is also described in Masys [36] with regards to crime scripting in support of counter-terrorism and in Strang and Masys [55] regarding intelligence analysis and visualization.

Reflection is a key element in the learning process. Hilden and Tikkamaki [17: 8] argue that reflection '…can be seen as an active and purposeful process of exploration and inquiry, where one becomes aware of the assumptions that govern one's actions, thinking and feeling' and is the '…bridge between experience and learning'. In so doing, reflection facilitates examination of new insights and challenges traditional mindsets and supports experimental actions [17: 80]. As such, intelligence agencies can be thought of as learning organizations '…where the process of intuiting, interpreting, integrating and institutionalizing are taking place in the processes of participating, constructing and sharing knowing, socially supporting and reflecting' [17: 79].

Reflective practices have been described in an analysis of Fukushima Daiichi nuclear accident [38]. As described in Masys et al. [38] '…reflective response is a 'dynamic developmental process' [5: 5] that occurs at the interface of individual, collective and critical levels that can be applied to support complexity management. It encompasses interrogating the assumptions integrated into the analysis along with reflecting on the 'norms' and 'appreciations' which underpin judgments and actions [45]'.

Weick and Sutcliffe [63: 2] highlight how black swan events can be '…considered as an abrupt and brutal audit: at a moment's notice, everything that was left unprepared becomes a complex problem, and every weakness comes rushing to the forefront'. The example of such a black swan as the Fukushima Daiichi Nuclear Accident of 2011 highlights how assumptions shaped the mindset that negated foresight and anticipation of such an event. DIET Report [12: 44] revealed:

> 'Underlying NISA's views was the conviction that, with regard to nuclear emergency preparedness, it was not necessary to anticipate an accident that would release enough radioactive material as to actually require protective actions, since (they believed) rigorous nuclear safety regulations, including safety inspections and operation management, were in place in Japan'.

Contributing to the organizational impediments in the Fukushima case was the lack of organizational learning. As described in the DIET [12: i3] 'Japan has itself dealt with a number of nuclear power plant accidents, small and large. Most of these were responded to, but without sufficient transparency; sometimes they were concealed by the organizations concerned… while maintaining that accidents could not occur in Japan'. What was foreseeable was not actioned leaving Japan unprepared for this accident. This reluctance to embrace organizational learning is reflected in the '… lack of commitment for reviewing laws and regulations, in diligent reflection of accident responses and safeguarding measures in other countries. Without the diligent reflection the industry focused on the promotion of nuclear energy but not public safety and health' DIET [12: 17]. As described in Masys et al. [38] unreflective practices perpetuated this mindset. The unintended consequence was the erosion of preparedness and a dysfunctional mindset that dismissed possibilities.

Reflective practice allows the members of the organization to slow down to critically evaluate their own thinking, but also to investigate the shared, collective assumptions and expectations, as well as institutionalized rules and routines [17: 82].

Reflection thereby emerges as a key attribute of the design thinking approach. In support of the comments of Minister Sajjan, it is about challenging assumptions and worldviews and enabling an organizational learning posture to counter-terrorism strategy development.

Dialogue is a key enabler supporting collaborative learning and reflection. It has been described as thinking and seeing together. As a process, dialogue aims to reveal the construction of meaning thus creating shared understanding. In so doing it aims at the questioning of assumptions, development of common language and a shared world view.

By creating an environment that supports diverse perspectives, it enables reframing of the problem space. In this sense through perspective, reflection and dialogue we move beyond the 'right answer' to a question, to reconsidering assumptions, goals and values to better frame the question through a process of continuous learning. As described in Ramirez and Wilkinson [44: 13] '…reframing-reperception cycles: enable the generation of new and shared knowledge; support tolerance for divergent thinking; and act to contain two pathologies that inhibit learning for people working in groups, namely fragmentation and groupthink'. Further, Ramirez and Wilkinson [44: 10] argue that 'Reframing is enabled by articulating plausible often emerging stories, and supports the learners to reperceive their world and to bring forth new options for action. … reframing occurs through strategic conversations that explore new territory, and that accommodate disagreement and render it a productive asset'.

Dialogue opens up a space to deal with complexity and is a key element of design and systems thinking. The application of Soft Systems Methodology and Rich Pictures [35, 37, 55] provides a methodology that combines dialogue, visual thinking and ideation.

Bringing Perspective, Reflection and Dialogue as part of the design thinking approach requires a mindset that embraces collaboration. Head and Alford [16: 16] argue that '…the presence of functioning cooperative networks increases the likelihood that the nature of *the problem* and its underlying causes can be better understood …This manifests itself as a shared understanding of the problems and overarching purposes'. In the collaborative space it is about involving a wider array of actors, offering more diverse insights into why a situation has arisen. Design thinking, as a social process, creates dynamic, innovative safe space from which reframing and reperception generate new knowledge and shared understanding. Further to the role of collaborative networks in the design thinking process, Head and Alford [16: 17–18] argue that '… collaborative networks can tap into a wider body of specific knowledge and skills than can unilateral decision makers. Communication among the parties increases the likelihood of them engaging in problem-solving behavior and finding ways forward'. The collaboration space suggested above in deriving counter terrorism strategies '…entails a degree of *trust and mutual commitment* among the parties'.

Perspective, Reflection and Dialogue as a framework to support design thinking emphasizes that system boundaries, dynamics, functions and outcomes are open to multiple framings that are contextual, positioned and embedded with assumptions, methods, forms of interpretation, values and goals.

As such from this framing, narratives, in particular those associated with terrorism and counter-terrorism strategies emerge that represent the storyline from a particular perspective. These narratives represent particular worldviews and power relations. Through Perspective, Reflection and Dialogue, we uncover the various narratives that are often omitted, deleted or black boxed.

6 Facilitating Design Thinking

What happens when creativity is lacking? We can draw upon the final reports from the Fukushima nuclear disaster and 9/11. In terms of the accident, lack of imagination and urgency translate as a lack of foresight informed action. Fukushima nuclear accident can thereby be considered as '…not a natural disaster but clearly man-made' DIET [12: 12]. The failure of imagination is a stark reminder of the findings of The 9/11 Commission Report [60: 336] that articulated '…a failure of imagination and a mindset that dismissed possibilities' as a key underlying factor.

The mindset that reified as a lack of imagination and inaction is rooted in the mental models regarded as incorporating ones biases, values, learning, experiences and beliefs about how the world works.

The application of systems thinking and design thinking will help to broaden the search for approaches that may enable the researcher to understand the nature of the wicked problem more clearly and/or to address it more comprehensively. As described in Tatham and Houghton [57: 27], through the use of a variety of systems analysis tools, it is suggested that one is better placed to understand the "tensions and contradictions" that 'are most likely to be the real sources of 'wickedness' of the problem [16, p. 16]'.

Considering the key elements of perspective, reflection and dialogue, the following approaches lend themselves to dealing with wicked problems:

- Visual thinking
  - Cause and effect diagrams [55]
  - Mindmaps [55]
  - Giga maps [50]
  - Layered scenario mapping: a multidimensional mapping technique for collaborative design [30]
  - Counter-terrorism architectural frameworks [36]
- Scenario Analysis [31]
- Red Teaming and AFD [31, 58]

Most design thinking models start with building a clear understanding of the problems. Visual thinking has been described in detail in Strang and Masys [55]. They argue that '…Within the context of intelligence analysis, visualization of knowledge…aim to create, assess, reference or transfer insights, experiences, attitudes, values, expectations, perspectives, opinions and predictions'. Sketches, diagrams, graphics, visualizations, external representations play significant roles in thought and communication supporting intelligence analysis. Exploring mental models and revealing assumptions, visual thinking helps to externalize internal thought, and serves as a methodology for inference, reasoning and insight'.

6.1 Scenario Analysis

As described in Masys [31: 321] 'Scenario planning represents a set of tools and methodology for foresight facilitating discourse with the purpose not to predict, but to facilitate a change in the mindset of the people who use them: in short to think the unthinkable [13, p. S13]'. For counter-terrorism strategy development, the use of scenarios stretch and refocus thinking and mental models thereby challenging the assumptions concerning linearity to recognize the fundamental interdependencies and complexity associated with the emergence of black swans. The OPSA Strategic

Reframing [44] leverages causally linked variables (system mapping) and stories (narratives) to create plausible scenarios. In this way, scenarios enable the strategic conversation. Through modeling in the scenario process, the design thinkers:

- Open up causal logics
- Build new and shared knowledge
- Facilitate decision support and option analysis
- Rehearse the future and visualize scenario implications
- Support engagement and communication [44: 77].

Scenario analysis examining the possibility and plausibility space, Ramirez and Selin [43] provides a context to support ideation and prototyping. It creates the 'stories' that rooted in the big data support design strategies making them 'real'. This immersion in big data is likened to 'empathy' in the design thinking paradigm. It is about feeling the data and context that establishes a vivid 'narrative' to describe the terrorists 'business model' [36]. This empathy helps frame both the problem and solution space in the iterative process of design. Conceptualizing terrorist organizations as complex adaptive networks rather than simple systems recognizes the inherent agility and complexity of their operations. In so doing this view argues against the static monolithic view of a terrorist organization thereby highlighting that single point vulnerabilities do not exist, nor do silver bullet solutions. This is presented in Masys [32] description of radicalization and recruitment from a CAS model perspective.

Scenario planning certainly is a key element in design thinking whereby challenging mental models can be seen as opening up perceptions, transitioning toward a learning organization. In other words, strategic interventions are based on a more full understanding of the implications, both internal and external to the organizations.

6.2 Red Teaming

As described in Masys [31: 323] the process of red teaming has long been used as a methodology to offer alternative interpretations and challenge established thinking in order to identify and reduce risks. Red teaming has a history of application within both industry and the military domain [8]. Red teaming is used as a way to challenge planning assumptions, established mental models, feasibility, vulnerability and risk and thereby helping to avoid the common fault of group think [20]. As such, red teaming can facilitate insights into otherwise "hidden mechanisms" that precipitate catastrophic surprises.

Greenemeier [15] describes how the Defense Advanced Research Projects Agency (DARPA) is developing an 'improv program' with the design thinking question 'To stop a terrorist, it helps to think like one'. DARPA wants to assemble the world's biggest "red team"—a group of outsiders that can help the Department of Defense get ahead of terrorists looking to attack military personnel, equipment or operations. This red teaming approach '… encourages engineers, entrepreneurs and tech enthusiasts to imagine how someone might repurpose commercially available devices as weapons'. The design elements of empathy, perspective, reflection and dialogue figure prominently in this approach.

Anticipatory Failure Determination (AFD) can be a key approach to enabling creative thinking in red teaming. Anticipatory Failure Determination (AFD) is an application of TRIZ (Theory of Inventive Problem Solving). It is an efficient and effective method for analyzing, predicting and eliminating failures in systems, products, and processes [22]. The AFD modeling process guides users in documenting the situation, formulating the related problem(s), developing hypotheses, verifying potential failure scenarios, and finding solutions to eliminate the problem(s).

AFD has two broad applications:

1. Failure Analysis: determination of the cause of a failure that has already occurred.
2. Failure Prediction: determination of possible failures that have not yet occurred.

Traditional failure analysis focuses on the question 'How did this failure happen?' In terms of failure determination, AFD poses the question 'If I wanted to create this particular failure, how could I do it?' In terms of failure prediction it poses the question 'If I wanted to make something go wrong, how I could do it in the most effective way?' Failure prediction thereby reflects an iterative application of failure determination to envision all the possible end states, mid states, initiating events and possible scenarios leading to these states. This methodology thereby views the failure as an intended consequence and thereby reflect a design thinking approach to problem framing. In support of counter terrorism, it opens up the black box of plausible scenarios to develop prevention/mitigation, preparedness, response and recovery strategies.

6.3 Enabling Design Thinking: Ideation

Deep Dive methodology, as practiced by IDEO, incorporates multiple perspectives, observation, interaction and iteration in the design process. The recent events in Paris, Turkey, Ivory Coast and Brussels capture salient indicators that can be leveraged in a design thinking approach. What is required in such wicked problems is greater understanding of the dependencies and interconnectivity that arises from a systems view of the problem space. Masys [32] highlights actor network theory (ANT) as a systems lens to begin to reveal and unravel the complexities associated with terrorism and radicalization to support the design of intervention strategies. As described in Masys [32], Bjelopera and Randol [4: 11] argue that as the terrorist threat becomes increasingly homegrown, a key way to fight it is to develop an understanding of how radicalization works and formulate ways to prevent the radicalization from morphing into violent extremism. Global systemic factors, state structural, and socio-cultural factors [7: xxiii] provide a macro-perspective on the problem space of terrorism. By opening these 'blackbox' (macro-perspectives) and 'following the actors' [24, 32, 36], ANT traces a complex, interconnected network that is best characterized as a rhizome [10].

Analysis conducted by Davis and Cragin [7], Jenkins [21] and TNO [61] highlight key causal factors with regards to terrorism. When we deploy an ANT approach to this body of work leveraging insights and data from recent terrorist events described, we recognize the extent to which the actors emerge from their associative (relational) interdependencies that characterize the process of radicalization. In particular, events in the past become embedded in the relational actor network and thereby become an actor in the process of translation. This relational 'mapping' is described in Masys [32] in terms of the translation process (problematization; interessement; enrolment; and mobilization) facilitating an examination of how 'actors enlist other actors into their world and how they bestow qualities, desires, visions and motivations on these actors [23]', as cited in Powell and Owen [42: 141]. It highlights key points of the network to support disruption of radicalization through the design thinking mindset.

Perspective, reflection and dialogue figure prominently in understanding the radicalization and counter violent extremism space.

To further examine the design principles associated with counter-terrorism, we can draw upon the application of the CTAF approach [36] that leverages interdisciplinary tools and approaches to support opening the black box of terrorist organizations. CTAF applies DODAF, crime scripting and Social Network Analysis.

CTAF [36], illuminates various dimensions of the problem space (Fig. 3).

By opening the black box of terrorism through the different views of DODAF, we gain 'empathetic' insights underlying the terrorist organization as described in Masys [36].

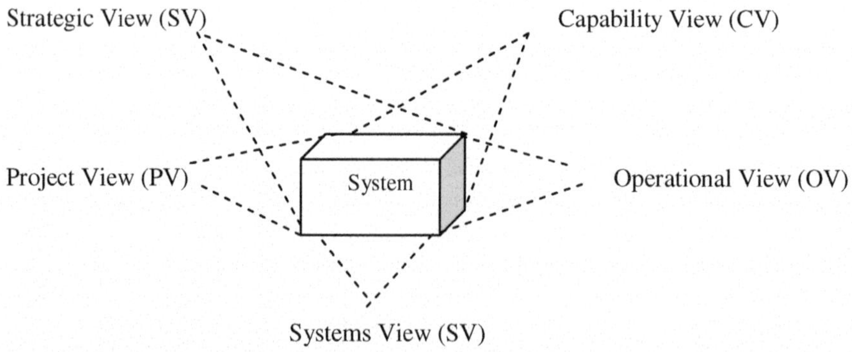

Fig. 3 DODAF views [36]

The recent attacks in Paris and Belgium highlight the distributed network of ISIS and affiliates within the EU. From an epidemiological perspective stemming from the disease model, the design thinking approach:

- Seeks to understand the underlying derivation, dynamics and propagation of radicalization and violent extremism.
- Recognizes that radicalization and violent extremism do not arise in a vacuum. Contextual 'pathogens' exist.
- Disruption of the complex, multifaceted phenomenon of radicalization and violent extremism requires a multipronged effort.

Bringing systems thinking and design thinking together as described through such approaches as CTAF and actor network theory can shed light on the shadows that characterize the dark networks that reside globally and that have emerged in the EU. What this does is help frame and contextualize our understanding of the terrorist organization through processes of design thinking. In effect we are examining ways to disrupt the business model of the terrorist organization. Perspective, reflection and dialogue are the foundation of design thinking to support counter-terrorism and reify through the application of ANT and CTAF.

7 Conclusion

Returning to the metaphor of the medical model, as noted in Lipsitz [28: 243] '… prevailing trends to use disease protocols, financial levers, and siloed programs to manage the health care system are fatally flawed and will lead to unintended consequences …. If health care is viewed as a complex rather than a mechanical system, several of its intrinsic properties can be exploited to influence its dynamic behavior and guide it in a more favorable direction'. This resonates with the counter-terrorism design strategy.

Problem structuring through perspective, reflection and dialogue is a critical aspect of the design process that takes into account the diversity of goals, assumptions and meanings. Tackling the wicked problem of terrorism requires reframing. As noted in Lockwood [29: xiii] '…by thinking like designers—being able to see the details as well as zoom out of the big picture—we can really add value by challenging the status quo'. We need new, transformative counter-terrorism strategies rooted in innovation. Being able to think across the counter-terrorism domain, to understand the interdependencies and interrelated systems, and being able to identify where to intervene to achieve a impact, is an important skill for a counter-terrorism intervention design.

Design thinking represents a radical shift in how to approach counter-terrorism strategy development. Design thinking addresses the fundamental assumptions, values, norms, and beliefs that reside within our organizational posture dealing with terrorism. As such, described by Lockwood [29: 86] 'the design thinking process involves a team approach, and the goal is to unlock the creative potential of the organization and its partners. The power of interdisciplinary teams is undeniable, and the ability of design teams to see both the big picture as well as the details is important and rather unique'.

This chapter highlights how 'design thinking' and systems thinking can facilitate scenario planning as an approach to reveal the business models of terrorist organizations and support 'design' of counter-terrorism strategies.

References

1. Ackoff RL (1974) Redesigning the future. Wiley, New York, NY
2. Atun R (2012) Health systems, systems thinking and innovation. Health Policy Plann 27: iv4–iv8
3. Berthiaume L (2016) Sajjan wary of 'ripple effects' from anti-terrorism policies. Ottawa Citizen. Published on: 29 January 2016, Last Updated: 29 January 2016, 8:16 PM EST. http://ottawacitizen.com/news/national/sajjan-wary-of-ripple-effects-from-anti-terrorism-policies#
4. Bjelopera JP, Randol MA (2011) American Jihadist terrorism: combating a complex threat. In: Perkins SJ (ed) Homegrown terror and American Jihadists: addressing the threat. Nova Science Publishers Inc., New York, pp 1–118
5. Bolton G (2005) Reflective practice: writing and professional development. Sage Publications, London, UK
6. Chernov D, Sornette D (2016) Dynamics of information flow before major crises: lessons from the collapse of Enron, the subprime mortgage crisis and other high impact disasters in the industrial sector. In: Masys AJ (ed) Disaster forensics: understanding root cause and complex causality. Springer Publishing
7. Davis PK, Cragin K (2009) Social science for counter-terrorism: putting the pieces together. RAND MG849
8. Defence Science Board Task Force (2003) The role and status of DoD red teaming activities. September 2003
9. Dekker SWA (2003) Illusions of explanation: a critical essay on error classification. Int J Aviat Psychol 13(2):95–106
10. Deleuze G, Guattari F (1987) A thousand plateaus: capitalism & schizophrenia. University of Minnesota Press, Minneapolis
11. Dennis K (2007) Time in the age of complexity. Time Soc 2007(16):139–155
12. DIET Report Executive Summary (2013) https://www.nirs.org/fukushima/naiic_report.pdf
13. Farber D, Lakhtakia A (2009) Scenario planning and nanotechnological futures. Eur J Phys 30:S3–S15
14. Flood RL (1999) Retinking the fifth discipline: learning within the unknowable. Routledge, London
15. Greenemeier L (2016) Pentagon paying techies to think like terrorists. Scientific American, on 11 March 2016. http://www.scientificamerican.com/article/pentagon-paying-techies-to-think-like-terrorists/?WT.mc_id=SA_SP_20160314
16. Head BW, Alford J (2013) Wicked problems: implications for public policy and management. Adm Soc 20(10):1–29
17. Hilden S, Tikkamaki K (2013) Reflective practice as a fuel for organizational learning. Adm Sci 3:76–95
18. Jackson MC (2003) Systems thinking: creative holism for managers. John Wiley & Sons Ltd, West Sussex, England

19. Jackson MC (2010) Reflections on the development and contribution of critical systems thinking and practice. Syst Res Behav Sci 27:133–139
20. Janis I (1972) Victims of groupthink. Houghton-Mifflin, Boston
21. Jenkins BM (2011) Stray dogs and virtual armies radicalization and recruitment to Jihadist terrorism in the United States since 9/11. RAND Occasional Paper 343
22. Kaplan S, Visnepolschi S, Zlotin B, Zusman A (1999) New tools for failure and risk analysis: Anticipatory Failure Determination (AFD) and the theory of scenario structuring. Ideation International Inc., Southfield, MI
23. Latour B (1996) On actor-network theory. A few clarifications. Soz Welt 47:369–381
24. Latour B (2005) Reassembling the social: an introduction to actor network theory. Oxford University Press, Oxford
25. Lee AJ (2016) Resilience by design. Springer Publishing
26. Levine S, Crosskey A, Abdinoor M (2011) System failure? Revisiting the problems of timely response to crises in the Horn of Africa. Humanitarian practice network, Overseas Development Institute
27. Liao Q, Shi L, Wang C (2013) Visual analysis of large-scale network anomalies. IBM J Res Dev 57(3):13:1–13:12 Paper 13. International Business Machines Incorporated
28. Lipsitz LA (2012) Understanding health care as a complex system: the foundation for unintended consequences. JAMA 308(3):243–244
29. Lockwood T (ed) (2009) Design thinking: integrating innovation, customer experience and brand value. Allworth Press, New York
30. Luras S (2015) Layered scenario mapping: a multidimensional mapping technique for collaborative design, CoDesign
31. Masys AJ (2012) Black swans to grey swans—revealing the uncertainty. Int J Dis Prev Manag 21(3):320–335
32. Masys AJ (2013) Human security—a view through the lens of complexity. In: Gilbert T, Kirkilionis M, Nicolis G (eds) Proceedings of the European conference on complex systems 2012, Springer proceedings in complexity, pp 325–335
33. Masys AJ (2014) Radicalization and recruitment: a systems approach to understanding violent extremism. Int J Syst Soc 1(2):51–65
34. Masys AJ (ed) (2014) Disaster management: enabling resilience. Springer Publishing
35. Masys AJ (ed) (2015) Applications of systems thinking and soft operations research in managing complexity. Springer Publishing
36. Masys AJ (ed) (2016) Exploring the security landscape—non-traditional security challenges. Springer Publishing
37. Masys AJ, Vallerand A (2014) Major event security planning: secure by design—through the strategic use of integrated modeling and simulation. M&S J
38. Masys AJ, Ray-Bennett N, Shiroshita H, Jackson P (2014) High impact/low frequency extreme events: enabling reflection and resilience in a hyper-connected world. In: 4th international conference on building resilience, 8–11 September 2014, Salford Quays, United Kingdom. Proc Econ Finan 18:772–779
39. Platts K, Tan KH (2004) Strategy visualization: knowing, understanding and formulating. Manag Decis 42(5):667–676
40. Pourbohloul B, Kieny M-P (2011) Complex systems analysis: towards holistic approaches to health systems planning and policy. Bull World Health Organ 89:242
41. Pourdehnad J, Wexler ER, Wilson DV (2011) Systems & design thinking: a conceptual framework for their integration. Working Paper #11-03, University of Pennsylvania. http://repository.upenn.edu/cgi/viewcontent.cgi?article=1009&context=od_working_papers
42. Powell JL, Owen T (2011) Actor network theory and social science: possibilities and implications. J Public Adm Gov 1(2):140–157. www.macrothink.org/jpag
43. Ramirez R, Selin C (2014) Plausibility and probability in scenario planning. Foresight 16 (1):54–74
44. Ramirez R, Wilkinson A (2016) Strategic reframing: the oxford scenario planning approach. Oxford University Press, Oxford

45. Reynolds M, Vince R (2004) Organizing reflection: an introduction. In: Reynolds M, Vince R (eds) Organizing reflection. Ashgate Publishing Ltd, Hampshire, UK
46. Rittel HWJ, Weber MM (1973) Dilemmas in general theory of planning. Policy Sci 4:155–169
47. Roam D (2008) The back of the napkin: solving problems and selling ideas with pictures. Penguin Group, New York
48. Rosenhead J, Mingers J (2001) A new paradigm of analysis. In: Rosenhead J, Mingers J (eds) Rational analysis for a problematic world revisited. John Wiley and Sons Ltd., West Sussex, England
49. Senge P (1990) The fifth discipline: the art and practice of the learning organization. Doubleday Currency, New York
50. Sevaldson B (2011) Giga-mapping: visualisation for complexity and systems thinking in design. In: Nordic design research conference, 2011, Helsinki. www.nordes.org
51. Shlain L (1991) Art and physics: parallel vision in space, time and light. Harper Perennial, New York
52. St. Clair RN, Jia YX (2006) Visual metaphors, visual communication and the organization of cognitive space. Int Commun Stud 15(1):157–167
53. Stares PB, Yacoubian M (2007) Terrorism as a disease: an epidemiological model for countering Islamist extremism. http://www.isn.ethz.ch/Digital-ibrary/Publications/Detail/?lang=en&id=46161
54. Sterman JD (2001) System dynamics modeling: tools for learning in a complex world. Calif Manag Rev 43(4):8–25
55. Strang S, Masys AJ (2015) Visual thinking for intelligence analysis. In: Masys AJ (ed) Applications of systems thinking and soft operations research in managing complexity. Springer Publishing
56. Styhre A (2002) Non-linear change in organizations: organization change management informed by complexity theory. Leadersh Organ Dev J 23(6):343–351
57. Tatham P, Houghton L (2011) The wicked problem of humanitarian logistics and disaster relief aid. J Hum Log Suppl Chain Manag 1(1):15–31
58. Taylor I (2016) Application of problem inversion to cascading critical infrastructure failure. In: Masys AJ (ed) Disaster forensics: understanding root cause and complex causality. Springer
59. Tetlock PE, Gardner D (2015) Superforecasting: the art and science of prediction. Penguin Random House
60. The 9/11 Commission Report (2004) http://www.9-11commission.gov/report/911Report.pdf
61. TNO (2008) Radicalisation, recruitment and the EU counter-radicalisation strategy. http://www.transnationalterrorism.eu/tekst/publications/WP4%20Del%207.pdf
62. Urry J (2003) Global complexity. Polity Press, Cambridge, UK
63. Weick KE, Sutcliffe KM (2007) Managing the unexpected: resilient performance in an age of uncertainty, 2nd edn. John Wiley & Sons Inc., San Francisco, CA

Economic Disruptions, Business Continuity Planning and Disaster Forensic Analysis: The Hawaii Business Recovery Center (HIBRC) Project

Jason Levy, Peiyong Yu and Ross Prizzia

Abstract It is proposed that modern disaster forensics can reduce supply chain disruptions, enhance disaster resilience and promote a more robust economy. This chapter examines the root causes of economic disruptions by presenting 'forensic analysis approaches' to disasters that impact the economy of the US island state of Hawaii. Supply chain disruptions and investigations pertaining to business disruptions are undertaken with an emphasis on modeling, understanding and characterizing the complex causality that defines them. In so doing this chapter uncovers creative, timely and important strategies for analyzing accidents and disasters that impact the economy of Hawaii. In order to promote business continuity planning and disaster forensics in Hawaii, the twenty-eighth Hawaii State Legislature enacted, and the Governor of Hawaii has signed, House Bill (HB) 1343 which provides funds for new state of the art Hawaii Business Recovery Center (HIBRC), a joint partnership between the State of Hawaii Emergency Management Agency (HIEMA), the State of Hawaii Department of Business, Economic Development & Tourism (DBEDT), the State of Hawaii State Procurement Office (SPO), and the University of Hawaii West Oahu (UHWO). This designated business recovery hub will provide both outreach and dissemination of business recovery resources in addition to serving as a center for presenting disaster forensics approaches to disaster investigations in Hawaii, thereby uncovering the complex causality that underlies them. The center will help inform businesses of the importance of disaster preparedness; assist with post-disaster business recovery efforts; and create a robust business recovery network that shares the highest-level of management and governance with business leaders and strives for best disaster management practices and continuous improvement.

Keywords Business continuity · Cybersecurity · Disaster forensics · Disaster risk reduction · Forensic disaster analysis (FDA) · Hawaii business recovery center (HIBRC) · Supply chain disruption

---

J. Levy (✉) · P. Yu · R. Prizzia
Honolulu, USA
e-mail: jlevy@hawaii.edu

A.J. Masys (ed.), *Disaster Forensics*, Advanced Sciences and Technologies for Security Applications, DOI 10.1007/978-3-319-41849-0_13

1 Introduction to Disaster Forensics and Business Recovery

Over one hundred years ago European criminal investigators began to use finger-printing and other identification techniques to solve crimes. As used in this paper, the forensics approach to disasters is used to "signify systematic, probing and dispassionate investigations" [1]. Supply chain disruptions and investigations pertaining to business disruptions are undertaken with an emphasis on modeling, understanding and characterizing the complex causality that defines them. While researchers in the field of disaster forensics have systematically examined business sector recovery for only a short time [2–11] they present a sobering picture of how disasters impact businesses, particularly smaller enterprises: a quarter of small businesses which experience a disaster never reopen—and a third of the remaining businesses can be expected to close within two years of the disaster [12]. As the most isolated island chain in the world, the US state of Hawaii has special needs for disaster preparedness and recovery. Transportation travel times and import costs to Hawaii are significant (nearly all food, fuel, and building materials are shipped by boat), the island's critical infrastructure is self-contained and much commerce and business activity occurs in low-lying areas subject to coastal risks and hazards. Accordingly a major disaster could precipitate a drop or stoppage of a majority of business activities in the islands. Moreover, small businesses (those with less than 50 employees) constitute a critical component of Hawaii's economy, accounting for 94 % of all businesses in the state: 86 % of Hawaii' businesses have less than 20 employees [13].

Particularly vulnerable are businesses in Hawaii with less than 20 employees and those that work from home [13]. According to Hawaii's Department of Business Economic Development (DBEDT) Natural Disaster Economic Recovery Plan [14] 38 % of businesses surveyed in Hawaii did not have a Business Continuity Plan: those Hawaiian companies without plans were primarily smaller companies of less than 25 employees. Larger businesses tend to fare better than smaller businesses in the event of a disaster, due to their increased access to resources and economy of scale. The smallest of Hawaii's businesses are not well informed about available disaster recovery resources or the benefits of disaster preparedness: some depend on land owners or property managers for recovery, disaster preparedness training is not common, most companies do not have emergency supplies at their office location, data back-ups are often on-site and are not regularly updated [14]. Specifically, a small business affected by an economic disruption has fewer financial resources (line of credit, cash reserves, capital assets, short term operating funds), lacks an extended network of employees and technical services located outside of the area of impact (to provide response assistance), and often depends upon a very small number of individuals to assist with emergency operations (and make other critical business decisions) in a crisis environment.

Interruptions of critical utility functions and the supply chain are the major impacts felt by Hawaii businesses. These are magnified by Hawaii's dependence on imported resources, including fuel and food. The loss of customers resulting from closures of airports and harbors could have a devastating effect, especially for the visitor industry. Neighbor islands are concerned about the vulnerability of O'ahu's infrastructure since their supplies come through harbors on O'ahu. There is a demand for reliable, credible, and accessible information on preparedness that is tailored to small businesses in Hawaii. There is also a need for localized, up-to-date information on road closures and the status of recovery efforts on all of the islands that make up the state of Hawaii. Businesses in high risk hazard areas need critical support in their preparedness, evacuation, response, and recovery. Many businesses in Hawaii anticipate insurance as their primary means of recovery assistance. However, less than half have business interruption insurance and many have incomplete or no disaster coverage. Some have difficulty finding recovery assistance and resources, or complying with the documentation requirements for insurance claims. This is intensified for those that do not speak English or that have other barriers to receiving information and support.

2 Disasters and Business Continuity Planning

In the 2001 Nisqually earthquake, large corporations in the US state of Washington such as Boeing were able to rely upon their extensive emergency plans, including the use of backup generators, activation of internal emergency operation centers, and the option to switch computing control to locations outside the region. Although Starbucks' US Seattle headquarters was evacuated, it was able to continue operations because of its multiple locations outside the area of strong ground motion [15]. Often smaller businesses do not have these same options and have great difficulty absorbing costs associated with structural mitigation and risk reduction strategies. However, businesses that develop and implement a disaster recovery plan are more likely to survive a disaster—and typically sustain less damage, loss, and downtime—than those that do not [16].

The single location of a small business leaves an owner's investments more vulnerable to total destruction when compared to a larger business chain, where risks are more spatially distributed [10]. Often, recovery aid guidelines put subtle pressure on business owners to remain at their same location, despite changes in the economic and risk landscape [7, 17]. Yoshida and Deyle [11] found that small businesses were less likely to be knowledgeable about hazard mitigation and specialized insurance. Furthermore, retail businesses that rent their floor space are more vulnerable to loss than those that do not rent [3]. In studying small businesses, researchers have also sought factors that increase the probability of small business recovery. In researching the 1994 Northridge earthquake, Tierney [10] found that business that rent their space were typically less able than building owners to engage in mitigation and preparedness activities. However, businesses that were relatively larger, older and financially stable, or had previous disaster experience were more likely to have engaged in preparedness activities prior to the earthquake. Following the earthquake, newer businesses and better-prepared firms were more likely to increase preparedness levels post-earthquake. The rising intensity and frequency of severe weather events and an increase in the value of business assets in vulnerable locations has direct implications for business continuity and broader objectives for any business enterprise. For example, the increasing frequency of heavy rain and rising sea levels will contribute to not only flood damage of key business assets and resources but also a disruption to corporate travel, supply chains or critical infrastructure as well as the possible mandatory evacuations (and the resulting lack of access to the business premises, etc.). Rising air temperatures can lead to a decline in productivity and profit through heat related illness, higher cooling costs, and the disruption of temperature sensitive processes or products. Drought risks also contribute to water scarcity and potentially higher business costs while low flows in rivers will also degrade the quality of water discharged under effluent consents.

3 Introduction to Disaster Forensics and Cybersecurity

By 2020 the world will generate fifty times the amount of information and seventy-five times the number of "information containers", far outpacing the growth of the information technology professionals required to manage it. Assuring the security of this data and the integrity of the supporting information systems are among the most challenging and urgent issues for businesses in the twenty-first century. Specifically, the number of cybersecurity incidents reported by US federal agencies has increased from 5503 in 2006 to 41,776 incidents in 2010 (an increase of over 650 %) targeting a host of assets from classified defense networks to critical infrastructure systems and the US Commerce and State departments [18, 19]. Cyber Security problems can be classified into six categories. First, there are security issues arising from human errors. An overwhelming majority of human error related security problems occur because of improper installation and mismanagement of hardware and software, updating of wrong files or entering incorrect data. Second, security problems may be due to analysis and design faults. Many security problems occur because of an ad hoc approach to the development of information systems (IS). This typically results in lack of correctness in specifications and a reactive approach to the design of security controls. Third, security problems occur because of the violation of safeguards by trusted personnel. It is a documented fact that nearly 70–80 % of security violations are a consequence of internal employees subverting organizational controls to gain unauthorized access. Fourth, security vulnerabilities may arise because of environmental damage. Many cyber security problems manifest themselves because of limited disaster recovery planning or other related unforeseen contingencies. The fifth source of cybersecurity problems are caused by system intruders. Both organized and unorganized hacking attacks consume considerable resources. Finally, given the over-reliance on information and communication technologies, organizational information assets have often been targeted through social engineering attacks or a range of malicious software, including phishing, bots, among others.

Organizations and national governments have reacted to the range of security problems by instituting security policies in their respective businesses or getting involved in establishing national and international security standards. In many instances such guidelines have also been transformed into state and federally mandated laws and regulations. In spite of these worthy initiatives only about 40 % of organizations in the US have a well defined security policy and many organizations fail to consistently incorporate the latest information security standards. Even in cases where such cyberstandards and policies have been adopted, organizational compliance has been reactive and sub-optimal, failing to keep pace with the evolving nature of cyber attacks. It therefore comes as no surprise that cybersecurity incidents continue to permeate and proliferate organizations across the country. It is hence essential to understand the nature and scope of deep seated structures that underlie the management of information security in organizations. Such a deep structural analysis also helps in understanding the linkage between the prevalent organizational norms and the actual policies and rules that are in place. A lack of concordance between the two certainly is a key ingredient in a cyber security failure.

Private sector organizations, governmental agencies and individuals in Hawaii increasingly rely on the security of their computer-based information assets. Despite the fact that Hawaii's information security professionals, disaster recovery experts, and business continuity practitioners have achieved a number of achievements in protecting businesses from cyber-threats, information security represents one of the most serious challenges to Hawaii's businesses. Information systems security, classically defined according to the requirements of confidentiality, integrity, and availability, is vital to protect Hawaii's private sector—and society at large. For example, a cyberattack on the maritime industry could affect the state's economic prosperity and quality of life as Hawaii's maritime industry is absolutely critical for importing food, fuel and other essential goods. Such an attack could also create risks to Hawaii's transportation, fishing and passenger cruise industries.

As commerce and business in Hawaii increasingly interacts, organizes, and create communities through cyberspace, businesses must devote more efforts towards protecting cyber networks and systems against intrusions and malicious activity. Hawaii's businesses are increasingly impacted by cyber-based risks including online crime, fraud and the theft of valuable intellectual property. These cyber-attacks have the ability to interrupt business operations, cause catastrophic social dislocation, severely degrade economic infrastructure, and threaten life-sustaining services. Hawaii's economy depends on networked information systems and critical infrastructure, which make them highly vulnerable to cyber-based risks including phishing, denial-of-service attacks, SQL injections, cross-site scripting, viruses, trojan horses and worms. While modest progress has been made towards preventing these sophisticated attacks and identifying attackers there is an urgent need to improve recovery to cyber incidents in Hawaii and to create dynamic, customized and robust information systems security solutions. Current cyber-challenges in Hawaii include a lack of cybersecurity leadership, ambiguous roles and responsibilities, technologic challenges, and workforce shortages.

Collective business cybersecurity and the protection of cyber-networks and computer systems in Hawaii can be achieved by leveraging fast-paced technological innovation, by establishing new cybersecurity frameworks to protect businesses from cyberthreats as well as by strengthening relationships among the private sector, governments, non-governmental organizations and the education sector. Cyberspace in Hawaii must be viewed as a critical business domain with associated responsibilities, implications and consequences for cyber-training, exercising, response and recovery. Serious concerns about cybersecurity and the vulnerability of Hawaii's communications networks remain with grave threats to business continuity, national security, civilian critical infrastructure to intellectual property potentially at risk: the need for increased information security is greater than any other modern technology [20]. One key challenge is that increasingly sophisticated cyberattacks now requires exponentially less intruder knowledge than in previous decades. Specifically, Fig. 1 reflects the disturbing trend toward increasingly sophisticated attacks on computer systems in Hawaii which can be launched by relatively unsophisticated intruders. The increasing complexity of a 'hyperconnected' world is exacerbating information security threats in Hawaii with mobile

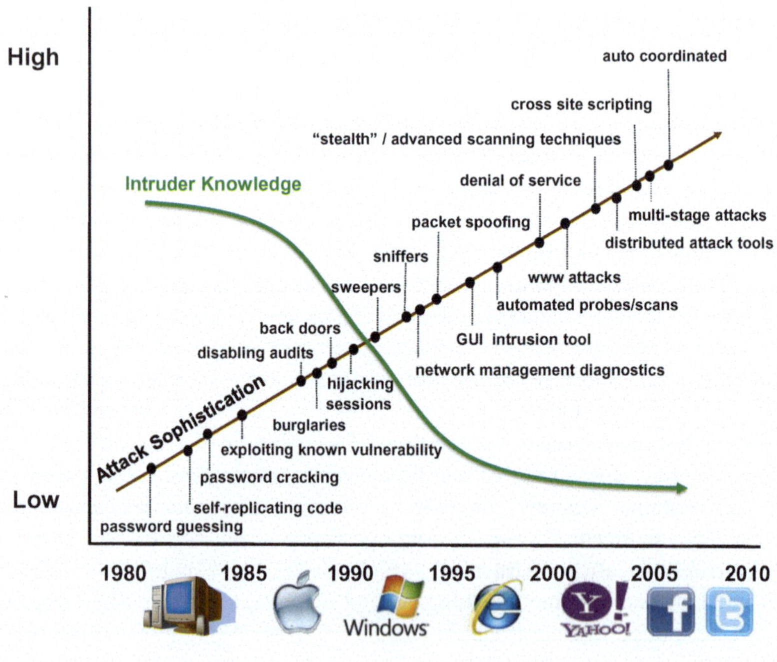

Fig. 1 Cybersecurity attack sophistication versus intruder knowledge (modified from CMU [37])

devices being a recent cause for concern. While flexible, remote and mobile work practices in Hawaii can lead to reduced fixed costs and overhead, mobile devices are easy to lose, difficult to secure, and have caused organizations to experience reputational damage and financial losses due to data breaches and theft (leading to downtime and replacement costs). Information security experts recently identified mobile devices, such as laptops, as a key threat to information security, in part due to the recent popularity of BYOD (Bring Your Own Device) policies, which can "extend the boundaries of the enterprise and blur the lines between what is and is not part of the network" [21].

In order to better understand the root causes of disasters in Hawaii a number of round panel discussions with business leaders and government officials over the past year have been held to determine key sources of risk in Hawaii. A systematic look at risks posed to businesses in Hawaii identified the following major and emerging threats: business continuity risks (supply chain disruption, etc.), cyber incidents (cyber crime, IT failures, data breaches), reputational and brand damage (via social media and other sources), natural catastrophes (storms, floods and earthquakes) as well as macroeconomic and market developments (volatility, intensified competition, commodity price increases). As cyberthreats constitute the most rapidly growing source of concern among a majority of small and large businesses in Hawaii it is herein explored in depth.

Based on surveys and in-person consultations with business stakeholders in Hawaii, five traditional approaches to improve cybersecurity for Hawaii's businesses are put forth: checklists, risk analyses, formal methods, soft systems approaches and bow-tie diagrams [22–26]. A large body of academic work has identified, classified, and evaluated these approaches to managing cybersecurity, and analyzed their prescriptive and descriptive implications. Often intended as security evaluation guidelines, checklists constitute the first category of information systems security. The earliest checklists, including IBM's 88 point security assessment questionnaire and the *Computer Security handbook* provided a straightforward taxonomy of threats and focused on a wide range of security issues from disaster planning and encryption to off-site backup and physical security. The American Federation of Information Processing Societies (ARIFP) provides a *Checklist for Computer Centre Self Audits*. This more sophisticated than previous tools in that a kernel style framework of hazards and associated defenses was developed. Computer center audits were often carried out with either the AFIPS and SAFE (*Security audit and field evaluation for computer facilities and information systems*) checklists [27].

Risk analysis methods were expanded in recent decades to overcome the problems involved with evaluating and implementing a large number of controls. Based on quantitative risk assessments, these risk-based approaches enable security professionals to select from a checklist only the most critical risks by identifying threats and hazards, articulating the relationship between risk and vulnerability, evaluating whether risks exceed accepted thresholds and selecting risk treatment strategies based on the expected probability distributions and consequences of loss. Popular risk analysis tools for cybersecurity include the Central Computer and Telecommunications Agency's (CCTA) CRAMM (CCTA Risk Analysis and Management Method) approach and the Methodologie d'Analyse des Risques Informatiques et d'Optimisation par Niveau (MARION). More details can be found in Broder and Tucker [28]. International standards and frameworks for enterprise risk management provide a broad overview of strategic and tactical risk management processes with particular focus on organizational governance (e.g. leadership and accountability), risk monitoring, education and training, risk assessment (risk identification, risk analysis, and risk evaluation), risk treatment, consultation and communication, and risk treatments.

A third approach for managing cybersecurity, formal security models and methods, are discrete event-oriented approaches. They were popularized in the 1970s and 1980s in part to overcome the weaknesses of information systems security testing: "Testing can only prove the presence of bugs, not their absence." Formal models were traditionally directed at the systems-software level of computer security (i.e. [29, 30]) and now include popular soft computing approaches such fuzzy logic [31] and neural networks [32].

4 Deep Structures and Bow Tie Method for Cyber-Disaster Forensics

This chapter seeks to develop a detailed understanding of cyber disaster forensics and to propose original and valuable tools capable of crafting robust, flexible, cohesive yet sophisticated cybersecurity strategies. Specifically we put forth two additional approaches that integrate multiple cyber-systems security approaches, methodologies and tools in a holistic fashion: bow-tie methodologies and deep structures. By combining and enhancing the aforementioned security modeling techniques (checklists, risk analyses, formal methods and soft models) in a holistic and systemic fashion this chapter seeks to enhance the resilience of Hawaii's businesses in the face of an external attack.

An important paradigm for understanding cybersecurity is the deep and surface structure perspective [33], which can be used to create meaningful, dynamic, custom-fit and flexible information security solutions. A key benefit of deep structure approach to cybersecurity is that it can successfully incorporate the aforementioned information systems security approaches (checklists, risk analyses, formal methods and soft modeling techniques) in order to reveal the underlying key requirements of robust and flexible security solutions. By integrating knowledge, competencies, methodologies, and applications, the deep structure approach to information systems security provides key insights to better interpret cybersecurity challenges and incorporates best-practice security concepts from a range of complementary disciplines. The deep structure framework incorporates three models: the representational model, which identifies subsystems within an organization; state-tracking model, which ensures that different states trace IS security in the real world; and the decomposition model, which defines specific external events that are a stimuli to changes in internal events [26].

Bow ties can also include information security insights found in checklists, risk analyses and formal models [34]. For example, Fig. 2 shows that bow ties are considered a combination of the traditional fault trees and event trees, whereby the fault tree constitutes the left hand side of a bow-tie (before the "top event") and the event tree the right hand side (after the event). As shown in Fig. 2, bow ties are based around a "top event" (an undesirable event or situation with the potential for harm or damage) where control over the process has been lost (e.g. the breach of an information system), possibly leading to profound adverse consequences. The bow-tie method also includes successive layers of defenses, barriers and safeguards known collectively as preventative 'controls' that preclude 'threats' from releasing a hazard. For example, in the information security field, intrusion detection and prevention systems, firewalls and anti-virus programs provide protection against known vulnerabilities. It is possible to use the bow-tie methodology in conjunction with other risk techniques to identify the most susceptible bow-tie branches. For example, in the commercial software BowTie Pro™ the "Swiss cheese model" of accident causation [35] is used to highlight the fact that any information system, no matter how well protected, is susceptible to threats. The magnitude of this threat is a

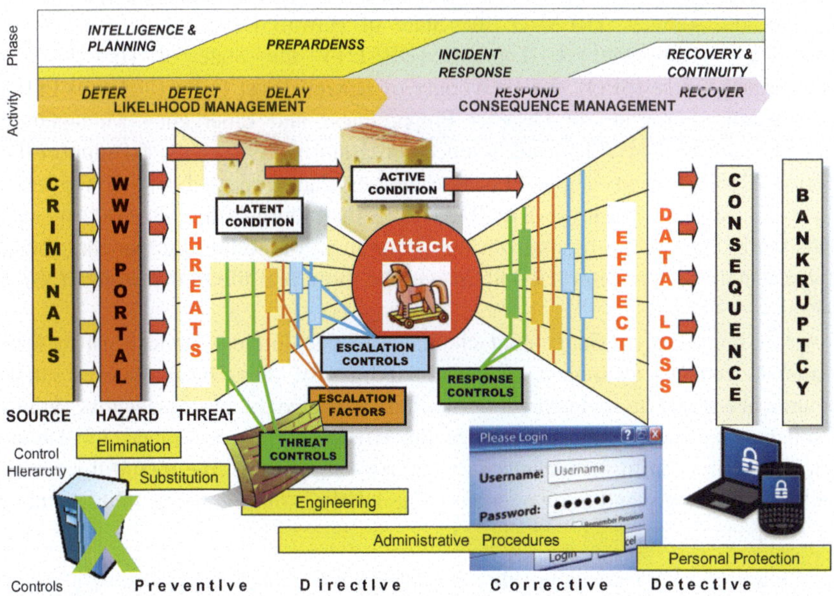

Fig. 2 Bow tie risk management framework for disaster forensics

function of which barriers contain the holes (i.e. gaps or weaknesses in system defenses) as well as the alignment and size of the holes. The bow-tie method of Fig. 2 (modified from [36]) captures the causal sequence of human failures that contribute to a disaster: failures are the product of either active factors that produce immediate events and involve operators (i.e. firewall installation errors), preconditions for unsafe acts (for example, fatigued staff or improper communication practices) or latent conditions, the underlying contributory factors that are inherent in the system and which may have lain dormant for extended periods until they synergistically contribute to an undesirable event. Latent factors include organizational influences (for example, a reduction in training expenditures and heavy workloads), unsafe supervision and environmental conditions (such as an unsafe or unhealthy work environment).

5 The Hawaii Business Recovery Center (HIBRC) Overview

The State of Hawaii Emergency Management Agency (HIEMA), formerly the "State Civil Defense" agency, supports County and State Agencies in their efforts to respond and recover from natural and man-made disasters. Modeled after the Louisiana Business Emergency Operations Center (LABEOC), the Hawaii Business Recovery Center (HIBRC) constitutes an innovative $1.6 million (US dollar) three year joint partnership between the State of Hawaii Emergency Management Agency (HIEMA), the State of Hawaii Department of Business, Economic Development & Tourism (DBEDT), the State of Hawaii State Procurement Office (SPO), the University of Hawaii West Oahu (UHWO) Public Administration program in Disaster Preparedness and Emergency Management (DPEM), and the UHWO Social Sciences program in Economics. To support this effort in 2015 the twenty-eighth Hawaii State Legislature enacted, and the Governor of Hawaii has signed, House Bill (HB) 1343 that appropriates $328,000 of state funds while US Federal funds from the Department of Commerce has contributed approximately $1.3 million US dollars over the three year life of the project (2016–2019).

A key emergency management challenge is that limited connections to local businesses reduces awareness of the capabilities of these businesses to satisfy resource needs and an understanding of what assistance the businesses may require to satisfy those needs. The mission of the HIBRC is then to engage local commercial and non-profit businesses in a mutually supporting role to fulfill resource needs of the County and State agencies. The HIBRC solution hence involves three key components:

- creation of an infrastructure that provides a robust and aware resource acquisition system that enables Emergency Mangers to fulfill needs from local businesses.

- development of a communications platform that informs local businesses about Emergency Manager needs; informs Emergency Managers about local businesses ability to fulfill requirements; and provides a means to match buyers and sellers using existing State
- on-going training and support capability.

The HIBRC business model involves working with participating trade associations, individual firms, and non-profits to identify locally available emergency resources (e.g. air cargo, heavy equipment, food distribution). Specifically, the HIBRC facilitates participant compliance with State procurement regulations and robust connections to the State's on-line procurement system, State of Hawaii eProcurement (HIePRO). Managed by UHWO with support from state agencies, the HIBRC maintains awareness of business capability status and operational needs and matches resource needs between Emergency Managers and local resources providers.

The goal of the HIBRC program will be to specifically enhance participation and access by commercial and non-profit organizations in disaster management response and recovery efforts through the development, deployment, and training, of public-private partnerships for the benefit of the State of Hawai'i. Key activities to be performed as part of the award are the following:

- The construction and equipping of a 2000 ft.[2] Business Recovery Center (BRC) Operations Center within HIEMA's Alternate State EOC located at HIEMA's Building 303 located adjacent to the State EOC at Diamond Head which is currently undergoing a $1.4 million hardening retrofit.
- The development and deployment of a web based BRC partnership portal
- The virtualization of Web EOC and the SPO's HiePRO (State of Hawaii eProcurement) system to permit on-line sourcing of executed Requests For Assistance (RFAs).
- The equipping and deployment of pre-positioned emergency communications satellite voice and data systems.
- The development of emergency procurement agreements and procedures with participating commercial and non-profit agencies.
- The development and execution of outreach, program training and exercise programs to support the BRC.

5.1 HIBRC Case Studies

The HIBRC project will significantly increase the resiliency of the State of Hawaii by providing for explicit partnership linkages between the State's Emergency Managers and Commercial and Non-Profit supporting enterprises. By developing and delivering a pre-defined means of exchanging resource requirements between Emergency Management agencies (Web EOC) and the State's on-line procurement system (HiePRO) participating business partners will be directly connected to the "market" for emergency goods. In return, BRC members will be able to communicate enabling requirements needed to deliver a specific good. For example, after a disaster HI-EMA Emergency Managers identify the need to identify and establish Points of Distributions (POD) in key locations and communicate with HIBRC: this is achieved by issuing a sourcing Request for Assistance (RFA) from WebEOC to create a solicitation, i.e. a Request for Proposal (RFP) on HiePRO. In other words, the HIBRC communicates with HI-EMA Emergency Managers via the WebEOC RFA process. The HIBRC also develops and manages a cloud based Sharepoint collaborative web portal to maintain a common awareness of HIBRC partner status between HIBRC partners and HI-EMA Emergency Managers.

The HIBRC then queries partnering local businesses to determine store availability (open, closed, potential to open). At this point, candidate stores communicate to the HIBRC any resources required to operate a POD beyond their capacity, including assistance needs with security, power, route clearance, and logistics at targeted locations. The HIBRC then calculates an optimal match between emergency needs and store needs, accepts the RFA, and executes coordinating support. This may involve employing Emergency Support Function (ESF) capacity to provide needed assistance to the HIBRC partner to enable them to satisfy the original POD requirement. In this way HIBRC partners gain resources needed to deliver essential emergency services.

As a second example consider the case in which Emergency Managers identify a need for ice and communicate to HIBRC which then issues an HIePRO solicitation (HIBRC uses HIePRO to issue solicitations to local businesses and non-profits). HIBRC queries partnering local ice producers about their availability (open, closed, potential to open). An ice plant operator may inform the HIBRC about the need for auxiliary power and spare parts to fulfill the solicitation. The HIBRC then coordinates the delivery of temporary auxiliary power and the prioritized shipment of spare parts. Finally, the ice plant operator responds to the HIePRO solicitation for ice. The BRC deploys auxiliary communications to partners as needed.

This designated business recovery hub will provide both outreach and dissemination of business recovery resources in addition to serving as a center for business continuity information, education and leadership. The center will help inform businesses of the importance of disaster preparedness; assist with post-disaster business recovery efforts; and create a robust business recovery network that shares the highest-level of management and governance and strives for continuous improvement.

The goal of the HIBRC program will be to specifically enhance participation and access by commercial and non-profit organizations in disaster management response and recovery efforts through the development, deployment, and training, of public-private partnerships for the benefit of the State of Hawai'i. Key activities to be performed as part of the award are the following.

6 Description of HIBRC Components

The Construction component of the project is to be the renovation of approximately 2000 SF of space sited within HI-EMA's Building 303 located at 4204 Diamond Head Road, Honolulu, HI within the Diamond Head State Monument. The property is owned fee simple by the State of Hawaii TMK:31042006:0000 and has no liens or encumbrances. The constructed Diamond Head space will house the HIBRC Operations and Support functions by providing hardened and redundant communications and information display capabilities for both real-time and training events. The constructability of this project is high due to the proposed re-use of an existing facility: by constructing within an existing structure no major permitting or site impact issues are anticipated and the site has adequate utilities and parking to accommodate the HIBRC project. The space will also provide on-going administrative, meeting, and equipment storage facilities for deployable HIBRC communications assets. HIBRC communications assets include a fixed Very Small Aperture Terminal (VSAT) satellite terminal at the HIBRC, a deployable VSAT satellite terminal for a mobile application, and ten data capable portable satellite phones capable of accessing the virtualized (cloud based) HIBRC information assets.

Virtualization of the HIBRC Web Portal, HIEMA Web EOC, and SPO HlePRO systems into the Government Cloud will provide a more resilient business recovery informatics capability (at lower cost) that can be accessed from any location equipped with internet access. HIEMA staff will serve as project point of contact providing oversight, compliance, financial accounting, and reporting. They will also provide the following program support and technical assistance elements:

- BRC liaison and staff support to State EOC functions (Business and Industry Emergency Support Function);
- Planning and implementation annual training exercises;
- Procurement and contracting services;
- Design and construction management services;
- Virtualization infrastructure (hardware and software) development and support;
- Communications asset management and support; and
- Repair and maintenance of the HIBRC facility

7 Program Support and Technical Assistance

The program support and technical assistance provided by the University of Hawai'i West Oahu's Disaster Preparedness and Management (DPEM) program faculty will be to act as the Principal Investigator of the HIBRC grant managing the program development with the responsibility for the following roles:

- Assemble and manage the Business Recovery Team;
- Establish best practice business recovery procedures in collaboration with all entities and strengthen public-private partnerships;

- Develop operational practices to engage the private sector in disaster recovery;
- Liaise with other State Agencies such as State Procurement (SPO) and the Department of Business, Economic Development and Tourism (DBEDT); and
- Conduct the quarterly industry meetings in partnership with HIEMA and perform on-going outreach, seminars/webinars and training support for disaster preparedness.

In summary, state and federal funding has been used to provide technical support for systems integration and liaison with other State Agencies such as State Procurement (SPO).

The BRC will enhance the capacity of Hawaii businesses to promote disaster preparedness by contributing their personnel, materials, technical and management advice to businesses, organizations, communities, and agencies. The BRC will provide both outreach and dissemination of business recovery resources, in addition to serving as a hub for post-disaster logistic support, technologic innovation and information sharing.

Specifically, the BRC will engage local commercial and non-profit Hawaii businesses in a mutually supporting role to:

- fulfill resource needs of the County and State agencies
- strengthen connections between local businesses and these agencies to satisfy resource needs
- better understand the capabilities and limitations of local businesses and what assistance businesses may require to satisfy those needs.

There is clearly a need to enhance ongoing, short-term (days), intermediate term (weeks-month) and long-term (months-years) recovery for Hawaii businesses. The UHWO DPEM program is examining comprehensive, all-hazards, integrated and risk-based business continuity planning strategies for Hawaii's businesses which involves preparing, mitigating, responding and recovering from all possible threats and perils including natural, health-related, human-induced, hazardous materials and other technologic hazards. The HIBRC work will carry out disaster forensics and business recovery work across all hazards and emergency management phases in order to shorten the recovery time in the event of a business disruption and minimize financial losses.

8 Assemble and Manage the Business Recovery Team (BRT)

UHWO DPEM will assemble a disaster recovery team with members from the private sector, trade associations, public sector and non-profit organizations with a primary focus primarily on economic recovery (most emergency operations plans are primarily focused on health and safety issues). There is a need to effectively engage and pro-actively recruit the appropriate local businesses representatives from across Hawaii in activities that will protect their business assets and expedite the recovery of the local economy in the event of a disaster. A mix of strategic public and private sector representatives will be invited to participate in the business recovery team. To ensure that economic recovery decision making contains the most experienced professionals the business recovery team will include economic development stakeholders (Economic Development Officers, chambers of commerce officials, business & trade association professionals, special districts), public and elected officials (including emergency management personnel) as well as a significant representation of the business community. Team members will be selected based on their articulated position of authority (to provide proper support to the team) and knowledge of the needs of the private sector. For example, the Community Economic Development Officer (EDO) will have understanding and knowledge of their community's long term economic development resources, goals and plans, zoning laws, permitting processes and any laws or ordinances around the financing of various economic development plans.

The economic impacts of each post-disaster recovery action should be carefully considered. Even the seemingly harmless decision of closing a few streets can slow recovery, force businesses to relocate or close, and drive residents and customers away—thereby affecting the resilience of the workforce and economy. Accordingly, the business recovery team will establish agendas, facilitate participation, discussion and information exchange from all representatives; delegate business recovery tasks and follow-up; and evaluate objectives and outcomes in an expedited yet comprehensive decision making process.

The business recovery team could specifically address the major pre and post-disaster economic recovery issues that businesses and communities face such as:

- understanding the capabilities and limitations of businesses;
- providing assistance for business continuity plans;
- selection of redevelopment priorities;
- identification and implementation of disaster mitigation strategies;
- development of post-disaster recovery financial and technical assistance programs; and
- expansion of response and recovery coordination between small businesses, the public sector and their surrounding communities

9 Establish Best Practices for Disaster Forensics and Business Recovery Procedures

Business surveys have been used to better understand the disaster services and capabilities that local Hawaii businesses provide and the type of equipment they have on hand in the event of a disaster to be used in cleanup, debris removal, emergency response, as well as long-term rebuilding efforts. Future preparedness and disaster response activities will be strengthened by documenting the capabilities, essential services, and equipment of local businesses. Understanding the vulnerabilities and weaknesses of Hawaii's businesses before disaster strikes will also help to identify emergency management gaps in the community. Local and state government agencies should also work closely with the HIBRC to effectively communicate and educate local businesses on disaster recovery and other preparation activities.

Pre-disaster planning is a critical element in the success of long-term economic and community recovery as many businesses will be negatively impacted by a disaster event (and some may never reopen). For example, a 2004 Gallop organization poll commissioned by the National Federal of Independent Businesses (NFIB) found that at least 30 % of the surveyed small businesses had been closed 24 h or longer at least once within the last three years while the U.S. Bureau of Labor Statistics reports that 93 % of businesses that suffer a significant data loss are out of business within five years. However, many small businesses lack a plan which helps them highlight the natural, health-related, technologic or human-induced threats that can disrupt their supply chains and viability. Business continuity plans constitute a *sine qua non* tool for disaster resilience because they help businesses to limit the effects from potential hazards and identify critical processes or operations that must remain open after disaster impact. Backing up and storing data off-site is very important but many small businesses fail to back-up important documents (such as legal information, key customer contacts, and financial records) in a remote location.

The HIBRC brings together leading business continuity professionals in Hawaii from both the public and private sector to share best practices and participate in continuing education and creating the business continuity professionals of tomorrow. The UHWO DPEM program will promote disaster resilience through regular networking and educational programs. As a recognized business recovery expert resource, UHWO DPEM will act in an advisory capacity to organizations and government institutions throughout Hawaii, helping to provide technical support for systems integration, to create industry disaster recovery standards and to promote business resiliency through public-private partnerships. These private-public sector partnerships will ensure that communication flows between these two groups to reveal any potential conflicts and/or duplication of effort in the recovery process. The public sector should be informed of the private sector's priorities for re-investment and redevelopment. Governments must also carefully consider the implications of decisions that could possibly delay the recovery effort. Trust and collaboration will be strengthened as stakeholders cooperate in the pre-planning phase so that these relationships can be relied on when the disaster strikes.

10 Improve Operational Practices to Engage the Private Sector in Disaster Recovery

Operational practices will be developed to help the private sector recover effectively and efficiently after a disaster event. The Business Recovery Team at UHWO thoroughly reviews existing business continuity operations, emergency management and mitigation plans that exist with a special focus on the impacts to the business community and the local economy. UHWO faculty are establishing a tiered system of re-entry for relevant individuals into the disaster stricken area in order to coordinate re-entry pass distribution to all businesses requiring clearance. It is important that the business re-entry plan prioritizes critical businesses which will restore essential services after a disaster. It is proposed that the HIBRC create an online website for qualifying businesses to apply for re-entry vehicle passes, learn about Hawaii's evacuation re-entry plans and remain connected with the HIBRC business continuity network (a main point of contact for businesses and clients in the event of an evacuation). After a major crisis, local business workers and owners may be restricted from returning to their property/operations if it remains in the evacuated area. However, any delays in allowing employees to return to their businesses can have devastating impacts for society including fewer employment opportunities for residents and a significant decline in the tax revenue base, thereby extending community vulnerability and recovery times. For example, following a mass evacuation, limited access to the area following the crisis event can lead to extensive delays in the reopening of businesses. This in turn may cause perishable inventory to spoil, employees to leave the community in search for work and the disruption of businesses operations needed to restore key resources and infrastructure, provide essential services and goods in the impacted area (from food and water to health services). For example, more than half of businesses within the police perimeter of the Oklahoma City bombing in 1995 closed permanently (even those without physical damage) due in part to the fact that business managers, owners and employees faced significant delays in returning to their workplaces.

A business re-entry plan should involve a tiered/phased re-entry system which accommodates business re-entry needs. In this way emergency managers can help local businesses to open in a timely manner, thereby enhancing local economic recovery. To respond to this issue, the HIBRC is working with state, local and county emergency management personnel to establish clear procedures of securing re-entry ID cards and developing a tiered system of re-entry following a disaster. The purpose of this tiered system is to allow for the expedited, safe, orderly return of re-entry of the following three tiers:

- tier 1 agencies/groups involved in emergency response and restoring normal operations following a disaster (search and rescue; emergency healthcare staff; utilities and infrastructure repair personnel; damage assessment teams; and pre-designated government staff and other critical personnel);

- tier 2 groups such as relief workers; healthcare agencies and suppliers; insurance agents (not allowed in tier 1); food and building material retailers, fuel distributors and stations, debris management, financial institutions, anchors for key industries, major employers with a large workforce, and/or a major tax revenue source for the community and businesses facing high risk circumstances (fragile inventory, hazardous waste, perishable goods, etc.);
- remaining tier 3 residents and business operators (not allowed under tier 2) from the restricted area as well as licensed contractors and other repair service providers.

A business re-entry plan must be sensitive to business considerations and should assist community leaders and emergency personnel in administering recovery efforts. This can be only achieved if emergency managers work in tandem with government officials and the private sector in advance of a major disaster to promote business continuity plans and procedures. Conditions on the ground, including damaged infrastructure (roadway, sewer, water and electricity lines) may preclude community members from returning promptly to their homes and businesses.

Once safety concerns are addressed, the business employees, managers and owners should be allowed to quickly return to their workplaces to secure property and re-establish business operations, thereby facilitating recovery and promoting the economic engines of their local communities. It has been shown that staggering business re-entry after a disaster promotes safety and ensures expedited re-entry for the critical businesses.

11 Conclusion

Over one hundred years ago European criminal investigators began to use fingerprinting and other identification techniques to solve crimes. As the field of forensics expanded in scope throughout the twentieth century from the criminal justice field to other areas its application became ubiquitous. It is shown that disaster forensics and business continuity planning can help us to understand the origins of these corporate disasters and ways to prevent them. Unsustainable business practices can lead to technologic, social and environmental disasters. A major disaster in the US state of Hawaii could precipitate a drop or stoppage of a majority of business activities in the islands. As the most isolated island in the chain in the world, Hawaii has special needs for business continuity planning and supply chain management. Accordingly, a disaster forensics and business continuity planning architecture is put forth to better understand how disasters impact Hawaii's businesses, how they recover, and how they might better prepare for unexpected, extreme events under conditions of climate change. In-depth interviews with logistics managers, security professionals, and government officials are undertaken to investigate how private sector organizations, government agencies and non-governmental organizations in

Hawaii perceive disaster risks and supply chain disruptions and what they are doing to respond and address them.

The proposed HIBRC will provide both outreach and dissemination of business recovery resources in addition to serving as a center for presenting disaster forensics approaches to disaster investigations in Hawaii, thereby uncovering the complex causality that underlies them. This designated business recovery hub center will help inform businesses of the importance of disaster preparedness; assist with post-disaster business recovery efforts; and create a robust business recovery network that shares the highest-level of management and governance with business leaders and strives for best disaster management practices and continuous improvement. It is shown that modern forensics can reduce supply chain disruptions, enhance disaster resilience and promote a more robust economy. This chapter examined the root causes of economic disruptions by presenting 'forensic analysis approaches' to disasters that impact the economy of the US island state of Hawaii. In so doing this chapter uncovered creative, timely and important strategies for analyzing accidents and disasters that impact the economy of Hawaii. It is proposed that future work examine supply chain disruptions and investigations pertaining to business disruptions with an emphasis on modeling, understanding and characterizing the complex causality that defines them.

References

1. Forensic Investigations of Disasters (FORIN) project (2015) Integrated research on disaster risk. http://www.irdrinternational.org
2. Alesch DJ, Holly JN, Mittler E, Nagy R (2001) When small businesses and not-for-profit organizations collide with natural disasters. Paper presented at the First Annual IIASA-DPRI Meeting on Integrated Disaster Risk Management: Reducing Socio-Economic Vulnerability
3. Chang SE, Falit-Baiamonte A (2002) Disaster vulnerability of businesses in the 2001 Nisqually Earthquake. Environ Hazards 4(2–3):59–71
4. Dalhamer JM, Tierney KJ (1998) Rebounding from disruptive events: business recovery following the Northridge earthquake. Sociol Spectr 18:121–141
5. Flynn DT (2007) The impact of disasters on small business disaster planning: a case study. Disasters 31(4):508–515
6. Furlong SR, Scheberle D (1998) Earthquake recovery—gaps between norms of disaster agencies and expectations of small businesses. Am Rev Pub Adm 28(4):367–389
7. Graham LT (2007) Permanently failing organizations? Small business recovery after. Econ Dev Q 21(4):299–314
8. Kroll CA, Landis JD, Shen Q, Stryker S (1991) Economic impacts of the Loma Prieta earthquake: a focus on small business. The University of California Transportation Center, Berkeley
9. Runyan RC (2006) Small business in the face of crisis: identifying barriers to recovery from a natural disaster. J Contingencies Crisis Manag 14:12–26
10. Tierney KJ (1997) Business impacts of the Northridge earthquake. J Contingencies Crisis Manag 5(2):87–97
11. Yoshida K, Deyle RE (2005) Determinants of small business hazard mitigation. Nat Hazards Rev 6(1):1–12

12. Insurance Institute of Business and Home Safety (2012) https://www.disastersafety.org/commercial_maintenance/commercial-vulnerability-assessment_ibhs/. Accessed 20 Aug 2015
13. DBEDT (2012) Department of Business, Economic Development and Tourism. Research & Economic Analysis
14. DBEDT. Department of Business, Economic Development and Tourism (2014). http://files.hawaii.gov/dbedt/op/spb/2014_nders_final.pdf
15. Freitag R (2002). The business of earthquakes—what we learned from the Nisqually earthquake. Paper presented at the CREW Conference
16. FEMA (2010) Businesses with preparedness plans more likely to survive disasters, Sept 2010
17. Vale LJ, Campanella TJ (eds) (2004) The resilient city: how modern cities recover from disasters. Oxford University Press, Oxford
18. Center for Strategic and International Studies (2011) Significant cyber incidents since 2006. http://csis.org/files/publication/110606_Significant_Cyber_Incidents_Since_2006.pdf
19. General Accountability Office (2011) Cybersecurity: continued attention needed to protect our nation's critical infrastucture and federal information systems. US Government, 16 Mar 2011
20. Goldman J (2010) CIO Insight, federal investment in cyber security to reach $13.3 billion by 2015. Government-CIO Insight. 29 Dec 2010. http://www.cioinsight.com/c/a/Government/Federal-Investment-in-Cyber-Security-to-Reach-133-Billion-by-2015-305281/
21. Deloitte (2011) Raising the bar: 2011 Technology, Media & Telecommunications (TMT). Global security survey—key findings. Deloitte Touche Tohmatsu Limited (DTTL) TMT Industry Group, Netherlands
22. Dhillon G, Torkzadeh G (2006) Value-focused assessment of information systems security in organizations. Inf Syst 293–314
23. Dhillon G, Backhouse J (2000) Technical opinion: information systems security management in the new millenium. Commun ACM 43(7)
24. Siponen MK, Baskerville R (2001) A new paradigm for adding security into IS development methods. In: Eloff JHP, Labuschagne L, Solms R, Dhillon G (eds) Advances in information security management & small systems security. Kluwer Academic Publishers, Boston, pp 99–111
25. Siponen MK (2005) An analysis of the traditional IS security approaches: implications for research and practice. Eur J Inf Syst 14(3):303–315
26. Thomas M, Dhillon G (2011) Interpreting deep structures of information systems security. Comput J Adv Access
27. Krauss L (1972) SAFE: security audit and field evaluation for computer facilities and information systems. Amacon, New York
28. Broder JF, Tucker E (2012) Risk analysis and the security survey. Elsevier, Waltham
29. Landwehr CE (1981) Formal models for computer security. ACM Comput Surv 13(3):247–278
30. McLean J (1990) The specification and modeling of computer security. Computer 23(1):9–16
31. Yang Y, Zhou Y (2011) Fuzzy logic based method for network information security risk assessment
32. Lian S, Gritzalis S, Nedjah N, Lin I-C (2011) Special issue on soft computing for information system security. Appl Soft Comput 4257–4259
33. Wand Y, Weber R (1990) Towards a theory of the deep structure of information systems. In: Proceedings of international conference on information systems, pp 61–71
34. Talbot J, Jakeman M (2009) Security risk management body of knowledge (Wiley Series in Systems Engineering and Management)
35. Reason JT (1990) Human error. Cambridge University Press
36. Talbot J, Jakeman M (2011) Security risk management body of knowledge (Vol. 69). John Wiley & Sons
37. Carnegie Mellon University (2006) CERT/CC vulnerability discovery project. CERT Coordination Center, Software Engineering Institute, Pittsburgh

An Event-Driven, Scalable and Real-Time Geo-spatial Disaster Forensics Architecture: Decision Support for Integrated Disaster Risk Reduction

Jason Levy

Abstract "An event-driven, scalable and real-time, geo-spatial disaster forensics architecture" uses advances in decision support systems to apply forensic theory, insight and analysis to disaster related research and practice. It examines water resources disasters and their impact on humans, the built environment and natural systems. The chapter also identifies, and describes timely and innovative decision support architectures to analyze climate related disasters, enhance emergency preparedness, reduce disaster risk, promote disaster resilience and improve disaster mitigation, adaption, and management. The root causes of water resources disasters are explored and a distributed, scalable and real-time disaster forensics architecture with event-driven messaging and advanced geomatics engineering capabilities is put forth. Emphasis is given to vigilant monitoring, assessment, response and recovery of floods and oil and molasses spills in the US state of Hawaii. The decision support and situational awareness advances found in this chapter complement the recent success of water resources disaster risk management and disaster forensics in Europe and elsewhere. The herein proposed disaster forensics architecture helps managers uncover creative, timely and important strategies for analyzing water resources accidents and disasters. In this manner, professionals have additional tools to model the complex causality of disasters and are better equipped to apply disaster forensics theory to the promotion of a more holistic, sustainable relationship between society and the environment. Specifically, this contribution provides theoretical insights and practical examples to manage water resources disasters under uncertainty.

Keywords Disaster forensics architecture · Decision support systems · Flood risk management · Disaster risk reduction (DRR) · Forensic disaster analysis (FDA) · Water resources hazards

---

J. Levy (✉)
Honolulu, USA
e-mail: jlevy@hawaii.edu

A.J. Masys (ed.), *Disaster Forensics*, Advanced Sciences and Technologies for Security Applications, DOI 10.1007/978-3-319-41849-0_14

1 Introduction: Towards a Next-Generation Disaster Forensics Architecture

A comprehensive, scalable and real-time disaster forensics architecture can help to uncover how natural hazards become—or do not become—disasters and to investigate the optimal strategies to respond to crises (once they emerge). This is consistent with advancing the objectives of the Integrated Research on Disaster Risk (IRDR) research programme, a program focusing on disaster risk associated with flood, drought and other natural hazards. Co-sponsored by the International Council for Science (ICSU), the International Social Science Council (ISSC), and the United Nations International Strategy for Disaster Reduction (UNISDR), the decades-long IRDR initiative promotes a timely, important and transdisciplinary approach to mitigating disaster impacts, and improving related policy-making mechanisms. The Forensic Investigations of Disasters project identifies five priority areas that can be addressed in an integrated manner in order to effectively apply the fields of forensics theory and analysis to disaster related research and practice: policy, management, scientific research and development and disaster risk reduction [9]. The development of a real-time, event-driven, scalable, distributed and geospatial disaster forensics architecture has the potential to address each of these five priority areas. Specifically, advances in geomatics engineering and provides insights into the root causes of disasters that go beyond superficial after action reports and disaster case studies. Finally, in keeping with the IRDR and FORIN research objectives, this chapter's activities are aligned with Goal 4 of the 2013-1017 IRDR Strategic Plan: "Reducing risk and curbing losses through knowledge-based actions".

Understanding the root cause and complex causality of disasters is essential: The impact of the 2011 Tōhoku earthquake and tsunami, the 2013 Super Typhoon Haiyan, the 2004 Boxing Day Indian Ocean tsunami and Hurricanes Katrina and Sandy provide a painful reminder of the vulnerabilities of communities to high impact, low probability events. The disaster forensics methodologies, architectures and frameworks herein proposed are built around natural and technologic disaster case studies in the state of Hawaii with a focus on hazard reduction policies, disaster science, risk management, and socio-economic vulnerability. Particular emphasis is placed on building a culture of disaster preparedness which fosters dialogue between the private sector, academics, NGOs, community leaders, practitioners and government officials. This process creates a common discourse about disaster forensics science, policy and management.

Natural disasters and crises are on the rise worldwide and their financial impacts have grown exponentially. Hurricane Sandy, for example, which struck the US northeast coast in 2012, caused more than US$60 billion in damages and losses, while the financial toll of Hurricane Katrina in 2005 was even higher (exceeding US $100 billion). In the face of such catastrophic threats, advances in forensics analysis approaches to disaster investigations have become disaster risk priorities in both the US and around the world. This paper constitutes a major attempt to address, comprehensively and in-depth, both a science and policy-focused approach to the many issues associated with understanding the root causes of disasters, building a culture of disaster preparedness and increasing societal resilience. Specifically a real-time disaster forensics architecture is developed to deal with the complex causality associated with natural and technologic disasters. Advances in decision support and geomatics engineering are incorporated in the disaster forensics architecture. The overall goal of the proposed disaster forensics architecture is to learn from previous incidents and manage disaster risk in ways characterized by adaptation, transformation, and resilience. Forensic analyses show that devastation and loss allow for new opportunities to identify innovative, timely and important strategies in analyzing crises and disasters. On the other hand, traditional plans and policies for mitigating disaster losses are inadequate and so disaster forensics scholars must better understand how hazards impact vulnerable communities and the policy frameworks in which they are managed.

2 Water Resources Disaster Forensics and Complex Causality

The proposed disaster forensics decision support architecture is designed to assess the causes and consequences of natural disasters, to model the interdependencies and complex causality of disaster risk and to provide guidance for disaster response and recovery policies. A large body of research exists on disaster mitigation, preparedness, response and recovery. However, most of that work has addressed the risk, vulnerabilities and capacities of families and households, not the complex causality and root causes of disasters at the community, regional and national level of analysis. This chapter sheds light on some of the challenges associated with building a disaster forensics decision support architecture. The goal is to build a culture of disaster prevention and to make communities safer and more resilient in the face of significant natural and technological hazards. It is essential to highlight the risk created by high impact, low probability events and to develop disaster risk reduction strategies and policies for managing the unexpected and cascading impacts of these Black Swans (extreme events), particularly transboundary catastrophes that cross policy domains, geographic, political, and sectoral boundaries. Since the disaster forensics field draws on a diverse range of paradigms and influences, this chapter includes case histories, scientific investigations, empirical studies, conceptual-theoretical work, policy perspectives, institutional analyses, and risk analyses, among others. Current gaps in knowledge about managing disasters is highlighted with the aim of opening new research avenues for disaster forensics research and integrated disaster risk management, thereby revealing the complex causality that characterizes disaster investigations and analyses. Accordingly, our cross-disciplinary and transnational case studies examine the causes and consequences of natural disasters, with a special emphasis on crafting comprehensive disaster policy solutions to effectively address them.

The use of a disaster forensics decision support system can help to provide insights into understanding the antecedent conditions and root causes of disasters and assist with implementation of disaster mitigation and preparedness policies. In 2011, for example, US President Obama issued Presidential Policy Directive 8 (PPD 8): National Preparedness, which is "aimed at strengthening the security and resilience of the United States through systematic preparation for the threats that pose the greatest risk to the security of the nation, including acts of terrorism, cyber attacks, pandemics, and catastrophic natural disasters." A key component of PPD 8 was the establishment of the US National Preparedness Goal which describes a vision for US preparedness nationwide and identifies the core capabilities necessary to achieve that vision across five disaster management phases—Prevention, Protection, Mitigation, Response and Recovery. While many urban jurisdictions are likely to meet the National Preparedness Goal, others, particularly the more rural and vulnerable regions, including local, state, territorial and tribal communities, will likely face greater challenges.

Understanding the root cause and complex causality of water resources disaster—and the adaptive management of crises once they occur—remains a significant challenge for disaster professionals. Accordingly a complex systems disaster forensics architecture is put forth to better understand how water resources disasters impact communities and businesses, how they recover, and how they might enhance disaster planning and management in order to cope with future, unexpected, extreme events under conditions of climate change. Water resources constitute the lifeblood of human civilization and environmental systems—they support a diverse array of ecosystem components, as well as providing a precious source of freshwater supply, both surface water and ground water. However, issues related to water resources disasters and sustainable watershed management are among the most urgent environmental priorities: watershed resources are being increasingly polluted and depleted due to population pressures, inadequate hazards management and economic activities. Humanity currently stands at a "watershed" moment concerning the sustainable management of disaster risk. Preparing for the impending shift to a sustainable future requires an enhanced capacity for disaster forensics and problem solving, by utilizing both disaster forensics analyses, market institutions and a transformational, participatory approach to community decision making. Despite several decades of intensive research, water resources hazards remain a challenge for robust, integrated, and adaptive environmental planning. Water resources disaster forensics seeks to understand the root causes of disasters while meeting socio-economic goals for current (and future) generations and maintaining the ecological and hydrological integrity of watersheds. However, disaster forensics is inhibited by a number of factors in the environmental domain. First, environmental laws and regulations often work at cross purposes. Second, incentives and policies too often emphasize technical and biophysical watershed aspects without integrating financial and social issues. Finally, devolution of responsibility for watershed planning to local agencies has occurred without the requisite financial, scientific, and human resources. Given the vital importance of water resources hazards and disaster forensics, this section addresses a number of important questions regarding sustainable watershed theory and practice. What are the best disaster management practices to protect and restore watersheds, including surface and ground water? How best can we integrate values, ethics, and engineering technology to minimize water resources hazards, and promote watershed conservation (including water quality and supply concerns)? How can we deal with overlapping jurisdictional boundaries to improve disaster risk reduction and improve the long-term welfare of riparian communities?

The proposed disaster forensics architecture will enhance the capacity of authorities, communities and businesses to promote disaster preparedness and facilitate the dissemination of knowledge pertaining to water resources hazards, in addition to serving as a portal for post-disaster logistic support, technologic innovation and information sharing. Specifically, real-time, distributed and geospatial disaster forensics architectures can be used to help disaster response organizations, tribal and first nations communities, government officials and the private sector (including both local commercial and non-profit businesses) work together in mutually supporting roles in order to:

- assist in disaster investigations and analyses
- understand disaster resource needs of businesses, communities and first responders
- better understand the root causes and complex causalities of disasters
- strengthen connections between local businesses and these agencies to satisfy resource needs
- improve disaster communications
- incorporate the socio-economic, human and technologic vulnerabilities of communities
- develop operational practices to engage the private sector in disaster recovery
- liaise with other disaster organizations to conduct the necessary outreach, seminars/webinars and training support for disaster preparedness.

3 Flood Disaster Risk

Floods, a temporary inundation of land from excessive rainfall or wave action, constitute a frequent, widespread, and increasing natural hazard in many regions of the world, jeopardizing important environmental quality, economic resilience and social development goals such as addressing poverty, ensuring adequate food, water, and sanitation, and protecting the environment. Statistically significant increases in global land precipitation have been observed over the twentieth century, such as wetter winters in Europe and wetter summers in the Asia monsoon region. The Inter-governmental Panel on Climate Change (IPCC) and global climate models predict larger inter-year and intra-year variations in precipitation and changing climatic conditions. In addition to climate change and rising sea levels, the extent and impact of flood events is also increasing due to human activities, including land-use changes (such as the alteration of vegetation along rivers and coastlines), growing populations (and denser concentration of people and infrastructure in flood-prone areas), and development in headwaters (altering natural hydrologic balance). The concomitant effect of these changes has led to a greater, more rapid runoff. While the National Flood Insurance Program (NFIP), established by the U.S. Congress in 1968, limits development in the floodway (the central portion of the floodplain), regulations do not constrain construction in the remaining floodplain areas as long as development occurs in areas that are either higher than the 100-year flood level, or are protected by levees with at least 100-year protection Although the NFIP has significantly mitigated flood damages in Hawaii, major flood challenges exist in older areas of the state that were developed prior to flood control regulations and building standards as well as in areas with flood control improvements that are inadequate to contain or control larger floods by present standards. Under the NFIP each US county has mapped flood hazard areas and established a permit system to regulate development within these flood hazard areas.

Flooding constitutes a powerful and prominent mechanism for the release of hazardous material, particularly in urban areas, leading to significant adverse impacts including ecological degradation and health effects [16, 30]. Globally, the number of "great floods" (floods with discharges exceeding 100-year levels from basins larger than 200,000 km^2) is on the rise. From 1990 to 1999, floods killed approximately 100,000 people and affected over 1.4 billion. Society is expected to become increasingly vulnerable to flood risks. For example, about 20 % of the world's population now lives in coastal ecosystems (within 100 km from the coastline) and the population density of coastal ecosystem zone (175 people per km^2) is the highest among all ecosystems. Urban flooding may also inundate businesses and residences, disable critical lifelines (gas, powerlines, etc.), submerge underground spaces, paralyze transportation networks and lead to hazardous chemical releases (as natural disasters lead to technologic problems, often referred to as "na-tech" disasters). For example, the 1972 flooding of contaminated areas in Western New York may be at least partially responsible for increases in leukemia and lymphoma rates [31]. Floodwaters from Hurricane Katrina flushed out hazardous materials across the U.S. Gulf coast, including pollutants from Superfund sites, oil and grease from flooded parking lots and roads, as well as petrochemicals from submerged refineries, industrial facilities, and underground gasoline tanks [21]. In the Netherlands, France, and Belgium, the 1993–94 Meuse River winter flooding led to heavy metal soil contamination, including the release of cadmium, zinc, lead, and copper [1]. The release of oil products into water bodies in Nagoya, Japan has negatively impacted water quality in the Tokai region of the country [11]. Hartmann et al. [10] use the Japanese Pollutant Release and Transfer Register (PRTR) database to identify the location of hazardous chemicals stored at industrial facilities using inundation scenarios based on meteorological and elevation data [33, 34] during the September 11–12, 2000 Tokai, Japan floods.

4 Flood Disaster Risk Management in Hawaii

Flooding is among the most common and devastating of all natural disasters in the US state of Hawaii: Hawaii is vulnerable to several types of flooding events including storm water runoff floods (i.e. flash floods, storm surge floods, riverine floods, urban floods, high surf floods), rainfall flooding (i.e. frontal storms, upper troughs, convective type storms, hurricanes or tropical storms) and dam failure (dams can exacerbate flooding should they fail). Direct losses from floods include drownings and injuries as well as damage to critical infrastructure and property, agricultural production, and sites of historical and cultural value; direct losses to property typically results from soaking, dislocation, and destruction of personal belongings. Indirect health problems often arise, such as water-borne infections, exposure to chemical pollutants released into flood waters, and vector-borne diseases.

Rising sea levels and climate change pose a risk for Hawaii and other coastal regions—disaster forensics research has uncovered that regional sea level rise contributes to frequent and more intense coastal flood events. Structural flood defense measures are also modeled in the proposed disaster forensics architecture as they prevent or mitigate damages from storm surges and play a major role in determining flood risk. By examining the interconnected complexity of human adaptation, climate change, adaptation, and flood damage the disaster forensics architecture provides a valuable assessment of potential economic consequences for appraising the efficiency of adaptation measures. Even if sea temperatures eventually stabilize, sea levels will continue to rise and threaten coastlines for generations. In order to improve adaptation, insights from disaster forensics suggest that additional mitigative and preventive measures should be implemented to help coastal regions adapt to an increase in flood risk and limit flood damage costs.

To help advise the public about flood risks, the US National Weather Service (NWS) issues alerts providing important information about dangerous flood-related conditions including:

- Flash flood Watch (means that a flash flood is possible in the area with the next 36 h and the public should remain alert)
- Flash flood Warning (means flooding is likely to occur within the next hour or is already occurring and immediate action should be taken)
- Urban and Small Stream Advisory (flooding of small streams, streets, urban storm drains, and low-lying areas).

The aforementioned flood alerts may occur frequently in Hawaii. For example, from February 19 through April 2, 2006 alone, the National Weather Service Forecast Office in Honolulu issued over 500 non-routine alerts including:

- 111 Flash Flood Warnings (typically there are 2 to 3 flash flood events each year during this same time period across Hawai'i)
- 88 Special Marine Warnings (for waterspouts and/or strong thunderstorms over the water within 40 miles of land that are capable of producing winds greater

than 40 mph or large hail. Typically there are about 30 special marine warnings in Hawaii per year)

- five Winter Weather Advisories (means snowfall of 2–5 in. is likely in the next 24 h).

5 Decision Support for Flood Disaster Forensics

In order to improve flood disaster planning and management a comprehensive, event-driven, scalable, real-time, distributed, and geo-spatial flood forensics architecture is put forth which includes relevant analytic techniques and decision support systems (Fig. 1). While water resources hazards are on the rise, advances in situational awareness, decision theory, multi-media facilities, and computational capabilities have provided new opportunities for the development and application of real-time, distributed and geospatial disaster forensics tools to improve the visualization, modeling and management of such hazards. For example, a number of innovative techniques currently exist for capturing the statistical and methodological uncertainties associated with flood modeling and improving long-term flood investment decisions. One such tool involves probabilistic forecasts with large multi-model ensembles (rather than a single simulation). Accordingly, the proposed

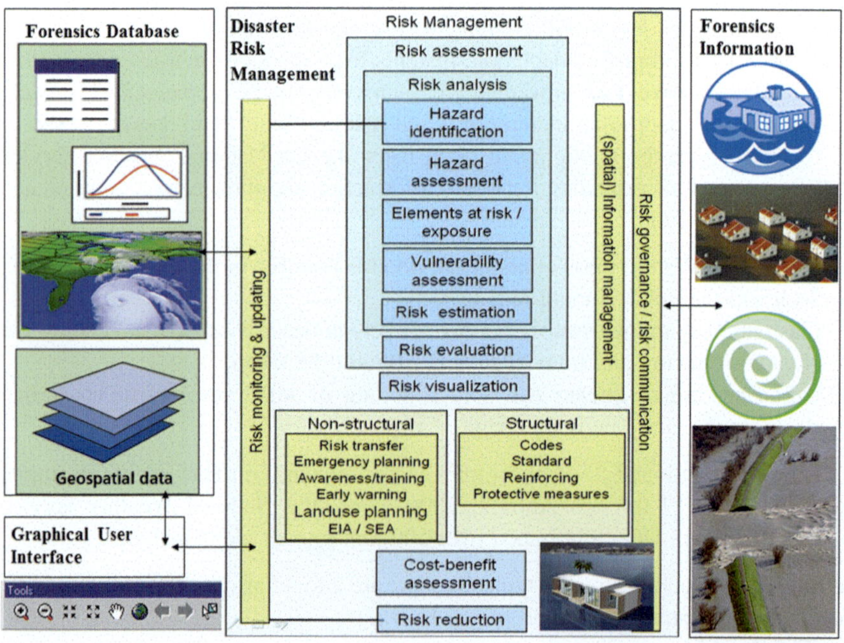

Fig. 1 A real-time, geo-spatial disaster forensics architecture for managing natural hazards

flood forensics architecture also involves numerical weather and climate models for predicting severe rainfall anomalies in Hawaii from several days to several months in advance. Such predictions are extremely valuable, allowing time for proactive flood protection measures to be taken. A dynamical climate prediction system is used in order to predict rainfall and improve flood investigations and planning; specifically ensemble prediction with dynamically conditioned perturbations is employed in order to reduce the uncertainty associated with seasonal climate prediction. The resulting disaster forensics information can be used to shape policy and guide future research and to reduce flood disaster risk.

The components of a real-time, geospatial disaster forensics architecture for managing natural hazards are now put forth (Fig. 1). Note that such a disaster forensics architecture involves three main components: a disaster forensics database, a disaster risk management system and a Graphical User-Interface. In the disaster risk management process key steps include risk analyses (hazard identification, hazard assessment, the identification of elements at risk, a vulnerability assessment), risk assessment (the evaluation of risk) and risk management (which includes risk governance/risk communication, spatial information management, risk monitoring and updating as well as mitigation activities). Finally, the disaster forensics decision support architecture involves a customized, flexible, and interactive GUI that interacts with analytical tools, databases, and models.

Remotely sensed and hydrological spatial data can help to understand the root cause and complex causality of disasters. There are three broad geospatial approaches to address flood forensics [5]: pre-processing data into a suitable format (such as spatial databases), direct GIS support (analysis, calibration, and other tasks) and post-processing data (such as creating flood-risk maps and performing cost-benefit analyses). Satellite imagery provides innovative and insightful flood management results by obtaining reliable, continuous, large-area, and synoptic coverage of key variables in a flood forensics investigation (including real-time estimates of rainfall, water levels, ocean waves, surface winds, and sea-surface heights), although calibration with ground-measured data is required. Multi-date satellite imageries is often used to reconstruct the history of previous extreme hydrological events and to monitor the effect of human activities and natural disasters on land use and land cover over time.

Decision support for flood forensics dates back approximately half a century, when the U.S. Army Corps of Engineer's (USACE) Hydrologic Engineering Center (HEC) first began developing algorithms and computer systems to evaluate hydrologic risk and flood damage reduction plans. More recent examples of flood forensics computer modeling and analysis tools include the European Union's River Basin Modeling Management and Flood Mitigation (RIBAMOD) project [3]. More specific application areas include flood control [26], flood evacuation [25], and flood risk mapping, forecasting and management [28]. Maxwell [17], Shim et al. [24], and Turban et al. [29] provide reviews on the theory and practice of decision support and disaster forensics for flood management.

Flood forensics typically involves examining challenges arising from both structural flood management approaches (i.e. codes, standards, reinforcements) and/or non-structural tools such as risk transfer, early warning, emergency planning, awareness/training, Environmental Impact Assessments (EIA) and Social Impacts Assessments (SIA). Until the 1960s, flood mitigation measures in the US, Canada and other Western countries were dominated by large-scale structural mitigation projects such as the state promotion of massive dams, dikes and diversions. For example, since the nineteenth century, flood control works by the US Army Corps of Engineers (USACE) have been used, with mixed long-term success, in an attempt to regulate large rivers such as the Mississippi. This emphasis on building flood embankments, constructing flood relief channels, and sometimes a series of flood control dams, constitutes the structural (so called "hard") approach to managing hydrologic systems. Although providing some benefits, disaster investigations have questioned the effectiveness of the structural approach. While the US continues to rely heavily on structural flood control measures, many countries around the world (particularly in Europe) the recommendations of disaster forensic analyses and investigations suggest that society shift toward a flood management policy of "living with floods", focusing on humanity's ability to co-exist with floods, rather than be protected from them. In particular, after the severe Rhine River flooding of 1993 and 1995, the Dutch government adopted a flood control policy of "more room for rivers" with an emphasis on establishing new storage and conveyance space. In this spirit, six European governments created the Meuse River High Water Action plan which provides billions of dollars for "longer storage and more release" for rivers. Here, the disaster forensic approaches have guided valuable flood risk management results. For example, in the German state of Baden-Wurttemberg, the addition of over 200 m^3 of floodplain storage has reduced flood stages to 1950 levels.

A real-time geospatial disaster forensics decision support architecture for flood disaster forensics is now put forth (Fig. 2). Note that the architecture involves similar components to the generic natural disaster forensics framework although it is specifically tailored to flood risk management and involves decision analysis tools for disaster decision support. Over the past two decades, advances in geomatics engineering have provided original and valuable data to accurately simulate complicated hydrologic and hydraulic phenomena (particularly in heavily urbanized watersheds). Key geospatial technologies for flood management include remote sensing, GIS, and mobile computing global positioning systems. For example, remotely sensed imagery has been used extensively for flood forecasting, primarily based on the presence of rain-bearing cloud patterns [8], and detailed flood mapping, which delineates inundated areas (detects flood boundary river inundation, stage and discharge) by comparing pre-flood and peak flood images (using clearly visible water levels). The mapping results can then be used to identify flood counter-measures and to create land development priority maps—a function of flood hazard and population as input for hydrologic models—and flood hazard assessment [12, 22]. Flood forensics is enhanced by the use of SPOT (Systeme Pour l'Observation de la Terre) multi-spectral imageries: they can identify inundated areas by capitalizing on the very low reflectance of water in the near infrared portion of the spectra [23]. At a larger scale, these medium resolution imageries are

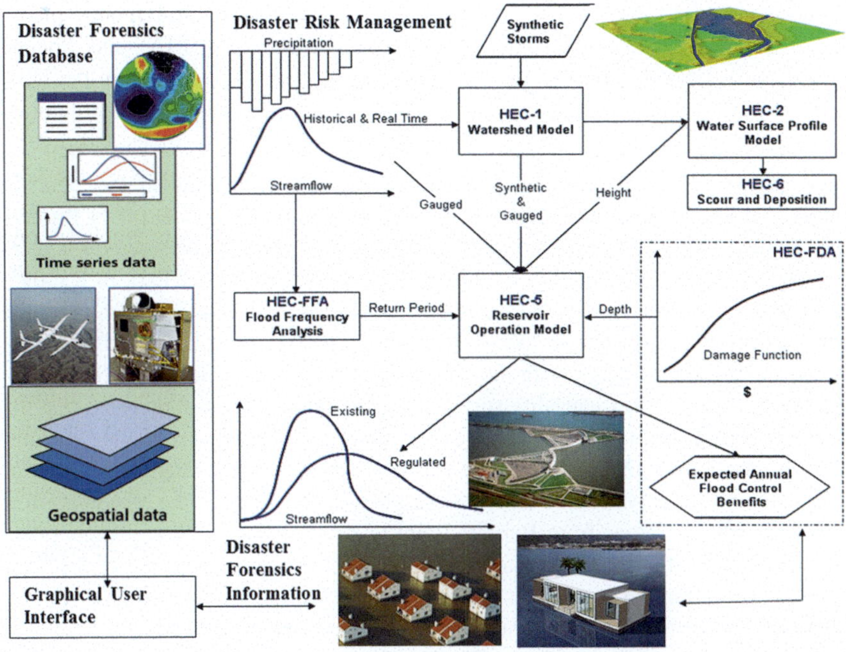

Fig. 2 A real-time, geo-spatial flood forensics decision support architecture

often supplemented by coarser resolution imagery such as the Normalized Difference Vegetation Index (NDVI) remotely sensed data from the National Oceanographic and Atmospheric Administration's (NOAA) Advanced Very High Resolution Radiometer (AVHRR). While LANDSAT, SPOT, and AVHRR imageries require cloud-free conditions, active (microwave) remote sensing, such as synthetic aperture radar (SAR), present on European Remote Sensing (ERS) and RADARSTAT satellites, is not sensitive to atmospheric conditions. The aforementioned tools can be used in concert to enable near real-time disaster forensics and flood risk monitoring although data obtained from different sensors often presents significant challenges [27]. However, a combination of optical and microwave remote sensing has yielded superior results for flood mapping in mountains.

The flood risk management components involve the U.S. Army Corps of Engineer's (USACE) Hydrologic Engineering Center (HEC) tools including the generation of synthetic storm data, and the use of watershed models (HEC-1), water surface profile tools (HEC-2), a damage function analysis (HEC-FDA), flood frequency analysis (HEC-FFA), reservoir operation models (HEC-5), scour and deposition models (HEC-6). These tools can be used to generate the expected annual flood control benefits from various structural and non-structural engineering approaches.

6 Disaster Forensics for Oil and Molasses Spills in Hawaii

The Hawaii Oil Spill Center (HSC), funded by the Member Companies of the Clean Islands Council and the Marine Spill Response Corporation, is dedicated to Oil Spill Emergency Response preparedness in Hawaii. In 2014, a mix of ship fuel and water spilled from a loading buoy anchored offshore from Hawaii Independent Energy; the single-point mooring is anchored approximately a mile off of Barbers Point Harbor and used to transfer crude oil and refined products between oil tankers and Hawaii Independent Energy's refinery in Campbell Industrial Park in Kapolei. The resulting release of oily water soiled the ocean for more than half a square mile near Barbers Point Harbor in West Oahu. A thorough forensic investigation by Hawaii Independent Energy with the help from the US Coast Guard uncovered the root causes and complex causality surrounding the oil spill. It was revealed that the release occurred during a hose change-out operation as part of a routine maintenance program, prior to which the hose was flushed twice with water. The source of the release was quickly secured and five vessels were involved in the response effort (such as inspecting areas from Barbers Point to Iroquois Point for any signs of oil and picking up oil from the ocean surface) including teams from the U.S. Coast Guard, Hawaii State Department of Health and Marine Spill Response Center together with Clean Islands, a first-response oil spill response vessel.

There have been other spills to affect Hawaii waters. For example, a September 9, 2013 accident spilled 1400 tons of molasses (produced from sugar cane at the Hawaiian Commercial and Sugar Co. on Maui) as Matson Navigation Co. (Matson) transported molasses from Honolulu Harbor to the mainland to be sold (and currently ships molasses about once a week). A forensics investigation highlighted that the culprit was a faulty pipe that discharged 233 thousand gallons (1400 tons) of molasses from a 16-hundred ton load meant for a California-bound Matson container ship. While molasses doesn't harm people directly, the pollutant killed a large number of fish and that can cause an increase in sharks and barracuda and eels which can in turn present a safety hazards to humans in the water. In fact, the molasses release was more hazardous to sea life than a conventional oil spill because the concentrated sugary material quickly sinks to the bottom of the harbor and caused widespread de-oxygenation. Various species of coral were injured or killed, and within a day, thousands of dead marine life quickly appeared near the Keehi Lagoon: more than 26,000 fish and other marine species suffocated and died. A molasses spill is also more difficult to remediate than an oil spill, which can be cleaned by skimming the surface. Natural currents and weather eventually diluted the molasses and flushed it out of the harbor and a nearby lagoon.

As an unregulated product, neither Matson nor government officials had a contingency plan to respond to a molasses spill since molasses was not considered a hazardous material (such as oil or gas) which require a spill response plan. Until recently, there was no requirement for businesses shipping molasses in Hawaii to pass routine pipeline inspections, or to develop and follow molasses spill response plans. However, on September 20, 2013 the Hawaii Department of Transportation determined that all businesses which pump products through State of Hawaii port pipelines must provide evidence of these response plans and pass routine inspections of molasses pipelines. Previously no such reporting had been required. In particular, the molasses response plan focuses on prevention and early detection, with regular inspections of pipelines and hourly monitoring of transmission operations. This is essential since molasses spills are often very difficult and expensive to clean up. To encourage compliance large fines are in place for discharging pollutants and violations of the Clean Water Act. For example, Matson could have to pay up to $25,000 US dollars a day for such violations. It is hoped that these regulations will reduce disaster risk, protect the environment, and promote safe handling and transportation procedures.

7 An Event-Driven, Scalable, and Real-Time Geo-spatial Oil Spill Decision Support Architecture

Figure 3 involves a real-time, geo-spatial oil spill decision support architecture. This event-driven and scalable oil spill monitoring and management system for disaster forensics differs from a traditional web-based request/response interaction model in two important ways. Specifically, the disaster architecture is scalable (i.e.

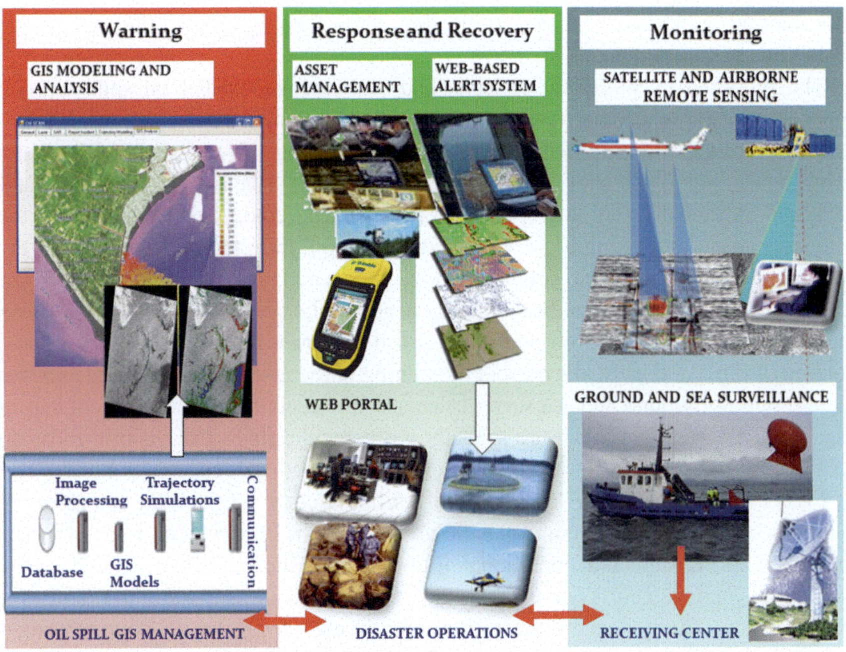

Fig. 3 An event-driven, scalable, and real-time geo-spatial oil spill decision support architecture

it can easily be extended to include new functions or to integrate new information services) and event-driven (has the capability to automatically match events with the subscriptions from the clients and push the event information to relevant decision makers in real-time). A publish/subscribe interaction model is used in event-driven system architectures (i.e. event-based computing) [18]. This powerful interaction paradigm includes a unique middleware component that facilitates the interaction between the distributed clients and is responsible for conveying oil spill content (i.e. information messages) from the producer clients (i.e. publishers such as the emergency management authority or oil spill management consortium) which generate and publish the oil spill event information to consumer clients (i.e. subscribers such as shipping companies, wildlife officials, recreational boaters), who are interested in receiving the events' oil spill information messages. This interaction procedure enables the clients to interact without direct knowledge of each other (i.e. they are loosely coupled from each other) and allows for the integration of heterogeneous, autonomous, and dynamic clients and sensor devices into the oil spill forensics architecture, thereby leading to better scalability and communication efficiency. Specifically, oil spill consumers receive only oil spill content notifications (i.e. events) that they have already registered for (subscriptions) while producers publish event messages (notifications) which might be of interest to the clients. The bulk of the computation and the communication processes occur inside the core of the middleware component (i.e. the event-notification service) which handles the published events, matches them with the registered subscriptions, and asynchronously pushes the events to matched subscriber clients in a timely manner. The internal architecture of the notification service has a major influence on the scalability of the system.

The proposed event-driven, scalable, and real-time geo-spatial oil spill decision support architecture also captures three main emergency management phases: oil spill disaster warning; response and recovery; and monitoring. First, the oil spill disaster management phase includes Geographic Information Systems (GIS) modeling and analysis. Specifically, the oil spill GIS management component incorporates the database, image processing tools, GIS models, oil spill trajectory simulations and the communication system (left column of Fig. 3). Second, the oil spill response and recovery component includes asset management and the web-based alert system. This incorporates not only the web portal but also the disaster operations system (middle column of Fig. 3). Finally, the oil spill monitoring phase involves satellite and airborne remote sensing, ground and sea surveillance and the receiving center (right column of Fig. 3). The system's middleware spatially processes the rules and algorithms for oil spill events and supports real-time data transaction from heterogeneous sources.

Figure 4 documents the proposed oil spill system configuration for resource acquisition, command and control and communication. In particular, Fig. 4a shows the schematic design and operational framework of the real-time oil spill monitoring and management system while Fig. 4b presents the software and hardware configuration. The system architecture of Fig. 4a includes the modules for resource acquisition (spill detection, assessment, forecasting, modelling, mapping and early warning), command and control (oil mission planning, coordination, decision making, mobilization and communication) and communication (message, voice, FTP, email, video, and internet). The software and hardware configuration of Fig. 4b involves the Central Repository System (CRS), Disaster Data Processing Modules (DDPM), a Command and Control System (CCS) and a Portal Management System (PMS). Each component of Fig. 4b is now described. First, the Central Repository System (CRS) is composed of computer servers and database storage servers. In particular the ArcSDE v9.3 workgroup geodatabase and Oracle 10 g database servers are used for storage and access management of spatial data. Second, the Disaster Data Processing Modules (DDPM) assist with oil spill monitoring and data modeling. Image analysis and processing is performed using

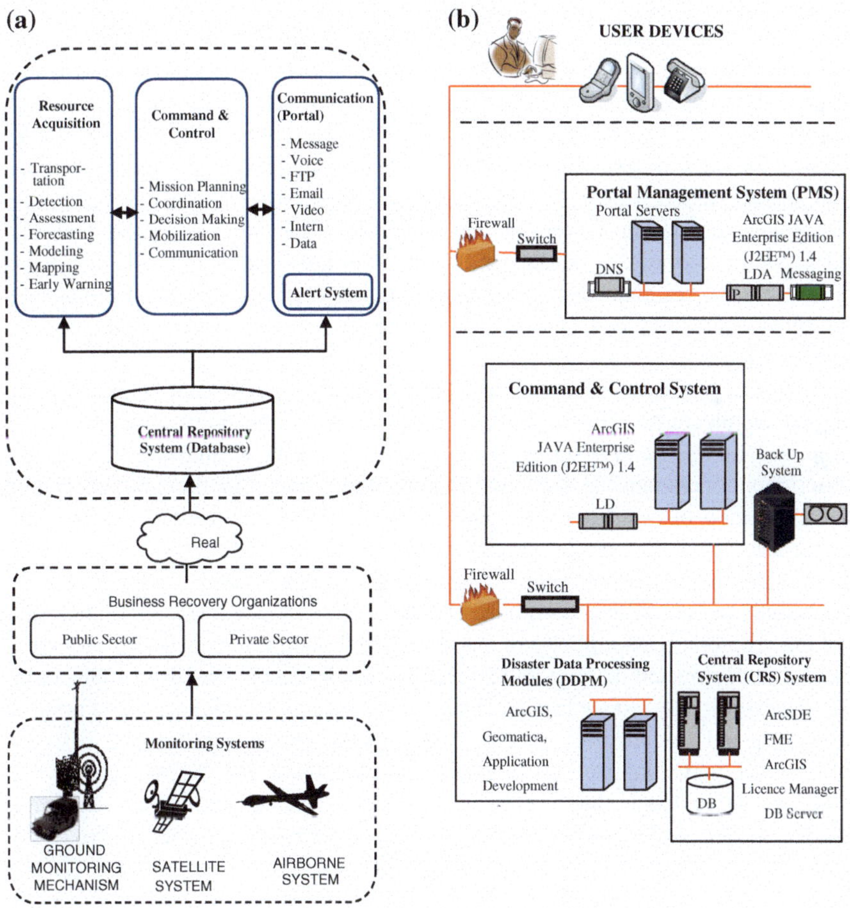

Fig. 4 Oil spill system configuration for resource acquisition, command and control and communication. **a** System architecture, schematics and operational framework. **b** Software and hardware configuration of engineering system design

Geomatica 10.1 while disaster models have been developed in ArcGIS 9.3. Third, the Command and Control System (CCS) serves as a bridge between the portal system, the data processing modules and the central repository system. Predesigned forms developed in JAVA Enterprise Edition (J2EETM) 1.4 link with the ArcGIS Server. Fourth, the Portal Management System (PMS) manages all incoming and outgoing data transactions through the CCS. The portal system is an internet based high-performance and secure messaging platform which facilitates communications between all decision makers. This secure communication system ensure the confidentiality, integrity and availability of digital communications through the use of session encryption, content filtering, and user authentication. The PMS was developed using Java and ArcGIS server and supports GIS data transactions.

The system receives information through satellite images, airborne data and ground surveys or devices. Specifically, high resolution (less than 3 m) imagery is essential for calculating important oil spill data that is used in the creation of oil spill maps. This architecture provides real-time geospatial oil spill data to key emergency management decision makers in order to improve oil spill modeling, management and overall situational analysis. Web-based technologies then link data directly to a central repository which includes all tabular and spatial data required for oil spill modeling (as well as all thematic output products generated from related disaster models). This real-time system also assimilates ocean and meteorological data into numerical computer models that simulate ocean dynamics and generate forecasts of ocean conditions. This information is useful for ocean vessels, natural resource managers, transportation companies, coastal researchers and other users that can benefit from oil spill forecasts, models and disaster simulations.

There are two types of notification service architectures: centralized and distributed architectures. The centralized architecture of the notification service is comprised of one central server component unit which is addressed by all the subscription operations as well as all the publications. As the first generation implementation of event-based systems, the centralized notification service architecture uses matching engine algorithms to match the publications (i.e., notifications) with the registered subscriptions. Consequently when subscriptions are successfully matched the system sends the published notifications to the subscribers. The relative simplicity of the centralized architecture makes it less complicated to deploy and manage and is ideal for small-scale event-based applications [4]; however, it may reduce the scalability of the system, introduce a single point of failure [6] and become less efficient when faced with a high volume of event publications, high subscriptions diversity and wide users distribution.

On the other hand, the distributed architecture of the notification service is comprised of several interconnected server components (known as event brokers or dispatching servers) which act as a centralized middleware between users [7, 19]. Connectivity to other subnets in the system network is due to the fact that event broker is connected to other brokers within the notification service. Every broker in the notification service not only serves the local users that are connected to it but also acts simultaneously as a publisher and a subscriber on behalf of the publications and the subscriptions that are directed to it from its subnet or the neighbour brokers. The role of a client, publisher or subscriber, ends by accessing the closest brokers in the system network and performing the required publish/subscribe operation. The distributed architecture of the notification service is ideal for achieving highly scalable event-based systems over wide-area or large-scale networks (e.g., the Internet). However, extensive efforts continue to examine key challenges facing the distributed notification service architecture including the security of publish/subscribe systems [2, 32], fault tolerance in reliable notification delivery and self-stabilizing strategies [14, 20], and automatic topology configuration and self-organizing publish/subscribe systems [13, 15]. Accordingly, improving the efficiency of the distributed notification service in publish/subscribe systems and achieving a reliable and efficient event-based system remain important research goals.

8 Conclusion

The chapter has shown that a forensics architecture for natural disaster management can reduce flood disaster risk, enhance disaster resilience and promote a more robust economy. The root causes of water resources disasters are examined and a real time, geo-spatial disaster forensics architecture is put forth. The disaster forensics architecture can be extended to address natural, technologic, human-induced and health-related threats. The decision support research found in this chapter complements the recent advances in water resources risk management and disaster forensics in Europe and elsewhere. It was shown that the proposed disaster forensics architecture can be used to promote situational awareness during floods and oil spills, improve resource allocation, identify hidden dangers, and manage unexpected hazards. The decision support systems also provide valuable insights into creative, timely and important strategies for preventing water resources accidents and disasters ranging from oil and molasses spills to floods. By so doing disaster forensics professionals have additional tools to model the complex causality of disasters and to apply disaster forensics theory, insight and analysis to water resources research and practice in order to promote a more holistic, sustainable relationship between society and the environment. This chapter provides theoretical insights and practical examples to show how advances in geomatics engineering, communication technologies and decision support can reduce the impact of water resources disasters. In summary, disaster forensic decision support architectures are *sina qua non* for improving real-time water resources hazard monitoring and response, improving crisis management, developing smarter flood and oil spill contingency plans, and improving flood prediction under uncertainty.

It has been shown that a robust flood disaster forensics architecture should include the systematic and comprehensive modeling and management uncertainty for five key reasons. First, extreme flood events are, by definition, rare, often requiring the use of synthetic data. Second, translating the physical hydrological system into a numerical model involves simplification. This in turn leads to model (conceptual representation) uncertainty, particularly when the structural flood model is inadequate or ambiguous. Third, predictive uncertainty arises in many contexts around the world, such as extrapolating flood models to predict inundation behavior under conditions of variability and change. Fourth, there are limitations in measuring hydrologic forcing variables (such as precipitation and temperature) due to inadequate spatio-temporal sampling densities. Fifth, significant uncertainty may arise due to insufficient understanding of socio-economic watershed characteristics (risk attitudes, infrastructure costs, etc.) and an inability to determine the 'best' model parameter set combination (based on available variables). The accurate parameterization of some hydrologic and meteorological processes is particularly difficult (cloud formation, soil moisture dynamics, land-surface interactions, etc.).

The proposed disaster forensics architecture should be especially useful for emergency managers, community leaders, government agencies, business leaders, tribal governments, indigenous communities, and businesses in coastal areas concerned with flood mitigation measures and disaster monitoring, including nonprofit organizations, research institutions, disaster management agencies, and other stakeholders. For example, the proposed flood forensics architecture can be used to meet the rising expectations of society for more timely and reliable heavy rainfall warnings. To apply the lessons of this paper a flood disaster forensics committee with members from the private sector, trade associations, non-profit organizations and the public has been assembled in the US state of Hawaii in order to apply the proposed disaster forensics architecture with a primary focus on early warning, disaster investigations, natural hazard management, emergency management resource allocation, business continuity planning and economic recovery.

References

1. Albering HJ, Van Leusen SM, Moonen EJ, Hoogewerff JA, Kleinjans JC, Human JC (1999) Health risk assessment: a case study involving heavy metal soil contamination after the flooding of the river meuse during the winter of 1993–1994. Environ Health Perspect 107:37–43
2. Belokosztolszki A, Eyers DM, Pietzuch PR, Bacon J, Moody K (2003) Role-based access control for publish/subscribe middleware architectures (trans: San Diego). ACM New York, California, pp 1–8
3. Casale R, Samuels P (eds) (1998) RIBAMOD river basin modeling management and flood mitigation. Concerted action. In: Proceedings of the first workshop on current policy and practice EUR 18019 EN. ISBN 92-828-2002-5
4. Cao F (2006) Architecture design for distributed content-based publish-subscribe systems. Unpublished thesis, Princeton University
5. Clark MJ (1998) Putting water in its place: a perspective on GIS in hydrology and water management. Hydrol Process 12:823–834
6. Cugola G, Jacobsen H-A (2002) Using publish/subscribe middleware for mobile systems. ACM SIGMOBILE Mobile Comput Commun Rev 6(4):25–33
7. Cugola G, Nitto ED, Fuggetta A (2001) The JEDI event-based infrastructure and its application to the development of the OPSS WFMS. IEEE Trans Software Eng 27(9):827–850

8. Feidas HN, Cartalis C, Cracknell AP (2000) Use of METEOSAT imagery to define clouds linked with floods in Greece. Int J Remote Sens 21(5):1047–1072
9. Forensic Investigations of Disasters (FORIN) project (2015) Integrated research on disaster risk. http://www.irdrinternational.org
10. Hartmann J, Levy JK, Okada N (2006) Managing surface water contamination in Nagoya, Japan: an integrated water basin management decision framework. Water Resour Manage 20:411–430
11. Isaji C (2003) Integrated water quality management for drinking water of good quality. Water Sci Technol 47(9):15–23
12. Islam MM, Sadu K (2002) Development of priority map remote sensing data for flood counter measures by geographic information system. J Hydrol Eng 7:346–355
13. Jaeger MA (2005) Self-organizing publish/subscribe (trans: Grenoble). ACM New York, France, pp 1–5
14. Jaeger MA, Muhl G (2005) Stochastic analysis and comparison of self-stabilizing routing algorithms for publish/subscribe systems. In: Proceedings of the 13th IEEE international symposium on modeling, analysis, and simulation of computer and telecommunication systems (MASCOTS'05), Atlanta, Georgia, pp 471–479
15. Jaeger MA, Parzyjegla H, Mühl G, Herrmann K (2007) Self-organizing broker topologies for publish/subscribe systems (trans: Seoul). ACM New York, Korea, pp 543–550
16. Lindell MK, Perry RW (1997) Hazardous material releases in the northridge earthquake: implications for sesimic risk assessment. Risk Anal 17(2):147–156
17. Maxwell DT (2004) Decision analysis: aiding insight VII. OR/MS Today 31(5):44–55
18. Muhl G, Fiege L, Pietzuch P (2006) Distributed Event-Based Systems, Germany: Springer
19. Pallickara S, Fox G (2003) NaradaBrokering: a distributed middleware framework and architecture for enabling durable peer-to-peer grids (trans: Rio Janeiro, Brazil)
20. Pallickara S, Bulut H, Fox G (2007) Fault-tolerant reliable delivery of messages in distributed publish/subscribe systems. In: Proceedings of the fourth international conference on autonomic computing (ICAC'07), Jacksonville, Florida, USA. IEEE Computer Society, p 19
21. Reible DD, Haas CN, Pardue JH, Walsh WJ (2006) Toxic and contaminant concerns generated by hurricane Katrina. J Environ Eng 132(6):565–566
22. Saatchi SS, Nelson B, Podest E, Holt J (2000) Mapping land cover types in the Amazon Basin using 1 km JERS-1 Mosaic. Int J Remote Sens 21(6):1201–1234
23. Sanyal J, Lu X (2004) Application of remote sensing in flood management with special reference to monsoon Asia: a review. Nat Hazards 33:283–301
24. Shim JP, Warkentin JF, Courtney DJ, Power R, Sharda R, Carlsson C (2002) Past, present, and future of decision support technology. Decis Support Syst 33:111–126
25. Simonovic SP, Ahmad S (2005) Computer-based model for flood evacuation emergency planning. Nat Hazards 34:25–51
26. Soscini-Sessa R, Castelletti A, Weber EA (2003) DSS for planning and managing water reservoir systems. Environ Model Software 18:395–404
27. Tholey N, Clandillon S, De Fraipont P (1997) The contribution of space borne SAR and optical data in monitoring flood events: examples in northern and southern France. Hydrol Process 11:1427–1439
28. Todini E (1999) An operational decision support system for flood risk mapping, forecasting and management. Urban Water 1:131–143
29. Turban E, Aronson JE, Liang T-P (2004) Decision support systems and intelligent systems, 7th edn. Prentice Hall Inc., New Jersey
30. White K (1993) Signs of an olive branch: confronting the environmental health consequences of the midwestern floods. Environ Health Perspect 101:584–588
31. Young S, Balluz L, Malilay J (2004) Natural and technologic hazardous material releases during and after natural disasters: a review. Sci Total Environ 322:3–20
32. Wang C, Carzaniga A, Evans D, Wolf AL (2002) Security issues and requirements for internet-scale publish-subscribe systems (trans: Island of Hawaii). IEEE Computer Society, Washington, DC, p 303

33. Zhang J, Hori H, Tatano H, Okada N, Zhang C, Matsumoto T (2003) GIS and flood inundation model-based flood risk assessment in an urbanized floodplain. In: International conference of GIS and remote sensing in hydrology, water resources and environment (ICGRHWE). Three Gorges Dam Construction Site, China, IAHS Publication 289, Three Gorges, pp 1–8
34. Zhai G, Fukuzono T, Saburo I (2005) Modeling flood damage: case of Tokai flood 2000. J Am Water Resour Assoc (JAWRA) 41(1):77–92

Advances in Economics and Disaster Forensics: A Multi-criteria Disaster Forensics Analysis (MCDFA) of the 2012 Kahuku Wind Farm Battery Fire on Oahu, Hawaii

Jason Levy and Peiyong Yu

Abstract The discipline of economics and its many sub- and closely related disciplines offer valuable modeling techniques to relate and apply forensic theory, insight and analysis to disaster related research. We herein propose advances in economics and disaster forensics to also reduce disaster risk and assess the direct and indirect impacts of disasters. This chapter constitutes a landmark attempt to address, comprehensively and in-depth, the many timely and important issues associated with using the field of economics to build a culture of disaster prevention and to understand the root cause and complex causality of disasters. In particular, advances in microeconomic, macroeconomic and forensic analyses are used to assess the causes and consequences of energy related disasters. A timely, original and valuable Multi-Criteria Disaster Forensics Analysis (MCDFA) approach for the forensic analyses of disasters is put forth and the 2012 battery room fire at the Kahuku wind-energy storage farm on Oahu, Hawaii is used as a case study to illustrate the proposed approach. Modeling identifies dynamic volt-amp reactive (D-VAR) technology as a preferred alternative over lead acid batteries for the Kahuku windfarm.

Keywords Disaster forensics · Energy storage system · Macroeconomics · Microeconomics · Multi-criteria disaster forensics analysis (MCDFA) · Political economic theory renewable energy · Windfarms

J. Levy (✉) · P. Yu
Honolulu, USA
e-mail: jlevy@hawaii.edu

A.J. Masys (ed.), *Disaster Forensics*, Advanced Sciences and Technologies for Security Applications, DOI 10.1007/978-3-319-41849-0_15

1 Introduction

Disaster losses are on the rise around the world in both developing and developed countries. A number of high impact and low probability events have occurred in the past decade alone: in the period from 2005 to 2014 more than six thousand natural and technologic disasters occurred, killing nearly one million people and causing more than one trillion dollars in economic losses [47]. Prior to the 20th century the discipline of economics was known as political economics (the convergence of politics and economics) and economic considerations continue to play a central role in our modern disaster forensics policies. However, recent accidents, injuries, fatalities in the energy sector require a systematic investigation of the root causes and complex causalities of energy related disasters—and serve as a barrier to achieving a sustainable energy future in general and increasing the penetration of renewables.

Better understanding the role of economics as it pertains to the root causes and complex causalities of disasters is consistent with the Forensic Investigations of Disasters (FORIN) project in order to guide disaster management policy, encourage coherence across essential disciplines and advance methodological diversity [29]:

- **Policy**: conduct disaster forensic analyses with inputs from multiple disciplines, stakeholders, and policy makers relevant to the renewable energy field.
- **Management**: highlight the link between energy disaster related research and practice and forensic theory, insight and analysis in order to improve policy formulation and application and share results online through high-quality publicly available case studies.
- **Scientific research**: build an interdisciplinary forensics capacity and implement science-based disaster forensics results as they apply to renewable energy systems.
- **Development**: determine the local manifestations of disasters, promote a culture of disaster risk reduction culture among all stakeholders, and foster wider dialogue between analytical researchers and implementing practitioners, building a common discourse in the process.
- **Disaster risk reduction**: illustrate risk drivers and promote sustainability through methodologically diverse science-based research, providing wider emphasis on reducing the risk of ecological damage, destroyed infrastructure and human accidents, injuries fatalities

Political considerations continue to influence the distribution of public and private relief money although the use of government funds to support disaster survivors is a relatively recent occurrence. For example, in 1887 US President Grover Cleveland noted that there was no constitutional basis for public funds to be used to reduce the suffering of individuals as a result of a disaster [9] and the public expected the costs of disasters being borne by individuals. Comprehensive economic assessments of disasters were not systematically completed before the twentieth century. An early economic study from 1920 estimated the impacts of the

Halifax ship explosion of December 1917 [63]. However, in the mid to late twentieth centuries, as public aid to victims became more socially acceptable and governments began investing in higher expenditures for disaster and emergency management, more resources in the US and other countries were allocated to estimating the economic impacts of disasters.

Understanding the economic preferences and the values of stakeholders can help disaster forensics professionals to master key competencies including administrative, management, and public policy skills—and assist disaster managers as they communicate with disciplinary experts and practitioners. Fortunately, the field of disaster forensics has become increasingly transdisciplinary, allowing forensics professionals to study topics across economics disciplines including microeconomics, macroeconomics, insurance, and political economic theory. This is because the disaster forensics profession analyzes disasters from a variety of perspectives and interacts with economics disciplinary experts and practitioners from multiple fields. Specifically, disaster forensics professionals with multidisciplinary knowledge are best qualified to facilitate meaningful dialogue with a range of experts and to incorporate advances in economics, science and management advances into disaster prevention, mitigation and planning efforts.

2 The Principles of Economic Loss Assessment: Understanding Complexity

Disasters constitute a complex chain of events that can derail sustainable development (especially in developing countries) and disrupt both the local, national and global economy. Calculating economic damages constitutes a complex, challenging and onerous task because the cost of a disaster is ultimately related the type of extreme event, the magnitude and duration of the event, the structure of the local economy, the time of onset, and the population density, critical infrastructure and natural resources of the geographic affected area. Rapid "back-of-the-envelope" type damage estimates calculated shortly afterward tend to significantly overstate a recent disaster for a number of reasons: buildings, infrastructure and crops may suffer less destruction than originally assumed; the media may sensationalize damages and political leaders may overestimate losses to maximize political leverage and obtain maximum disaster relief assistance. While the direct physical damage caused by disasters (such as debris removal, property damage to buildings and infrastructure, etc.) is often inspected, analyzed and estimated by engineers, architects, and construction specialists [45] economists focus on indirect losses and the losses associated with employment income. Indirect losses typically include commuter disruptions, loss of local tax revenues, reduced tourism and activities at damaged firms, loss of income in secondary and tertiary employment, and business disruptions not directly attributable to damage. For example, when a utility plant is damaged there are indirect impacts to businesses (utility customers) since the customers are not able to operate [50]. As another example, severe flooding occurred during the 2011 monsoon season in Thailand negatively impacted the production of vehicles in Europe and Japan and led to global supply chain disruptions, including a shortage of Thai parts and idled production of Prius hybrids and Camry sedans. Japanese automakers were also struggling with a record high yen and trying to recover from the aftermath of the March 11, 2011 Tohoku Earthquake in east Japan that led to a tsunami, a nuclear facility malfunction, and global economic impacts. As the biggest Southeast Asian manufacturing hub for Japanese auto-related companies, Thailand's worst floods in half a century spread disruptions to Honda's car plants from Malaysia and Japan to the U.S. and Canada. This contributed to a correction in the Tokyo Stock Price Index (TOPIX) transport index and hindered plans by Japanese automakers to revive production and ramp up parts supplies and assembly in overseas plants.

It is important to emphasize that in the globalized twenty-first century, a significant disaster in one country can create socio-economic disruptions around the world. For example, the 2004 Boxing Day Sumatra-Andaman earthquake and tsunami killed more than 230,000 people in 14 Indian Ocean countries including more than 5000 people in Thailand. This led to the deaths of 543 Swedish citizens across the Indian ocean region; shockingly, no other natural or technologic disaster in Swedish modern history has led to more deaths (not even the Baltic ferry MS Estonia catastrophe, the deadliest European shipwreck disaster to have occurred in peacetime).

For example, as previously mentioned, the 2011 Thai floods damaged Honda's automotive plants near Bangkok, significantly disrupting production. As a result Honda did not require the same volume of trucking services to deliver raw materials or pick up finished Honda vehicles, which are carried by truck, railroad or large ship. This has the potential to impact the employment of truck drivers, railroad workers, shipyard workers.

Economically, the cost of a disaster and the losses that stem from it are two separate terms. First, costs are incurred when damaged or destroyed tangible assets (capital) are replaced, repaired or reinforced. This includes the use of structural mitigation approaches such as the construction of seawalls in flood prone areas, or the reinforcement of bridges or buildings in earthquake prone areas. In contrast, *losses* occur principally through the destruction of wealth—an economy's wealth are the physical assets such as roads, bridges, utilities, factories, buildings or natural resources that generate income (Table 1). To avoid double counting these losses, either the lost income that these physical assets help generate must be calculated or the decline in the assets' values.

There are high levels of uncertainty in estimating disaster losses since they manifest themselves in numerous ways and one must take into account several factors that are often overlooked, interconnected or notoriously difficult to measure. For example, how to determine an accurate value of a public utility, a hydroelectric project, a public bridge or a sewage treatment plant? The economic value of a physical asset is its present discounted value, but this is a subjective judgment. Determining an asset's true market value is an alternative approach, but this may be

Table 1 Definitions and concepts for calculating the economic effects of disasters

Term	Definition	Example
Losses	Change in health to an individual (including loss of life) or to the wealth of a business or individual caused by damage to structures or other physical assets	Injury, disability, or death of humans. Damage to levees, roads, bridges, utilities, factories, homes, buildings, infrastructure, equipment, buildings and structures, destruction of crops and forests
Direct versus indirect losses	Direct losses are those resulting from damages to humans, agricultural property, infrastructure and resources. Indirect losses follow from the physical damages	Direct losses: building damages, bridge collapse, loss of lives. Indirect losses: commuter disruptions, loss of local tax revenues, reduced tourism
Market versus non-market effects	Market effects are those that are reflected in national income accounts data; non-market effects do not appear in the national income accounts data	Market effect: loss of income due to disaster-caused destruction. Nonmarket effects: loss of leisure time due to longer commute as a result of the disaster
Costs	Highest-valued of foregone alternative use of a resource	Mitigation expenditures undertaken before the disaster occurs (for example, construction of levees or seawalls or reinforcement of buildings) and reconstruction of buildings, etc. during recovery period
Redistribution	Transfer of wealth between individuals or governments	Federal disaster relief, but also includes transfers that occur because resources or production are moved to a new region
Wealth	Present value of the income stream from the productive assets of society	The value of a forest or farmland is the sum of the flow of monetary benefits (income from sales of timber or crops) and non-monetary benefits (vistas and recreational benefits of a forest)

Adapted from Saint Louis Reserve [66] and Brookshire and McKee [14]

extremely difficult since many physical assets are not traded in the marketplace. Without accurate information, the asset's replacement cost is often used as a default. Other difficult questions typically arise. For example, assume that a major flood inundates a community that contains both a residential and agricultural area. How to measure the decline in property values that may occur in the vicinity of this disaster area? Is it possible to estimate losses associated with livestock that died, became injured or were unable to gain weight during the extreme event? What prices and production levels should be attached to crops that were destroyed or damaged in the floodwaters (before they could be harvested)? Finally, what about the farmers and other workers that died or suffered serious injuries? Can their expected lifetime earnings be accurately estimated?

With the growing size and political influence of the US middle class in the 1950s, survivors of disasters after World War II perceived the distribution of disaster relief as "a corrective to a naturally induced injustice" rather than a form of welfare [41] and expected larger amounts of public assistance regardless of their own resources. There continues to be general acceptance for the US federal government to provide generous economic support for disaster mitigation, preparedness, response and recovery activities, including grants and low interest loans to communities living in hazardous areas that are damaged by disasters, even at risk to increasing the "moral hazard" of such development. However, this increased federal intervention and support is increasingly leading to economic distortions at the local level such as decisions not to purchase private sector insurance and a failure to seek other own-source solutions to disaster needs [39]. Local and state governments are willing to encourage development in high risk areas for tax base growth, knowing that any potential disasters will be offset by federal assistance. As a result, more than half of the US now resides in coastal counties—an area that comprises less than one-fifth of the mainland US land mass [23]. About 40 % of the world's total population lives within 100 km of the sea and most of the world's *megacities* with more than 10 million inhabitants are in the coastal area. The coastal landscape around the world is increasingly filled with significant development intrusions (from primary residences to shopping malls, vacation homes, hotels, and resorts) in sensitive areas that is creating an unsustainable situation. Of course, disasters that affect densely populated areas such as Southern Florida endanger not only large numbers of people but also have the potential to magnify loss to homes, businesses, highways, roads, bridges and utilities. Decision makers and government leaders (including entrepreneurs and developers, water authorities, local government, and waste disposal officials and housing authorities) often prioritize short term economic development goals over ecological integrity and sustainable urban development, thereby increasing disaster risk. Specifically, development in barrier islands, sensitive estuaries and ecologically important wetlands is destroying the very resources that buffer society from disasters (and make these locations attractive for human habitation in the first place).

3 Macroeconomic Analysis of Disasters

Macroeconomic analyses of disasters consider the extent to which disasters cause regional and national impacts. The total national cost of a disaster is the replacement value of the property damaged [24]. There is an ongoing debate on the extent to which disasters pose a potential impediment to economic development at a national or regional scale. There are two opposing perspectives on the significance of disasters from a macroeconomic perspective (Fig. 1). While natural disasters lower US GDP by only 0.052 % per year [65] the same may not hold true for smaller nations with more specialized economies. Accordingly the first macroeconomic view considers disasters as a grave setback for national economic growth,

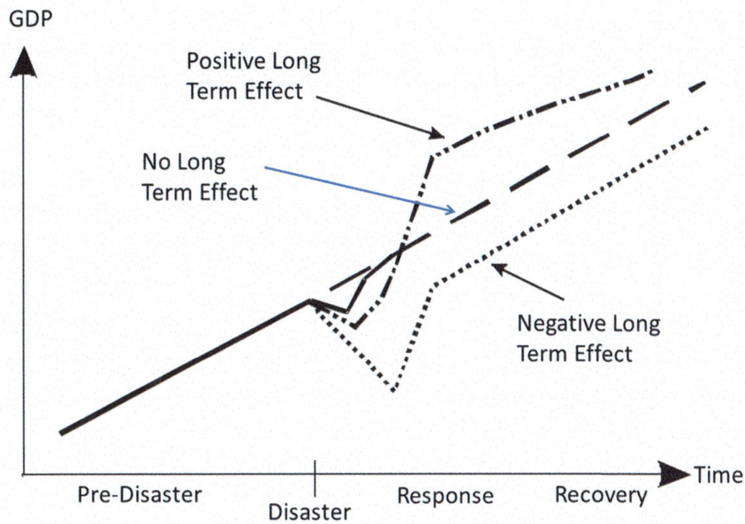

Fig. 1 Possible GDP trajectories after a disaster. Modified from Hochrainer [36]

especially sustained droughts in countries with an agrarian-based developing economy (Albala-Bertrand, cited in [25]) and countries heavily dependent on tourism or a single sector. This view is well represented by the following citation in regard to developing countries:

> Government decisions should be based on the opportunity costs to society of the resources invested in the project and on the loss of economic assets, functions and products. In view of the responsibility vested in the public sector for the administration of scarce resources, and considering issues such as fiscal debt, trade balances, income distribution, and a wide range of other economic and social, and political concerns, governments should not act risk neutral [48].

Accordingly, governments have a responsibility to address the economic and social challenges that result from disasters since it is proposed that they have a long term impact GDP suggests.

For example, the island nation of Haiti has yet to fully recover from the catastrophic magnitude 7.0 Mw earthquake that occurred on January 12, 2010. The earthquake, with an epicenter near the town of Léogâne (Ouest), approximately 25 km (16 miles) west of Port-au-Prince (Haiti's capital), occurred on January 12, 2010 and affected an estimated three million people with over 100,000 dead and approximately 300,000 collapsed or severely damaged residences and commercial buildings. Despite a substantial international aid response, Haiti's weak economic base was severely damaged and the nation has yet to recover. A 2013 survey disclosed that of the 1.5 million Haitians living in camps, more than a quarter million remained in 352 camps with limited services (for example, approximately half of the remaining camps have no health services for residents).

An empirical analysis of the impact of disasters on household and aggregate welfare in sixteen Caribbean and Latin American countries over a quarter century (1975–1999) shows that disasters lead to a worsening of the current account of the balance of payments, a moderate decrease in consumption growth and a substantial decline in the growth of investment and output [6]. As tourism-based economies many countries of the Caribbean and Latin American region are subject to market responses to hurricanes and disaster events—or predictions of crises—over which they have limited control. Exposure to shocks generated by catastrophic events is a particular concern to the poorest segments of the population in the Caribbean region (although all households are affected to some extent). Despite strong consumption growth in the Caribbean from 1975 to 1999, it is proposed that consumption volatility arises primarily due to ineffective (or underdeveloped) risk management mechanisms which allow production shocks to be transformed into consumption shocks [6].

A second macroeconomic school of thought argues that natural disasters do not negatively affect GDP and "if anything, GDP growth is improved" following large, external shocks [1]. This school of thought sees disasters as entailing little growth implications and consider disasters and their reduction as an opportunity for economic development (e.g. [1, 2, 17]). Even when many costs are included, it is argued that disaster impacts are typically too small to have a meaningful impact on a national economy: capital markets are simply too large to be disturbed beyond a short period of time by natural disasters [46]. For example, despite the enormous human toll extracted by the pandemic flu of 1918–1919 it had no major effect on the world economy [12]. While the 1995 Great Hanshin earthquake and subsequent fires led to the destruction of more than 100,000 businesses and killed 6500 people in Kobe, Japan with total damages estimated at $114 billion [38]—about 2.5 % of Japanese gross domestic product (GDP) in 1995—within 15 months the manufacturing sector was operating at 98 % of the pre-earthquake trend. Moreover, all department stores had reopened within 18 months and the port of Kobe was operating close to pre-earthquake levels within a year [38, 40]. With reference to the Kobe earthquake Horwich [38] suggests it is more telling to consider the impact of a disaster on economic potential as opposed to economic activity.

There are other global examples one can consider. One week after the 2004 Boxing Day Sumatra-Andaman Indian Ocean earthquake and tsunami, the Indonesian and Malaysian stock markets had gained value from the pre-disaster level, while the Thai and Sri Lankan stock markets declined only slightly [12].

Hurricane Katrina, the costliest natural disaster in US history, provides another salient case study that highlights post-disaster economic resilience. In 2006, the Bush Administration sought $105 billion for repairs and reconstruction in the Gulf Coast region and the total economic impact of Hurricane Katrina to Louisiana and Mississippi likely exceeded 150 billion $US (with over $50 billion in insured losses) [16]. While the hurricane affected swathes of Louisiana, Florida, Texas and Mississippi in August 2005, killing 1836 people and destroying a sizable proportion of the economic capacity of the Gulf Coast, US GDP rapidly rebounded. For example, in 2005, US Gross Domestic Product (GDP) growth went from 3.8 % in the third quarter (July–September, when Katrina hit) to 1.3 % in the fourth quarter (October–December, when production losses were felt) and by the first quarter in 2006 GDP bounced back to 4.8 % [10]. The disaster took 230,000 jobs from directly affected areas as documented by over-the-year changes in employment. Specifically, employment in St. Bernard Parish, Louisiana, was down by nearly forty percent in September 2005 from a year earlier [11]: Louisiana's St. Bernard, Orleans, and Jefferson Parishes had the largest percent declines in employment in the Gulf Coast region between September 2004 and September 2005. However, national US employment for the month of September, 2005 declined by a miniscule 35,000 [8].

The disaster had a major impact on Louisiana's chemical plants and sugar industry as well as oil production, importation, and refining across the Gulf Coast region; before the storm, one-tenth of all US crude oil consumption and nearly half of US gasoline production came from Gulf Coast refineries and Hurricanes Katrina and Hurricane destroyed or sunk 113 offshore oil and gas platforms and damaged 457 oil and gas pipelines, spilling nearly as much oil as the Exxon Valdez oil disaster. This impacted fuel prices across the US: oil prices increased by $3 a barrel while gasoline prices reached nearly $5 a gallon. In a bid to reduce gasoline prices oil from the Strategic Petroleum Reserves stockpile (also located in the Gulf Coast region) was released. Significant damage was done to the City of New Orleans, especially its tourism industry which generated nearly $10 billion annually before Katrina. While the port of New Orleans alone suffered $260 million in damages, it was open to ships a week later.

Given the starkly contrasting positions pertaining to possible GDP trajectories after a disaster, new analyses have attempted to reach a compromise by identifying circumstances under which disasters have the potential to cause significant medium-term economic impacts. For example, in a medium-term analysis, natural disasters seem to lead, on average, to adverse macro-economic consequences on GDP. While the negative effects may be small, they can become more significant depending on the size of the shock. The adverse macroeconomic consequences appear to be reduced by "higher aid rates as well as higher remittances" while "capital stock loss is the most important predictor" for negative economic impacts [37].

4 Microeconomic Analysis of Disasters

Microeconomic analysis of disasters provides key insights into the economic changes that are caused by a catastrophic event and the total economic impact. Key questions include: Which industries and businesses are most significantly impacted by the extreme event? Which are most likely to see increases in their business activities resulting from disasters? For example the construction and engineering sector of the economy may see massive growth in business after a disaster. Answers to these questions together with predictive microeconomic modeling helps decision makers to more efficiently allocate pre-disaster activities (disaster prevention, preparedness and pre-event mitigation strategies) [32, 46] and post-disaster response (including short term relief efforts and long-term recovery). In this way survivors and businesses most in need can receive the lion's share of the emergency management and disaster response resources.

Six data analysis techniques are identified to assess the indirect and income effects of disasters [18, 21, 71]: surveys, econometric models, Box-Jenkins methods (time series analyses), input-output models, general equilibrium models, and economic accounting models. First, surveys provide direct information from business owners and individuals directly or indirectly impacted by disasters. Surveys can be used to carry out basic disaster data (actual or expected costs to rebuild) or answer sophisticated questions (Which operations and activities were most affected by the disaster? How much revenue was lost because of information technology down time? Do you expect to lose customers because of reputational damage for your firm? Did you receive any government assistance to finance your rebuilding? Are you expecting to reopen in the same location? Do you intend to create redundant facilities at alternative locations in the future?).

While surveys offer an excellent tool for obtaining direct, relevant data about disaster impacts, they suffer from a number of challenges including the non-response bias (are respondents representative of the broader population of disaster victims?), a lack of response reliability (will survey answers change at a later date due to the emotional distress associated with the crisis), logistical problems (can the disaster area be accessed immediately?), strategic responses (do survey respondents strategically exaggerate economic losses to elicit maximum disaster assistance?) and a lack of time and resources to carry out surveys (are in-person interviews too expensive and time consuming to carry out? Can survivors be located after the disaster while remaining on time and within budget constraints?).

The lead author of this chapter designed and carried out surveys to assess the economic impacts of the December 2007 flooding in southwest Washington on businesses. The floods caused widespread damage to more than two hundred businesses and farms. The surveys were part of a Quick Response Grant from the Natural Hazards Center at the University of Colorado to research the effects of the 2007 floods on businesses in Centralia and Chehalis, Washington. Working with local Offices of Economic Development, a 28-question survey was administered to 63 businesses (37 were flooded businesses and 26 were unflooded businesses) in order to assess risk perception and disaster preparedness, together with the economic impacts of disasters and long-term recovery strategies. The surveys documented the direct and indirect economic impacts to local businesses. For example, poor sales were reported by both flooded and unflooded businesses in the weeks that followed the flood (some companies experienced a loss in revenue up to two months after the flood) highlighting the need for sector wide protective actions through workshops, trainings, mentorship, and networking. Responses to survey questions also highlighted that prior to the disaster, business' perceived the flood risk to be low and few companies were prepared for the flooding, although disaster preparedness was highest among those that had experienced previous flooding. Moreover, two-thirds of the respondents who had engaged in business continuity planning ranked it as very helpful [34]. Accordingly, it seems counterintuitive that businesses' perceived disaster planning assistance to be one of the least useful activities to support recovery. This suggests that more effort should be devoted to educating businesses about the value of engaging in disaster planning activities. Additional longitudinal follow-up surveys with affected business were carried out.

The second microeconomic analysis approach to study the economic impacts of disasters is econometric modeling. This technique is ideal when large quantities of data are readily available for the affected region. For example in the case of forest fire analyses typical data of interest might include forest acreage burned, human casualties, number of fire responders, historic weather data, the distribution and size of forest fires, the cause of wildfires (i.e. lightning, human activity, etc.), demographic data pertaining to the affected residential communities, reported economic damages, emergency preparedness logs, and social perceptions of wildfires. Economic modeling contains a number of challenges such as data availability (relevant economic data may be gathered and published with substantial lags), problems accounting for product substitution and rapid changes in imported goods (particularly in an region that experienced a significant shock), and the non-linear nature of production functions in a post-disaster economy.

Disaster scholars have offered credible analyses using regression techniques, see for example, the classic studies of Ellison et al. [25], Cochrane [20] and Guimaraes et al. [23]. A number of regression techniques are often used for disaster management including ordinal regression, logistic regression and multinomial regression. For example, logistic regression (also called logit regression or logit modeling) is a statistical technique allowing researchers to create predictive models for modeling in situations for which there is a binary response variable. The logit modeling technique is most useful for understanding the influence of several independent variables on a single dichotomous outcome variable. In a logistic regression the predictor variables can be numerical or categorical (including binary). Multinomial (aka polychotomous) logistic regression can be used when there are more than two possible outcomes for the response.

Logistic regression models have been shown to provide powerful fire prediction results when used in sophisticated fire occurrence prediction systems and models. Fire occurrence expressed in terms of a fire day is a Boolean event, meaning that only one of two outcomes is possible. For example, one or more fires occur over the course of a fire day [1] or no fires occur (0). In fire modeling, logistic regression models are often preferred over linear regression models since the logistic distribution is applicable to the fire occurrence problem and the approach is both flexible and easy to use. Since the probability of a fireday can be expressed as a logistic function [44] we now consider a binary response version of the model. Letting Y be the binary response variable, it is assumed that $P(Y = 1)$, the probability that one or more fires occur over the course of a fire day is possibly dependent on $\vec{x}$, a vector of fire predictor values. The goal is to model

$$p(\vec{x}) \equiv P(Y = 1|\vec{x}).$$

Since Y is binary, modeling $p(\vec{x})$ involves modeling $E(Y|\vec{x})$, the approach carried out in ordinary least squares, with a numerical response. If we model $p(\vec{x})$ as a linear function of predictor variables, e.g.,

$$\beta_0 + \beta_1 x_1 + \cdots + \beta_p x_p,$$

then the fitted model can result in estimated probabilities which are outside of [0,1]. To avoid this challenge we assume that

$$p(\vec{x}) = \frac{\exp(\beta_0 + \beta_1 x_1 + \cdots + \beta_p x_p)}{1 + \exp(\beta_0 + \beta_1 x_1 + \cdots + \beta_p x_p)},$$

where $x_1, \ldots, x_p$ is the original set of untransformed explanatory variables, but the predictors may include transformed and constructed variables. Note that

$$\log\left(\frac{p(\vec{x})}{1 - p(\vec{x})}\right) = \beta_0 + \beta_1 x_1 + \cdots + \beta_p x_p,$$

where $\log\left(p(\vec{x})\Big/\left[1 - p(\vec{x})\right]\right)$ is called the *logit*. Fortunately, the model for the logit is linear in the predictors while the model for $p(\vec{x})$ is more complicated, Also regardless the value of

$$\hat{\beta}_0 + \hat{\beta}_1 x_1 + \cdots + \hat{\beta}_p x_p$$

the corresponding estimate of $p(\vec{x})$ will be between 0 and 1. The unknown parameters (the coefficients, $\beta_0, \beta_1, \ldots, \beta_p$) are typically estimated by maximizing the likelihood,

$$\prod_{i=1}^{n} \left\{p(\vec{x})^{y_i}\left[1 - p(\vec{x}_i)\right]^{1-y_i}\right\},$$

which constitutes an expression for

$$P(Y_1 = y_1, \ldots, Y_n = y_n|\vec{x}_1, \ldots, \vec{x}_n).$$

where the maximum likelihood estimates are determined numerically, by maximizing the log likelihood.

While logistic regression models often work well for predicting categorical or multinomial disaster outcomes they suffer from a number of challenges in the disaster economics field including the following problems: it is challenging to identify the correct set of independent variables (if researchers include the wrong independent variables, the model will likely not provide useful or accurate predictions); an analyst cannot include all types of outcome variables (logistic regression cannot predict continuous outcomes and converting a continuous scale like the measurement of surface fire temperatures into discrete categories like "high" or "low" sacrifices the precision of the data set); each data point be independent of all other data points (logistic regression requires that independent observations are required because if observations are related to one another, then the model will tend to overweight the significance of those observations. This is a major disadvantage in disaster modeling because experimental disaster science and social psychology research often involves multiple observations of the same individuals); logistic regression models can "overstate" the accuracy of its predictions (in other words, a logit model can appear to have more predictive power than it actually does as a result of sampling bias. Hence a logistic regression would therefore be "overfit" which can lead to overconfidence in the logit models by overstating their accuracy). During a regression analysis non-disaster factors affecting the economy are statistically controlled for so that regression estimates relate to only those costs associated with the disaster; in other words, fully-partialed effects of a system shock can be modeled as an intrusion on a dataset.

Variations of hedonic pricing models (derived from the term hedonism) consider preferences in purchasing decisions and constitute valuable econometric modeling approaches for disaster studies. Classic hedonic estimation has been traditionally been used for the purpose of making inferences about non-observable values of different attributes like air quality, airport noise, and access to transportation [27]; examining price gradients based on seismic safety attributes for residential housing [15]; and assessing the impact of flood hazards on real estate values [42]. These models are most commonly used in real estate and disaster research to describe why some houses are more desirable (higher priced) even when other factors are the same; examples of house attributes include physical attributes, community attributes and attributes observed by the consumers but not econometricians [7]. Weaknesses found in traditional hedonic models include information asymmetry, measurement validity of explanatory variables, market limitations, multicollinearity, price changes and problems obtaining a priori parametric specifications. While the use of Box-Cox transformations [31] involves many parametric assumptions a more robust way to accurately predict the dependent variable is to employ non-parametric models which do not impose an a priori parametric specification. Several techniques can be used to estimate such models including 'cubic splines', 'nearest neighbors', 'series approximators' and 'kernel estimates' [3]. Unobserved variables usually have some structural constraints associated with them which can be used to infer their values from the data [19]. Hedonic modeling approaches can add valuable insights into consumer behavior in disaster prone areas especially if housing price models are modified for longitudinal analyses to explore changing hedonic factors after a disaster, such as housing prices in California after multiple major earthquakes or home values in Japan after hazardous material releases.

The 1960s can be considered as the golden age of large-scale econometric models with the rise of the Office of Business Economics (OBE), FRBMIT-PENN (FMP), Brookings, and the Wharton econometric model of the U.S. economy. Preliminary success of these massive, complex models in forecasting the short-run state of the US economy contributed to overconfidence about their general applicability to a wide range of problems in the analysis of economic theory and policy. By the 1970s econometricians began to express more doubts in the success of econometric models when faced with poor forecasts. Box-Jenkins time series analyses—an alternative to econometric models—constitutes a third analytic microeconomic technique for assessing the economic effects of disasters that soon eclipsed econometric models in popularity. AutoRegressive (AR) models were introduced by Yule [70] and supplemented by Slutsky's [64] Moving Average (MA) schemes. Wold [68] combined both AR and MA schemes and showed that ARMA (p, q) processes can be used to model all stationary time series as long as the appropriate order of p (the number of AR terms) and q (the number of MA terms) was appropriately specified. Accordingly any time series x_t can be modeled as a combination of past x_t values and/or past e_t values as shown in Eq. (1):

$$\begin{aligned} x_t = \phi_1(x_{t-1}) + \phi_2(x_{t-2}) + \cdots + \phi_p(x_{t-p}) \\ + e_t - \theta_1 e_{t-1} - \theta_2 e_{t-2} - \cdots - \theta_t e_{t-q} \end{aligned} \tag{1}$$

Utilizing the theoretical results suggested by Wold [68] in Eq. (1) for time series modeling requires four key steps. First, the original time series x_t must be transformed to be stationary around its mean and variance. Stationarity is essential to ascertain since if the time series is non-stationary then all the typical results of the classical regression analysis are invalid (i.e. regressions with non-stationary series may be "spurious"). In simplest terms, a time series process (x_t) is said to be stationary (i.e. exhibits covariance stationarity) when it has the following three characteristics: $E(x_t)$, the mean ofx_t, is constant for all t (i.e. the time series exhibits mean reversion in that it fluctuates around a constant long-run mean); $Var(x_t)$ is constant for all t (i.e. the time series has a finite variance that is time-invariant); and $E(x_t)$, $Var(x_t)$ and $Cov(x_t)$, the covariances of x_t, remain constant over time (i.e. the time series has a theoretical correlogram that diminishes as the lag length increases). Second, the order of p and q must be specified. Third, a non-linear optimization algorithm must be used to estimate the value of the parameters ϕ_1, ϕ_2, ..., ϕ_p and or θ_1, θ_2, ..., θ_q that minimizes the sum of square errors (or another loss function). Fourth, and finally, one must establish a practical way of modeling seasonal series (and establish the appropriate order of such models).

A Box-Jenkins analysis uses previous values of the variable under consideration to predict the next value as shown in Eq. (1). However, implementing Eq. (1) for actual time series in practice requires the availability of sufficient computing power in order to optimize the parameters of (1). Box and Jenkins (1976, original edition

[13]) popularized the use of ARMA models by providing guidelines for making a series stationary (in its mean and variance); using autocorrelations (and partial autocorrelation coefficients) for determining appropriate values of p and q (and their seasonal equivalent P and Q when the series exhibited seasonality); estimating the parameters with computer software; proposing a diagnostic check to confirm the residuals e_t are white noise (i.e. random residuals), in which case the order of the model was valid and the model could be used for forecasting purposes; on the other hand, if the residuals are random, then another model should be considered and the latter two steps repeated.

The above approach (proposed by Box and Jenkins) came to be known as the Box-Jenkins methodology for ARIMA models, the term deriving from autoregessive (AR), integrated (I) and moving average (MA). In the 1970s, empirical studies (for a survey see [4, 5]) showed that the Box-Jenkins methodology outperformed the large and complex econometric models, in a variety of situations. This led to their increased popularity, especially among academics. The order of an ARIMA (autoregressive integrated moving-average) model is usually denoted by the notation ARIMA (p,d,q), where p is the order of the autoregressive part, d is the order of the differencing and q is the order of the moving-average process. If no differencing is done ($d = 0$), the models are usually referred to as ARMA (p, q) models. Even though an original series of data may not be stationary, differences between successive observations may be stationary: $\mathrm{d_t = x_t - x_{t-1} = (1 - B)x_t}$ where B is the *Backshift operator*. An ARIMA model results when an ARMA model is applied to the $\mathrm{d_t}$. If the first differences are not stationary, the second differences might be, i.e. $\mathrm{d'_t = d_t - d_{t-1} = (1 - B)(1 - B)x_t}$. An ARIMA (1) process, with the middle number referring to the number of differences that are taken, can be described as

$$\begin{aligned} d_t &= \phi_1 d_{t-1} - \theta_1 e_{t-1} + e_t \\ x_t - x_{t-1} &= \phi_1 (x_t - x_{t-1}) - \theta_1 e_{t-1} + e_t \\ x_t &= (1 + \phi_1) x_{t-1} - \phi_1 x_{t-2} - \theta_1 e_{t-1} + e_t. \end{aligned} \tag{2}$$

Taking a Box-Jenkins approach to time series analysis, the ARIMA data analysis methodology has become an extremely valuable modeling approach in disaster research for a number of reasons: relative simplicity, analysis power and cost effectiveness (data does not need to be obtained for large numbers of relevant variables). Consider how ARIMA data analysis can be used to estimate a hurricane's impact on retail sales in South Florida: ARIMA can be used to account (control) for trends (such as an overall rise in retail sales over several years), to examine the correlation between current and previous observations (if a business is successful one month, they will likely be successful the following month), to account for seasonal variations (prices rises for Christmas, Thanksgiving, and other especially busy times) and to identify secular trends in the data (are overall consumer prices rising after a disaster?). By comparing predicted retail sales (without a hurricane) with actual sales after the hurricane, an ARIMA model provides an estimate of the hurricane's impact on retail sales. In a similar manner, ARIMA models have been used to assess losses in the tourism industry due to human-caused threats [26] and to estimate the impacts of disasters on stock markets [69]. The popularity of the Box-Jenkins methodology for ARIMA models faded during the 1970s as empirical studies (using real data) showed that Box-Jenkins (the short name attributed to the methodology proposed by Box and Jenkins) was often less accurate than simple modeling methods when post-sample comparisons were made [28, 30, 35, 44]. Another weakness of the ARIMA approach is a failure to account for confounding accompanying occurrences, such as changes in the Chinese labor markets causing a retailer to close in the midst of an active hurricane season (although the concomitant event of Chinese regulatory and political change are totally unrelated to the hurricane).

Input-output (I/O) economics represents an expansive collection of data describing our economic systems, and/or as an analytical technique for explaining and predicting economic system behavior. Accordingly, I/O models, when applied correctly, constitute a fourth powerful technique for assessing economy-wide disaster impacts. In the 1930s Wassily Leontief used transaction matrices to capture the flow of goods across industries. Regional I/O models provide multipliers that are based on behavioral assumptions related to how individuals adjust their labor supply, saving, and consumption decisions when there is an initial change affecting their income. While regional I/O multipliers can be used to estimate the economy-wide effects that a disaster has on a regional economy they are not a substitute for national-level macroeconomic (Keynesian) multipliers which are based on behavioral assumptions related to how individuals adjust their labor supply, saving, and consumption decisions when there is an initial change affecting their income. Economic surveys shed light on the raw materials, goods, and services purchased as inputs in the production process by various industries.

I/O models provide a description of how demand-satisfying production creates upstream and downstream economic activities. The I/O models produce three types of impact assessments: direct effects/impacts (captures the direct purchases by the industry being described); indirect effects/impacts (captures purchases by related companies in the supply chain) and induced effects/impacts (captures the economic activity created by employees spending a portion of their earnings in the local economy for goods and services). The economic multiplier (the total effect) constitutes the sum of the direct, indirect, and induced impacts, expressed as coefficients (a total effect greater than 1.0 is possible). Multipliers generated by the Regional Input-Output Modeling System (RIMS II) developed by the Bureau of Economic Analysis are widely used. Most I/O models include the economic activities (purchases) of households, though not all.

The US Bureau of Economic Analysis (BEA) of the US Department of Commerce and the Minnesota IMPLAN Group (the IMPLAN model) produce relatively easy to use off-the-shelf I/O models that provide cost effective and rapid initial estimates of economic activity and modeling capabilities at the county level: IMPLAN offers highly disaggregated estimates of economic activity (as many as 528 different industry categorizations) and more flexibility in adjusting regional purchasing coefficients. While I/O models have the major drawbacks of being static, linear and require sophisticated assumptions for calculating highly disaggregated impacts, they provide reasonable estimates for a disaster researcher. I/O models have been used to simulate the indirect regional economic effects of an earthquake event that damages electricity generating infrastructure [49]. Although a limitation of I/O models is that they are unlikely to accurately reflect facts on the ground after a disaster as they invoke fixed coefficients and do not easily account for product substitution. However the use of social accounting matrices (SAM) can expand the I/O model calculations to more realistically calculate potential disaster impacts by examining transfer payments, value-added accounting, and the distribution of disaster impacts across households at various income levels. A SAM I/O model was used to estimate potential impacts of damage to the electric industry in upstate New York to aid regional disaster planning [22].

Computable general equilibrium (CGE) models represent one of the most popular tools for estimating the economic impacts of disasters in recent years. CGE models overcome some of the weaknesses of I/O models because different elasticities of supply and demand can be applied across different tiers of economic activity [51]. Moreover, in the post-disaster environment, when a given input in a production process may be no longer available—but can be imported from another region—the CGE model can incorporate a range of input substitutions and more accurately estimates the direct, indirect, and induced effects of this change. However, CGE models rarely cover more than a few industrial sectors since this level of flexibility is data intensive. Disaster case studies include the use of dynamic regional CGE models to simulate the effects of a disease outbreak [67], to estimate the value of pre-event mitigation [51] and to analyze disaster impacts and policy responses [50]. Drawbacks of CGE approaches to disaster impact modeling include its intensive data requirements, practical limitations on the number of industries that can be effectively analyzed at once and an emphasis on equilibrium states—an unrealistic situation in the aftermath of catastrophe although Rose [50] offers a partial solution. Hence CGE tools are best used for disaster planning purposes and a priori assessments of potential impacts. The direct and indirect economic impacts of disasters can also be modeled with the Federal Emergency Management Agency's (FEMA) HAZUS-MH software which uses a combination of I/O, hybrid-I/O, and CGE modeling approaches. This flexible tool integrates GIS and allows users the choice of using preprogrammed assumptions about the local economy (for a quick and simple analysis) or elicit detailed local data (typically with the services of seasoned professionals). HAZUS-MH calculates indirect economic disaster impacts with IMPLAN data matrices followed by adjustment algorithms similar to those employed by hybrid-IO and CGE models.

Finally, the economic accounting approach constitutes a sixth microeconomic approach for estimating the impacts of disaster events. This technique explicitly estimates: (a) business losses using other methodologies such as case based analysis (surveys), GDP estimates (econometric) or I/O models; and (b) the value of human life and injuries. However, the latter task is highly controversial and subjective. For example, the value of a life lost ranges from \$20,000 (US National Safety Council) to \$5.8 million (Environmental Protection Agency) to \$250,000–\$7 million

(Special Master for the US Department of Justice overseeing claims related to the terrorist attacks on the World Trade Centers). These two elements are then added to estimates of physical losses to estimate the total economic disaster impacts [71].

The aforementioned six approaches highlight a number of challenges for studying the economic impacts of disasters: microeconomic disaster impact studies only include losses that can be measured in transactions rather than also attempting to quantify issues from socio-cultural assets to the cost of human psychological trauma [46]; over causality of post-event loss, using restrictive time horizons, and double counting losses [21] and the opportunity costs of disaster volunteers. However, advances in the analytic approaches to estimating the economic impacts of disaster events at the macro- and micro-economic levels have helped to overcome some of these barriers and obstacles. The challenge is to improve the accuracy and reliability of economic methods while provide timely information to disaster professionals, community leaders and elected officials.

5 Incorporating Economics in the Multi-criteria Disaster Forensics Analysis (MCDFA)

After obtaining the economic impacts of a disaster it is valuable to use the economic results in a disaster forensics multiple criteria analysis framework. Accordingly a unique group multiple criteria approach for decision making for disaster forensics has been developed that allows for decision making under uncertainty and integrates with the Analytic Network Process [52–60]. The proposed approach assumes that N decision makers, DM_k ($k = 1, \ldots, N$), evaluate m alternatives, A_j ($j = 1, \ldots, m$), under n criteria, C_i ($i = 1, \ldots, n$). First, the aggregation of local individual alternative weights into local group weights is considered. Let a_{ij}^k denote the local weight of decision-maker DM_k evaluating alternative A_j under criterion C_i. A unique aggregation method is developed in which the group weight for alternatives $x_i = (x_{i1}, \ldots, x_{im})^T$ under a criterion i is obtained as the solution x_i^* to the following quadratic programming problem:

$$\begin{aligned} &\min_{w,x} \frac{1}{2}\sum_{k=1}^{N}\sum_{j=1}^{m}\left(a_{ij}^k w^k - H\left(a_{ij}^k\right)x_{ij}\right)^2 \\ &s.t. \sum_{j=1}^{m} x_{ij} = 1, \quad w^k \geq 0\ (k = 1,\ldots,N), \quad x_{ij} \geq 0\ (j = 1,\ldots,m), \end{aligned} \tag{3}$$

With application to disaster forensics, we propose a Multiple Criteria Disaster Forensics Analysis (MCDFA) approach. As shown in Fig. 2, the general problem of deriving priorities from a matrix of pairwise comparison judgments is to solve for the derived priorities from the matrix $A = (a_{ij})$ where the a_{ij} are judgments from the

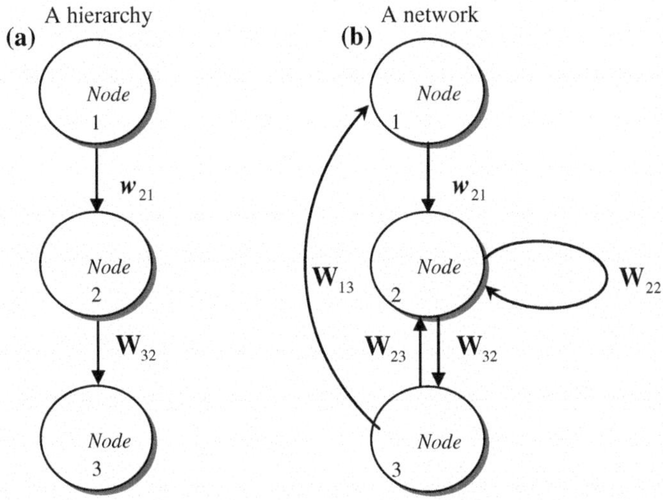

$$\mathbf{W}_h = \begin{matrix} \\ Node\,1 \\ Node\,2 \\ Node\,3 \end{matrix} \begin{matrix} Node1 & Node2 & Node3 \\ \end{matrix} \begin{bmatrix} 0 & 0 & 0 \\ w_{21} & 0 & 0 \\ 0 & \mathbf{W}_{32} & \mathbf{I} \end{bmatrix} \qquad \mathbf{W}_n = \begin{matrix} \\ Node\,1 \\ Node\,2 \\ Node\,3 \end{matrix} \begin{matrix} Node1 & Node2 & Node3 \\ \end{matrix} \begin{bmatrix} 0 & 0 & \mathbf{W}_{13} \\ w_{21} & \mathbf{W}_{22} & \mathbf{W}_{23} \\ 0 & \mathbf{W}_{32} & \mathbf{I} \end{bmatrix}$$

Fig. 2 **a** Linear hierarchy and **b** nonlinear network

fundamental 1–9 scale as discussed by Saaty [42–60] in the Analytic Hierarchy Process (AHP) and Analytic Network Process (ANP). If $a_{ij} = a_{ik}/a_{kj}$ for all i, j, and k, the matrices in Fig. 2 are consistent. A priority vector derived from paired comparisons represents the impact of a given set of elements in a component on another element in the system. When an element has no influence on another element, its influence priority is assigned a zero. The priority vectors derived from pairwise comparison matrices are each entered as a part of a column of a "supermatrix", which represents the influence priority of an element on the left of the matrix on an element at the top of the matrix [61, 62].

In order to model complexity (dependence, feedback and non-linear interactions) our MCDFA research uses a nonlinear hierarchical structure and its corresponding supermatrix (Fig. 2b) rather than a linear hierarchy (Fig. 2a, which constitutes a simple and degenerate case of a network). A node represents a component (or cluster) with elements inside it; a straight line/or an arc denotes the interactions between two components; and a loop indicates the inner dependence of elements within a component. An arrow from *Node1* to *Node2* shows that the elements of a component in *Node1* depend on another component *Node2*. Figure 2a, b illustrates the corresponding supermatrix of a hierarchy and a network with three levels of clusters: where $w21$ is a vector that represents the impact of *Node1* on *Node2*; **W13** is a matrix that represents the impact of *Node 3* on each element of *Node 1*; **W**22 is a matrix that represents the impact of *Node2* on itself (inner loop); **W**23 is a matrix that represents the impact of *Node 3* on each element of *Node 2*; **W**32 is a matrix that represents the impact of *Node2* on each element of *Node3*; and **I** is the identity matrix.

6 A Multi-criteria Disaster Forensic Analysis (MCDFA) of the 2012 Kahuku Wind Farm Battery Fire on Oahu, Hawaii

First Wind first broke ground on the Kahuku Wind Farm, west of Kahuku town on Oahu's rural North Shore, Hawaii, in March 2011. Located above the hills of Kahuku on the island of Oahu, the Kahuku Wind Farm began operation in early 2011 as the first utility-scale project on Oahu, Hawaii. Initially operated by First Wind, the 30 MW (megawatts) wind farm project consists of twelve 2.5 MW Clipper Liberty turbines and creates enough power to supply 7700 homes with renewable energy. Housed in a 9000-ft.^{2} metal warehouse with 12,000 lead acid batteries mounted in racks more than six feet high, the 10-MW battery storage system (installed by Xtreme Power) received a $117 million (US dollar) loan guarantee for construction from the U.S. Department of Energy (DOE) in 2010. However, the new battery storage technology also created additional disaster risk. Since operations started in 2011, three fires have occurred at the battery (energy) storage building where the energy collected by the 12 turbines is stored. A probe of the first two Kahuku fires—in April and May 2011—blamed manufacturing defects in the capacitors made by Electronic Concepts Inc (ECI) and contained in inverters by Dynapower. The first two fires burned themselves out or were extinguished by the Honolulu Fire Department (HFD) before causing extensive damage. In the pre-dawn hours of August 1, 2012 the warehouse containing 12,000 individual battery packs caught fire for a third time, burning for more than 13 h, forcing the wind turbines to shut down, and causing $30 million dollars in damages (Fig. 3) although no injuries were reported.

After determining no one was inside, the HFD response team launched a "defensive fire attack" from outside the warehouse focused on keeping the blaze from spreading to other buildings at the site. Parts of the battery storage building collapsed, but crews kept the flames from spreading to other structures. Firefighters did not enter the storage facility for seven hours due to the confined nature of the warehouse (rows of racks of batteries with small aisles in between) and the risks posed from thick smoke, toxic fumes, scalding heat, a collapsing structure and the potential for battery explosions. Firefighters avoided using water to extinguish the fire directly out of concerns for electric shock and risks of creating toxic chemical runoff. They later unsuccessfully used dry chemicals from the Hawaiian Electric

Fig. 3 Kahuku wind farm fire on August 1, 2012

Company (HECO) to try to extinguish the flames. While the cause of the fire remains unknown technical experts and fire inspectors identified that the fire started in or near the battery banks and then spread.

When the Kahuku windfarm project re-opened in February 2012 the authors consulted with First Wind and the local utility, Hawaiian Electric Co (HECO), to carry out a thorough forensic analysis and to determine the optimal course of action after the battery energy storage system fire. To assist in the forensics decision making, a group Multiple Criteria Disaster Forensics Analysis (MCDFA) model with dependence and feedback was developed (Fig. 4). Two disaster forensics alternatives are considered (nodes in the alternatives cluster C_5). Four additional clusters (economics, training, social, and information assurance) are shown: dynamic volt-amp reactive (D-VAR) technology to regulate voltage for safe, reliable integration of energy from the Kahuku wind farm into the grid and lead acid batteries. Note that the economic and assurance clusters (C_2 and C_3) each consist of four nodes. An arrow from one cluster to another means that at least one element in the cluster is connected to elements in the other to form a pairwise comparison set, with comparisons being done for importance or preference with respect to the source element. There is also feedback since an arrow goes in both directions between several pairs of clusters. The MCDFA model resulted in the following scores: dynamic volt-amp reactive (D-VAR) technology (0.61) and lead acid batteries (0.39). In 2014, operators restored the Kahuku wind farm to full capacity, minus the electricity storage component.

In 2015, the Hawaii Public Utility Commission approved another wind turbine installation at Kahuku which would send energy directly to power lines instead of buffering through a battery system.

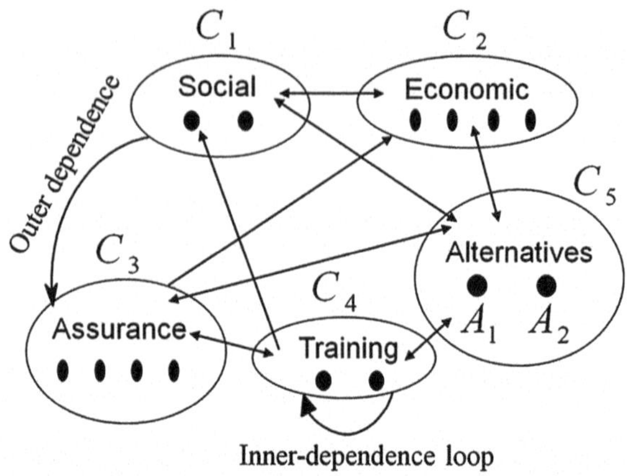

Fig. 4 Clusters, nodes and connections in the MCDFA

7 Conclusions

The discipline of economics and its many sub- and closely related disciplines offer valuable modeling techniques for reducing disaster risk, improving disaster forensics investigations and assessing the impacts of disasters. Economists have developed valuable theories of development for understanding choices that increase exposure and vulnerability to disasters, such as the location of residential and business locations in hazardous areas. This chapter addressed, comprehensively and in-depth, key issues associated with using the field of economics to build a culture of disaster prevention and enhance disaster forensic analyses. It is shown that advances in microeconomic and macroeconomic analyses can make a valuable contribution to the related broad fields of sustainable development and disaster risk reduction. This chapter assessed the causes, perspectives, and consequences of energy related disasters, as well as providing a new Multi-Criteria Disaster Forensics Analysis (MCDFA) approach for the forensic analyses of disasters. The 2012 Battery Room Fire at the Kahuku Wind-Energy Storage Farm on Oahu, Hawaii was used as a case study to identify measures for preventing loss from renewable energy storage system disasters. A key objective is to minimize the incidence rate and the scope of potential fire damage in battery-based energy storage systems. It is shown that in addition to special fire protection measures for detecting, fighting and preventing fires, procedural safety measures and comprehensive control technologies/systems for monitoring procedural operations and conditions are required. There is an urgent need to effectively engage accident investigation and forensics analysis professionals on Oahu, Hawaii in order to promote a culture of disaster resilience in the energy sector.

References

1. Albala-Bertrand JM (1993) Political economy of large natural disasters with special reference to developing countries. Clarendon Press, Oxford
2. Albala-Bertrand JM (2006) The unlikeliness of an economic catastrophe: localization & globalization. Working Papers 576, Queen Mary, University of London, Department of Economics
3. Anglin P, Gencay R (1996) Semiparametric estimation of a hedonic price function. J Appl Econom 11(6):633–648
4. Armstrong JS (1978a) Forecasting with econometric methods: folklore vs. fact. J Bus 51:549–564 (JSA) (SM)
5. Armstrong JS (1978b) Long-range forecasting: from crystal ball to computer. Wiley, New York (AA) (ESG) (SM) (PN)
6. Auffret P (2003) High consumption volatility: the impact of natural disasters. World Bank Policy Research Working Paper 2962. The World Bank, Washington, DC
7. Bajari P, Kahn ME (2008) Estimating hedonic models of consumer demand with an application to urban sprawl. Hedonic methods in Housing markets, Springer, New York, pp 129–155
8. Balls A, Swann C (2005, Oct 8) Little sign of Katrina damage as employment figures hold up. Financial Times, 3
9. Barnett B (1999) US government natural disaster assistance: historical analysis and a proposal for the future. Disasters 23(2):135–155
10. Bureau of Economic Analysis (2006) National Economic Accounts, Department of Commerce. http://www.bea.gov/national/Index.htm
11. Bureau of Labor Statistics, Quarterly Census of Employment and Wages (2005)
12. Becker G (2005, Jan 4) And the economics of disaster. Wall Street J A-12
13. Box GEP, Jenkins G (1970) Time series analysis, forecasting and control. HoldenDay, San Francisco
14. Brookshire DS, McKee M (1992) Other indirect costs and losses from earthquakes: issues and estimation. In: Indirect economic consequences of a catastrophic earthquake. FEMA, National Earthquake Hazards Reduction Program, Washington, DC
15. Brookshire D, Thayer M, Tschirhart J, Schulze W (2001) A test of the expected utility model: evidence from earthquake risks. J Polit Econ 93(1):369–389
16. Burton ML, Hicks MJ (2005, Sept) Hurricane Katrina: preliminary estimates of commercial and public sector damages. Marshall University: Center for Business and Economic Research
17. Caselli F, Malhotra P (2004) Natural disasters and growth: from thought experiment to natural experiment. IMF, Washington, DC
18. Chang S (2003) Evalutaing disaster mitigations: a methodology for urban infrastructure systems. Nat Hazards Rev 4:186–196
19. Chopra S, Trampy T, Leahy J, Caplin A, Lecun Y (2007) Discovering the hidden structure of house prices with a nonparametric latent manifold model. In: Proceedings of the 13th ACM SIGKDD international conference on knowledge discovery and data mining
20. Cochrane H (1974) Social science perspectives on the coming San Francisco earthquake: economic impact, prediction, and reconstruction. Natural Hazard Research Working Paper No. 25. Institute of Behavioral Science, University of Colorado, Boulder, CO
21. Cochrane H (2004) Economic loss: myth and measurement. Disaster Prev Manag 13(4):290–296
22. Cole S (2004) Performance and protection in an adaptive transaction model. Disaster Prev Manag 13(4):280–289
23. Crossett K, Culliton T, Wiley P, Goodseep T (2004) Population trends along the coastal United States: 1980–2008. National Ocean Service, Washington, DC. Accessed 23 Jan 2006 at http://www.oceanservice.noaa.gov

24. Dacy D, Kunreuther H (1969) The economics of natural disasters: implications for federal policy. The Free Press, New York
25. Ellison R, Milliman J, Roberts R (1984) Measuring the regional economic effects of earthquakes and earthquake predictions. J Reg Sci 24:559–579
26. Enders W, Sandler T, Parise G (1992) An econometric analysis of the impact of terrorism on tourism. Kyklos 45:531–554
27. Espey M, Lopez H (2000) The impact of airport noise and proximity on residential property values. Growth Chang 31(3)
28. Fildes R, Hibon M, Makridakis S, Meade N (1995) The accuracy of extrapolative forecasting methods: additional empirical evidence. INSEAD Working Paper
29. Forensic Investigations of Disasters (FORIN) project (2015) Integrated research on disaster risk. http://www.irdrinternational.org
30. Geurts MD, Ibrahim IB (1975) Comparing the Box-Jenkins approach with the exponentially smoothed forecasting model application to Hawaii tourists. J Mark Res 12:182–188
31. Goodman A (1978) Hedonic prices, price indices and housing markets. J Urban Econ 5 (4):471–484
32. Gordon P, Moore J, Richardson H, Pan Q (2005) The economic impact of a terrorist attack on the twin ports of Los Angeles-Long beach. A report by the center for risk and economic analysis of terrorism events. University of Southern California, Los Angeles
33. Guimaraes P, Hefner F, Woodward D (1993) Wealth and income effects of natural disasters: an econometric analysis. Rev Reg Stud 23:97–114
34. Green R, Miles S, Gulacsik G, Levy J (2008, Aug) Business recovery related to high-frequency natural hazard events. Quick Response Report 197. Natural Hazards Center, Boulder, CO
35. Groff GK (1973) Empirical comparison of models for short-range forecasting. Manage Sci 20 (1):22–31
36. Hochrainer S (2006) Macroeconomic risk management against natural disasters. DUV
37. Hochrainer S (2009) Assessing the macroeconomic impacts of natural disasters: are there any?. World Bank Policy Research Working Paper Series, vol (2009).
38. Horwich G (2000) Economic lessons of the Kobe earthquake. Econ Dev Cult Chang 48 (3):521–522
39. Kunreuther H (1998) Introduction. In: Kunreuther H, Roth R (eds) Paying the price: the status and role of insurance against natural disasters in the United States. Joseph Henry Press, Washington, DC, pp 1–16
40. Landers P (2001,Oct 9). Kobe disaster offers clues on rebuilding. Wall Street J. A19
41. Leitko T, Rudy D, Peterson S (1980) Loss not need: the ethics of relief giving in natural disasters. J Sociol Soc Welf 7(5):730–741
42. MacDonald D, Murdoch J, White H (1987) Hazards and insurance in housing. Land Econ 63:361–371
43. Makridakis S, Hibon M (1979) Accuracy of forecasting: an empirical investigation (with discussion). J R Stat Soc Ser A 142(Part 2):79–145 (lead article).
44. Martell DL, Bevilacqua E, Stocks BJ (1989). Modelling seasonal variation in daily people-caused forest fire occurrence. Can J Forest Res 19(12):1555–1563
45. McEntire D, Cope J (2004) Damage assessment after the Paso Robles (San Simeon, CA) earthquake: lessons for emergency management: quick response research report #166. Natural Hazards Center at the University of Colorado, Boulder, CO
46. Mileti D (1999) Disasters by design: a reassessment of natural hazards in the United States. Joseph Henry Press, Washington, DC
47. National Academies of Sciences, Engineering and Medicine (2016) "G-Science" Academies call for action to strengthen global disaster resilience, understand and protect the brain, and nurture future scientists. http://www8.nationalacademies.org/onpinews/newsitem.aspx?RecordID=04192016
48. OAS (1991) Primer on natural hazard management in integrated regional development planning. Organization of American States, Washington DC

49. Rose A, Benevides J, Chang S, Szczesniak P, Lim D (1997) The regional economic impact of an earthquake: direct and indirect effects of electricity lifeline disruptions. J Reg Sci 37 (3):437–458
50. Rose A (2004) Defining and measuring economic resilience to disasters. Disaster Prev Manag 13(4):307–314
51. Rose A, Liao S (2005) Modeling regional economic resilience to disasters: a computable general equilibrium analysis of water service disruptions. J Reg Sci 45(1):75–112
52. Saaty TL (1976a) Hierarchies, reciprocal matrices, and ratio scales. In: Modules in applied mathematics, mathematical association of America, Cornell Univ., Ithaca, New York
53. Saaty TL (1976b) Interval scale from pairwise comparisons. In: Lucas WF, Roberts FS, Thrall RM (eds) Discrete and system models. Springer-Verlag, New York, pp 248–250
54. Saaty TL (1980) The analytic hierarchy process. McGraw Hill Company, New York
55. Saaty TL (1982) How to structure and make choices in complex problems. Hum Syst Manage 3(4):255–260
56. Saaty TL (1986) Exploring optimization through hierarchies and ratio scales. Socio Econ Plann Sci 20(6):355–360
57. Saaty TL (1987) Risk—its priority and probability: the analytic hierarchy process. Risk Anal 7 (2):159–172. Research of Japan 31(8):479–481
58. Saaty TL (1989) Group decision making and the AHP. In: Golden B, Wasil E, Harker P (eds) The analytic hierarchy process. Springer, Berlin, Heidelberg, pp 59–67
59. Saaty TL (1990) Multicriteria decision making: the analytic hierarchy process. RWS Publications, Pittsburgh, PA
60. Saaty TL (1992) New perspectives on decision making in complex environments. Anheuser-Busch distinguished lecture, SEI Center of Advanced Studies in Management, The Wharton School of the University of Pennsylvania
61. Saaty TL, Vargas LG (2007) Dispersion of group judgements. Math Comput Model 46:918–925
62. Saaty TL, Vargas LG (2012) Models, methods, concepts & applications of the analytic hierarchy process vol 175. Springer Science & Business Media
63. Scanlon J (1988) Winners and losers: some thoughts about the political economy of disaster. Int J Mass Emerg Disasters 6(1):47–63
64. Slutsky E (1937) The sommation of random causes as the source of cyclic processes. Econometrica 5:105–146
65. Tavares J (2004) The open society assesses its enemies: shocks, disaster, and terrorist attacks. J Monet Econ 51(5):1039–1070
66. The Economics of Natural Disasters—Federal Reserve Bank of St. Louis (1994) https://www.stlouisfed.org/publications/regional-economist/april-1994/the-economics-of-natural-disasters
67. Wittner G, McKirdy S, Wilson R (2005) Regional economic impacts of a plant disease incursion using a general equilibrium approach. Aust J Agricult Resour Econ 49(1):75
68. Wold H (1938) A study in the analysis of stationary time series. Almgrist & Wiksell, Stockholm
69. Worthington A, Valadkhani A (2004) Measuring the impact of natural disasters on capital markets: an empirical application using intervention analysis. Appl Econ 36:2177–2186
70. Yule GU (1926) Why do we sometimes get nonsense-correlations between time series? A study in sampling and the nature of time series. J R Stat Soc 89:1–64
71. Zimmerman R et al (2005) Electricity case: economic cost estimation factors for the economic assessment of terrorist attacks. A report by the Center for Risk and Economic Analysis of Terrorism Events. University of Southern California, Los Angeles

Complexity and Disaster Forensics: Paradigms, Models and Approaches for Natural Hazards Management in the Pacific Island Region

Jason Levy

Abstract "Complexity and Disaster Forensics: New Paradigms, Models and Approaches for Natural Hazards Management in the Pacific Islands Region" assesses and applies complex systems theory, modeling and analysis to disaster forensics policy and research. To better understand the root cause and complex causality of disasters, complex systems theory, with roots in the fields of statistical physics, information theory, and non-linear dynamics, and systems analysis, is applied to help communities and nations achieve important social development goals, reduce institutional brittleness and increase disaster resilience by promoting positive transformations in the co-evolving and mutually dependent human-environmental condition, and by capitalizing on opportunities provided by human creativity, diplomatic openings, technologic capacities and environmental change. The case studies, investigations and models outlined in this chapter collectively demonstrate the quality, breadth and depth of complex systems and disaster forensics methodologies. Game-theoretic ("Small World") decision analyses, "Large world" (complex systems) models of mutually interactive game design are put forth to capture the complexity of climate related disasters and to reduce the threat of climate refugees in the Pacific Island region. The resulting risk management lessons learned were applied to communities in the Pacific Island of Vanuatu, the most natural disaster prone country in the world.

Keywords Complexity · Climate change · Disaster forensics · Forensic disaster analysis · Game theory · Large world decision making · Paradigms risk analysis · Root cause analysis

J. Levy (✉)
University of Hawaii, Honolulu, USA
e-mail: jlevy@hawaii.edu

A.J. Masys (ed.), *Disaster Forensics*, Advanced Sciences and Technologies for Security Applications, DOI 10.1007/978-3-319-41849-0_16

1 Introduction: Complexity and Disaster Forensics for Pacific Island Nations

Pacific island communities are faced with a unique set of environmental, economic and cultural issues pertinent to the management of natural disasters and disaster forensics investigations. In 2011, 80 % of global disaster-related economic losses occurred in the Asia and Pacific region [60]. These large disaster losses include not only destroyed property and critical infrastructure, but also the large scale loss of life and damage to ecological systems in the Asia-Pacific. For example, there have been several major coastal storms to affect Pacific islands in recent decades: Hurricane Iniki (central North Pacific) hit the island of Kauai in Hawaii in 1992, leading to $2.5 billion in physical damages while Super Typhoon Pongsona (western North Pacific) struck Guam on December 8, 2002 and caused $700 million in damages. Other notable historical storm "event anatomies" in the Pacific Ocean region include Typhoon Chata'an (western North Pacific) and Cyclone Heta (central South Pacific). The strong winds, heavy rains, and high seas (storm surge, etc.) that accompany these disasters pose a direct threat to the well-being of Pacific communities, with island communities disproportionately affected by natural disaster.

A central problem in natural disaster management is that our expansive species of primate, *Homo sapiens* (so-called "man the wise"), now enjoys a remarkable, although possibly temporary, dominance of its host ecosystem because of unprecedented neurological development. Christensen [7] points out that natural disaster management is more about "human problems" than environmental problems. Over the past four decades the scope and scale of environmental crises has expanded considerably, from local pollution and resource depletion issues to regional and global problems including soil erosion, climate change, and ozone depletion [24, 47]. In light of the novelty, urgency, and complexity surrounding the environmental problems of modernity, decision makers are forced to learn more about their worldviews and beliefs towards natural hazards, risks and disasters

Consider the issue of global climate change. Climate change threatens the fabric of life for people around the world—it affects key health, environmental and social dimensions including access to clean water, food production, and the sustainability of ecological systems and the urban built environment. The countries of the Asia-Pacific region, both developing and developed, are particularly exposed to human-caused climate-related hazards. While scientists have developed complex computer models of the earth's atmosphere [29] and the signs of global warming are increasingly apparent, humans continue to neglect sustainable patterns of energy use, food production, forest management, and transportation policies that could reduce the risk of climate-change related disasters. The human perception of environmental risks, hazards and disasters depends on one's "view of the world," or mental model (*Weltanschauung*). These psychological models "predetermine how we perceive reality" [2, 37] and influence how we perceive disasters, respond to them and undertake mitigation activities and disaster forensics investigations. Psychological models of disasters are constructed from past experiences [1] and help to simplify, structure, and 'make sense' of the deluge of information decision makers receive [10, 28, 31]. Many authors have categorized these mental models: Harvey [23] considers epistemological types; Maruyama [39] explores various "mindscapes" or "metatypes"; Timmerman [59] considers "myths and paradigms"; and Holling [26] describes prevailing environmental "belief systems". Mental models are the mechanisms through which we interpret reality, and hence it is these, not reality, that guide our behavior. Accordingly, mental models hold supreme significance in political, economic, and ecological decision making. For example they may help to explain why humanity continues to increase disaster risk and create a less sustainable future by polluting the planet, overharvesting natural resources and releasing greenhouse gases into the atmosphere on a large scale (several billion tons of carbon per year).

Information that confirms our existing mental models about disasters is readily accepted, while information contradicting existing disaster constructs is commonly ignored, reinterpreted, or even changed to suit the model better. Although less common, sometimes humans modify their mental models in order to accommodate new experiences about hazards, risks and disasters and conflicting information. When mental models do change, "they tend to do so rapidly, because the psychological re-structuring permits new kinds of information to enter" [1]. When mental models change across an entire scientific discipline, it is a "paradigm shift" [32, 64]. Extending Kuhn's concept, one can consider mental models across an entire culture or "cultural paradigms" [1, 31].

Often, mental models yield completely opposite conclusions, even among experts using the same data. For example, rangeland scientists disagree over whether rangelands are fragile, extensively degraded and facing disaster [12], intrinsically stable [56], unstable [14], highly resilient [52] or all of these, depending on rainfall variability [9]. An original categorization of human mental models and natural hazard/disaster paradigms is developed in Fig. 1 to illustrate four mental models and paradigms that have driven political debate, scientific research, and public concern about hazards and disasters: the Nature Constant (Engineering) Paradigm which dominated until the middle of the twentieth century; the Nature Random (Behavioral) Paradigm that was popular from 1950 to the 1970s; the Nature Ephemeral (Social Vulnerability) Paradigm that was widely adopted in the last quarter of the twentieth century; and Nature Resilient (Complexity) perspectives that are currently being explored by leading scholars, organizations and community leaders. For example, the Nature Resilient (Complexity) exhibits high paradigm complexity, incorporates biocentric (rather than anthropocentric) ethics, and emphasizes sustainability, disaster resilience, monitoring and adaptation.

Ethics constitute an important part of 'mental models' [1] since society must decide which system components are valued and should be preserved: how many resources should society dedicate to fostering disaster resilient communities? How highly should we value sustainable patterns of consumption and production? How many resources should we dedicate to disaster forensics analyses? Anthropocentric values tend to emphasize the use of natural resources for economic purposes and the well-being of humans whereas biocentric values deal with the intrinsic, non-utilitarian values of natural resources. The debate between John Muir and

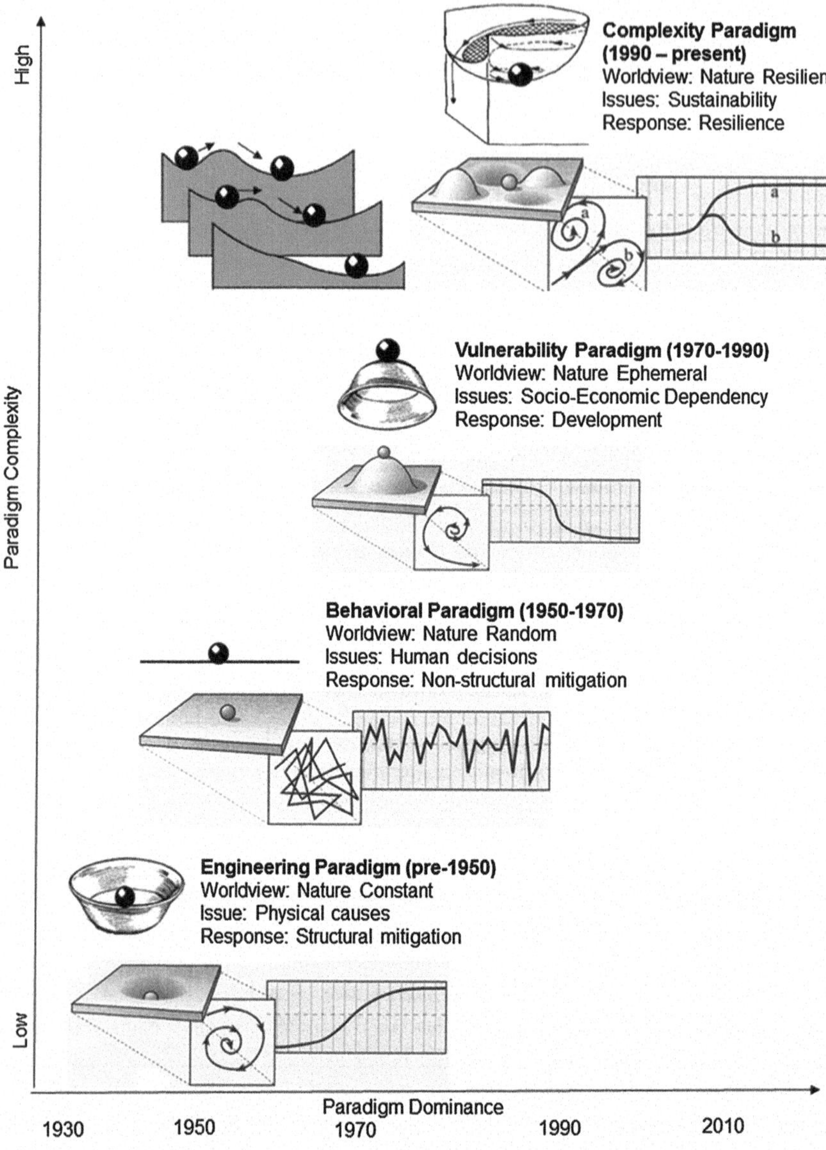

Fig. 1 Natural hazard and disaster paradigms (adapted from [27, 48, 51])

Gifford Pinchot serves to highlight this distinction: Pinchot (an early Conservationist/Utilist) argued that resources exist in part for human consumption and ecological resources should be managed to "reap maximum potential benefit for human life" [35]. On the other hand, Muir, an 'Inherentist/Preservationist', emphasized the intrinsic value of living systems.

The Western anthropocentrism which separates humans from nature and emphasizes domination and control of natural systems is not universally accepted. In China and Japan, for example, nature means "what is by itself." Needham [43] speaks of the irony with which the Chinese greeted the Jesuits' announcement of the triumphs of modern science. For them, the idea of 'managing' nature seemed a wonderful example of anthropocentric foolishness. According to Chinese tradition, nature is spontaneous harmony: "What can be controlled is never completely real; what is real can never be completely controlled" [41]. Disaster forensics professionals, community leaders, elected officials, corporate executives scientists and public policy experts often hold competing perspectives on how organizations should respond to low probability, high consequence extreme events and natural disasters, such as earthquakes, tornadoes, and earthquakes [22, 57]. A tension exists between command and control processes that seek to exploit nature, thereby increasing disaster risk (see the Engineering Paradigm in Fig. 1) [46, 50] and processes of innovation and discovery that are capable of learning from errors and thoroughly investigating disaster events to understand root causes and complex causalities [8]. This is consistent with the sustainable management approaches for resilience labeled under the Complexity approach. Theories of risk reduction through redundancy [26, 33, 54] have been rejected in practice as too costly for low probability events, at least in most situations.

Note that each worldview in Fig. 1 takes a different approach to the salient disaster issues and responses (thereby making use of different socio-economic and ecological indicators) and is the result of different historical and cultural influences. The remainder of the introduction describes how key disaster concepts of this chapter are organized according to Fig. 1. To understand the aforementioned mental models better, it may be helpful to picture nature as a ball on a 'dynamical landscape': Depending on the topography of the landscape, human activities will slightly oscillate the ball, or permanently dislodge it into the domain (basin) of another "attractor" (often referred to an equilibrium (or bowl). The ball represents the current position of the system (e.g. by the current amounts of grass, shrub and livestock) while the cup represents the basin of attraction. Here, the notion of ecological system stability is crucial: do systems tend to a single unique stable state from all initial conditions and disturbances? Or do large perturbations carry these systems into a new region of state space. In the former case, historical accidents are unimportant; in the latter, chance events can be of "overriding significance" [40].

2 Engineering Paradigm of Disasters: The Nature Constant Perspective (Pre-1950s)

A unique combination of events—the scientific-technological revolution, European domination of newly discovered lands, and seventeenth century laissez-faire—gave rise to the Western worldview that nature provides an endless supply of ecological resources for humans. The term ‘frontier economics’ was coined by Kenneth J. Boulding to describe this approach which prevailed in most countries until at least the late 1960s [6] and was adopted in many academic and business communities until at least the mid twentieth century: even if ecological limits are reached, they can be overcome by product substitution and other technological innovations. Examples include the ‘developmental’ paradigm that has dominated rangeland management [63]. In an influential paper, Arrow [4] argue that technological optimism and cornucopian beliefs continue to dominate economic policy since national and international economic agreements “usually ignore the environment”. In areas where the environment is beginning to infringe on economic policy, such as the General Agreement on Tariffs and Trade (GATT) and the North American Free Trade Agreement (NAFTA), it remains a “tangential concern”. While there was much fanfare over the 2015 international agreement at the UN Paris climate summit, the 21st Conference of Parties (COP21) which seeks to cap the increase in global average temperatures to 2 °C above pre-industrial temperatures it is unclear whether the current “Intended Nationally Determined Contributions” (INDC) will be sufficient to reduce greenhouse emissions. The goal is for all involved nations, both developed and developing nations—and those in between—to collectively lower greenhouse gas emissions, mitigate the extent of climate change and adapt to global temperature rise.

The Nature Constant perspective focuses on global stability and the supposed linear response of ecological systems to all human disturbances, regardless of intensity and scale of the shocks. This concept is well represented by the “ecosystem linearization” models of Patten [44] in which interactions between components are assumed to be linear. Clearly, Nature Constant engineering paradigm plays a central role in both understanding the causes of, and designing policy solutions to, hazards and disasters, including human-induced climate change. For example, if the Nature Engineering paradigm (Nature Constant worldview) prevails in a majority of countries there may not be sufficient resources and motivation to tackle the problem even though the COP21 UN Paris climate change agreement commits developed nations to raise $100 billion annually beginning 2020 in order to empower developing countries to fight global warming by strengthening their technical capacity to transition to renewable energies and sustainable development. Proponents of the Nature Constant view point out that as per capita income goes up, there is increasing environmental degradation up to a point, after which environmental quality ostensibly improves [5, 53]. This “inverted-U” relationship [55] is provided as evidence that economic growth will lead to a healthier environment.

A half century ago, a strong bias towards the Engineering perspective encouraged the development of large scale civil engineering and public works projects. While the US continues to rely heavily on structural flood control and other engineering measures for water resources disaster management, many countries around the world (particularly in Europe) have shifted towards a flood management policy of “living with floods”, focusing on society’s ability to co-exist with flood hazards, rather than be protected from them (see the Nature Resilient worldview in Fig. 1). In particular, after the severe Rhine River flooding of 1993 and 1995, the

Dutch government adopted a flood control policy of "more room for rivers" with an emphasis on establishing new storage and conveyance space. In this spirit, six European governments created the Meuse River High Water Action plan which provides "longer storage and more release" for rivers. The Netherlands alone has allocated three billion US dollars for a broad array of levee alternatives. Fortunately, these flood risk management efforts have delivered impressive theoretical insights and practical examples for the "Nature Resilient" advocates. For example, in the state of Baden-Wurttemberg, Germany, the addition of over 200 m^3 of floodplain storage has reduced flood stages to 1950 levels, thereby reducing the impact of extreme flood events under uncertainty and promoting a more holistic, sustainable relationship between society and the environment.

Recently, dams have been removed in increasing numbers in the US and around the world as they have filled with sediment, become unsafe or inefficient, or otherwise outlived their usefulness [3]. For example, the recent removal of the 64-m-high Glines Canyon Dam and the 32-m-high Elwha Dam in northwestern Washington State were among the largest yet, releasing over 10 million cubic meters of stored sediment. A major finding is that rivers are resilient, with many responding quickly to dam removal (see again the Nature Resilient Paradigm in Fig. 1). It is shown that most river channels stabilize within months or years, not decades, particularly when dams are removed rapidly; phased or incremental removals typically have longer response times [20]. The rapid physical response is driven by the strong upstream/downstream coupling intrinsic to river systems. Substantial fractions of stored reservoir sediment—50 % or more—can be eroded within weeks or months of breaching [20]. Sediment eroded from reservoirs rapidly moves downstream [38, 45]. Many rivers soon trend toward their pre-dam states [13, 38]. Dam size, river size, reservoir size and shape, and sediment volume and grain size all exert first-order controls on physical and ecological responses to dam removal. Larger dam removals have had longer-lasting and more widespread downstream effects than the much more common small-dam removals [20]. The total number of U.S. and international removals are, however, more than offset by dam construction for hydropower and other purposes and in regions with emerging economies, such as Southeast Asia, South America, and Africa [65]. Clearly, decisions regarding dam removals require difficult tradeoffs and the balancing of risks, continued economic function, and the potential for ecologic restoration. Climate change adds another level of uncertainty and is likely to increase the demand for fresh-water storage, both as a low-carbon energy source and for consumptive use.

3 Vulnerability Paradigm of Disasters: The Nature Ephemeral Perspective (1970–1990)

The Nature Ephemeral view argues that the environment cannot safely tolerate human activities (fundamental instability is the rule) with prescribing mandatory rules and regulations to ensure sustainable patterns of consumption and production:

survival is only deemed possible by applying stringent safety factors (avoiding large scale irreversible damage); ensuring maximum biological diversity (in structure and over space); curtailing human population growth (thereby preserving future options and ecological possibilities); and abandoning technological innovation (due to unknown associated hazards and risks). Accordingly, decentralized governance with fine-scaled local autonomy is necessary. The Earth's resource base, it is argued, cannot support increased economic growth: draconian legislation is necessary to replace existing transportation, energy, agricultural and technologic systems with more environmentally sustainable ones.

According to the International Union of Geological Sciences (IUGS), we are officially in the Holocene ("entirely recent") epoch, which began 11,700 years ago after the last major ice age. However, advocates of the Nature Ephemeral approach argue that the Holocene should be replaced by the "Anthropocene" era because human beings have permanently changed the planet through mass extinctions of plant and animal species, pollution of the oceans and alterations to the atmosphere, among other lasting impacts. The Anthropocene represents a new geological epoch in which humans are a dominant geophysical force producing previously unimaginable impacts on the earth. The theories and practice of disaster forensics and sustainability in the Anthropocene can provoke new understandings of hazards, values, crises, spatiality, politics, temporality, ethics, and responsibility. A new focus on disaster resilience becomes possible when society considers that the stakes involve irreparably damaging the world we inhabit. In the Nature Ephemeral worldview, the (biocentric worldview) was summarized by Chief Seattle during the Treaty of Walla Walla negotiations: "every part of the earth is sacred." It follows that every organism has a right to exist and should be preserved. Early advocates of this "inherentist, preservationist" ethic include John Muir, a nineteenth century naturalist who successfully campaigned for forest preservation in the United States. Ludwig [36] promote the Nature Ephemeral perspective in a provocative article on conservation policy: they argue that despite claims of sustainable environmental management, ecological resources are "inevitably overexploited, often to the point of collapse or extinction." The Nature Ephemeral perspective is often contrasted as a refreshing antidote to the anthropocentric, cornucopian perspective. Tropical rain forests are often held to fit this Ephemeral description [19]. Perhaps the Nature Ephemeral perspective has the most inherent validity as economic pressures are a reflection of human desires, and both human population growth and the per capita consumption of resources are increasing in an unsustainable fashion. In addition, even well-meaning specialists often cannot detect initial signs of resource overexploitation until ecological damage is severe. Worse still, consensus among scientists is seldom achieved, even after the total collapse of a resource: humans seem unwilling to take prudent environmental measures, even when there is a good scientific understanding that certain practices are ultimately destructive. The result of human greed, as we have seen too often, is the misuse, and subsequent destruction of resources.

An excellent example is the use of irrigation in arid regions. It is well-known that in ancient Mesopotamia the once highly productive wheat crop had to be replaced by more salt-resistant plants. The increased soil salt was a result of irrigation [62]. While many scientists warned of similar consequences in California due to large-scale planned irrigation, pleas from local biologists, some more than a century ago [25] fell on deaf ears [18]. Thus, 3,000 years of experience and sound scientific knowledge may not be sufficient to overcome shortsightedness and ignorance.

4 Complexity Paradigm of Disasters: The Nature Resilient Perspective (1990-Present)

This view is frequently described as a synthesis of the Nature Constant and Ephemeral perspectives (between the extremes): the environment is forgiving of most shocks, but large perturbations can knock variables into new regions of the landscape. In this view, the response of a living system to stress will be largely linear until a critical threshold is crossed, at which point a radical change (called a discontinuity or catastrophe) occurs [30, 58]. Here, severe environmental conditions (for example pest out breaks, fires, and windstorms) are used to test the survivability of system components or eliminate weak ones. In the Nature Resilient view, human culture is seen as embedded in nature; dependent on it, and capable to harm it [48]. By emphasizing the resilience of systems [21], insights from the Nature Resilient view may help societies to overcome pathological behavior, including institutional rigidity, social dependencies, political hegemony, and ecological degradation.

In the complexity paradigm it is helpful to distinguish state variables (ecosystem characteristics/attributes) from parameters (which establish the landscape of the ecosystem of interest) as shown in Fig. 2. State variables often change rapidly in response to feedback from system dynamics while parameters are often external to the system and typically independent of feedback from state variables within the model (or subject to only very slow feedback). One could also consider variables of a system to be measurable ecosystem state characteristics (e.g. population density, vegetation area, wind speed and sand flux) while parameters are the drivers of these state variables (e.g. birth/death rates, migration, etc.). Regime shifts (times when the ball on the landscape moves into the domain of another attractor) can arise due to changes in either variables or parameters.

Understanding the root causes and complex causalities of natural hazards, risks and climate-related disasters is also consistent with the Forensic Investigations of Disasters (FORIN) project which seeks to guide disaster management policy, encourage coherence across essential disciplines and advance methodological diversity [16]:

- **Policy**: conduct disaster forensic analyses with inputs from multiple disciplines, stakeholders, and policy makers relevant to the renewable energy field.
- **Management**: highlight the link between energy disaster related research and practice and forensic theory, insight and analysis in order to improve policy formulation and application and share results online through high-quality publicly available case studies.

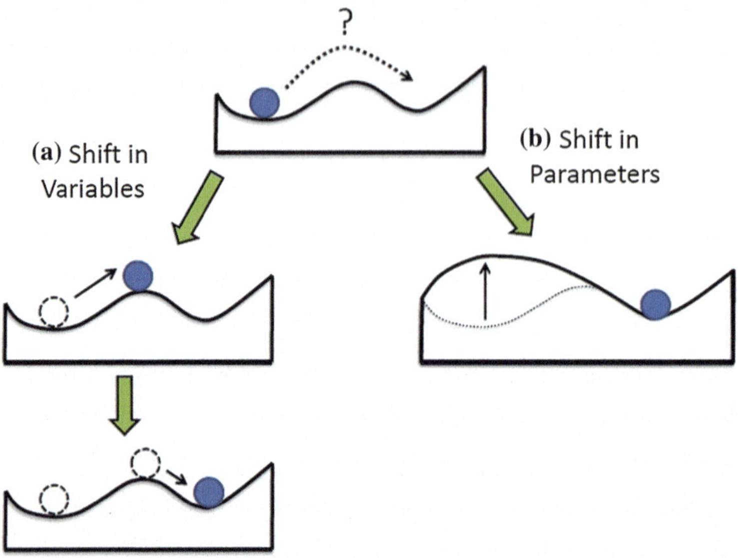

Fig. 2 Two-dimensional ball-in-cup diagrams: (*a*) a shift in state variables causes the ball to move laterally, (*b*) but a shift in parameters causes the landscape itself to change, moving the ball vertically

- **Scientific research**: build an interdisciplinary forensics capacity and implement science-based disaster forensics results as they apply to renewable energy systems.
- **Development**: determine the local manifestations of disasters; promote a culture of disaster risk reduction culture among all stakeholders, and foster wider dialogue between analytical researchers and implementing practitioners, building a common discourse in the process.
- **Disaster risk reduction**: illustrate risk drivers and promote sustainability through methodologically diverse science-based research, providing wider emphasis on reducing the risk of ecological damage, destroyed infrastructure and human accidents, injuries fatalities

5 The World's Most Vulnerable Country to Natural Disasters: Vanuatu

Situated on the 'Pacific Ring of Fire' with a cyclone season from November to April, the developing island state of Vanuatu is an archipelago of more than 80 islands (located northeast of Australia) with a population of around 263,000. It is the world's most vulnerable country to extreme natural events: exposure of its population to natural disasters has been ranked at 63.6 % [61]. Moreover, a recent global risk analysis study [42] found that Port Vila, Vanuatu's capital and main urban center, situated on the southwest coast of Efate Island is the world's most exposed city to natural disasters: Port Vila's 44,000 residents are more at risk to natural disasters than 1300 other cities around the world because the city is highly exposed to multiple hazards including earthquakes, tsunamis, storm surge, flooding and tropical cyclones.

Over the last decade Vanuatu maintained strong economic growth, at an average of 4 % GDP per annum (in real terms) which exceeds the growth of many other countries in Melanesia. Over the medium term, this GDP growth will continue to rely on the expansion of services, especially tourism, and industry. While stable macro-economic management, key microeconomic reforms and social stability helps to attract investment and tourists; however, a number of challenges have potential to undermine Vanuatu's economic and social stability. Specifically, Vanuatu has a widely dispersed population and limited investment in provincial areas: 70 % of people live in rural areas across 65 of the country's more than 80 islands. Widespread violence against women and children results in nearly three-quarters of women experiencing physical and sexual violence in their lifetime [11]. Other issues include access to justice more broadly, population growth, unemployment (the number of graduates entering the job market far outstrips the average number of jobs created in the formal economy), urbanisation leading to increased land pressures, poverty and insecurity, climate change and frequent natural disasters.

The annual rate of urbanization is 4 %, exceeding Vanuatu Government's ability to provide sufficient urban infrastructure: the urban percentage of the population of Vanuatu is expected to increase from 25 % today to 50 % by 2050, as more citizens move to urban areas in search of jobs and services. One-third of Vanuatu's population lacks access to basic services and more than 12 % live below the poverty line while literacy and numeracy, immunisation and nutrition are stagnating and, in some cases, declining. With insufficient revenue to adequately fund education and health services the country is struggling to meet its human development goals. Schools and health clinics often lack basic resources due to constraints on service delivery capacities and system inefficiencies. Moreover, poverty together with sub-standard housing, inadequate basic services and overcrowding pose a risk to social stability.

With limited financial resources, citizens in Port Vila often build improvised dwellings with salvaged or cheaply bought materials, such as timber, corrugated iron, tin and fabric: makeshift dwellings make up 27 % of all residences in the capital and 30–40 % of Port Vila's residents live in informal settlements such as Freswota and Seaside.

6 Vanuatu and Tropical Cyclone Pam

Island communities such as Vanuatu are especially vulnerable to a warming and more energetic climate system: climate-related sea level rise is contributing to storm surge, king tides, coastal floods and the degradation of groundwater resources.

Vanuatu is vulnerable to a range of hazards from storm surge to earthquakes, volcanic eruptions, floods, cyclones and tsunamis. Located in a tropical climate zone south of the equator, Vanuatu has been affected by at least 20 damaging cyclones in the past quarter century including Cyclone Lusi which impacted thousands of people across northern and central regions of Vanuatu in March 2014, destroying villages and crops. Approximately a year later, in mid-March 2015, severe tropical cyclone Pam, the most powerful cyclone to ever hit the Pacific swept through the South Pacific Ocean. Affecting most of Vanuatu's islands, tropical cyclone Pam was the worst natural disaster in the history of Vanuatu: the country was more directly affected by Cyclone Pam than any other nation. Vanuatu received the brunt of the Hurricane Pam although the storm also significantly impacted Tuvalu, Kiribati, the Solomon Islands and Papua New Guinea. With winds gusting up to 320 km/h tropical cyclone Pam had the highest 10-min sustained wind speed of any South Pacific tropical cyclone on record and was the third most intense storm in the Southern Hemisphere as measured by atmospheric pressure (after tropical cyclone Zoe of 2002 and tropical cyclone Gafilo in 2004). The storm killed 15 people directly and left 75,000 homeless, with many others killed indirectly or injured.

Cyclone Pam deepened hardships faced in Port Vila's urban settlements: heavy winds and torrential rain destroyed homes and bridges, cut off essential public services and flooded communities already faced with inadequate power, water and communication service, deteriorating roads, poor drainage and sanitation. During the first week after the cyclone more than four thousand displaced citizens were supported by forty-three evacuation centers in the urban and peri-urban areas of Port Villa. Post-cyclone Pam disaster recovery in Port Vila's settlements remains slow and is hindered by land tenure issues, finance and resource constraints. It is recommended that foreign governments work with Vanuatu to focus on the most cost-effective interventions to meet human development priorities and improve system efficiencies (Fig. 3).

More than a year after Cyclone Pam, Vanuatu still faces many challenges: subsistence crops were decimated (it is estimated that approximately 90 % were lost). The almost near complete crop decimation by the cyclone was followed by a recent drought: the entire region of Melanesia faced an El Nino-driven drought which exacerbates the damage created by Cyclone Pam as water sources dry up, crops fail, and residents struggle with food security. The financial cost of Cyclone Pam is approximately $590 million, more than half of Vanuatu's yearly GDP. The Government of Vanuatu estimates the economic impact of Cyclone Pam is equivalent to 64 % of GDP, primarily due to losses in the country's tourism and agricultural sectors; the tropical cyclone severely damaged about 80 % of the country's buildings and 90 % of the country's crops. Specifically, houses, roads, health centers, schools, bridges, sanitation systems, the power grid and other critical infrastructure were devastated.

The cyclone affected 63 % of the population (approximately 166,000 people spanning 22 islands) and thousands received emergency shelter and other forms of humanitarian assistance; thousands of people across the country continue to need food, water and sanitation as they wait for assistance to rebuild their homes. The

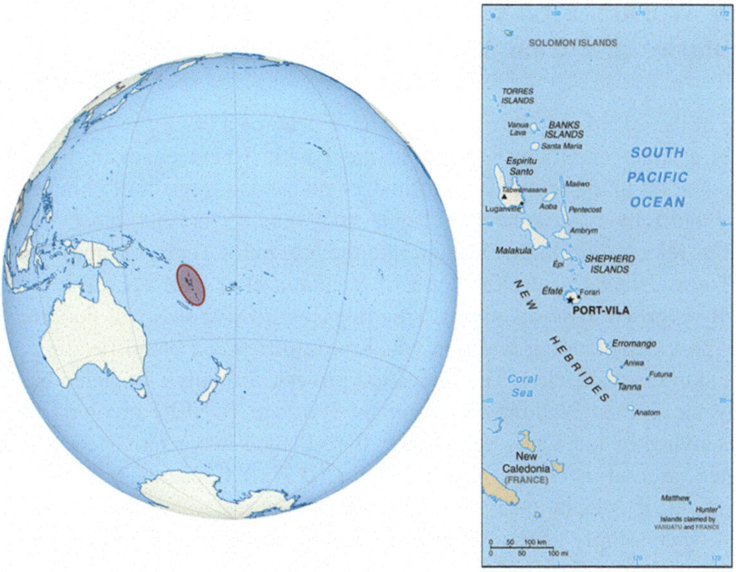

Fig. 3 Location of the island nation of Vanuatu. By TUBS—own work. This vector graphics image was created with Adobe Illustrator. This file was uploaded with Commonist. This vector image includes elements that have been taken or adapted from this: Polynesian triangle.svg (by Gringer), CC BY-SA 3.0, https://commons.wikimedia.org/w/index.php?curid=15127570

Government of Vanuatu and partnering aid organizations are helping to provide cyclone affected communities with access to safe drinking water and to restore their livelihoods through the distribution of seeds, fresh planting materials and livestock. Fortunately Vanuatu received assistance from its Pacific military partners and NGOs on the ground who attempted to help with the nation's recovery.

The Vanuatu government's National Disaster Management Office efficiently and effectively coordinated the Cyclone Pam aid effort including distributing aid to around 100,000 people across islands without airstrips, roads or functional telecommunications in the initial weeks. The International Federation of Red Cross and Red Crescent Societies (IFRC) working with National Red Cross Societies in the region, reached more than 40,000 people across Vanuatu, Tuvalu, Kiribati, the Solomon Islands and Papua New Guinea with direct humanitarian assistance, ensuring the distribution of essential supplies to people in need. However only in Vanuatu did the local Vanuatu Red Cross lead recovery efforts and repair and rebuild their homes and communities: nearly 30 % of humanitarian assistance was provided through the Vanuatu Red Cross [49].

7 The World's First Climate Refugees: Pacific Island Canaries in a Coal Mine

Citizens of the island nations of Oceania have become a canary in a coal mine. For example, a small community living in the Pacific island chain of Vanuatu has become the first in the world to be formally relocated as a result of climate change. The tiny and remote settlement of Tegua, in Vanuatu's northernmost province of Torba, was relocated to higher ground into the interior of Tegua (Tagua) Island, one of the Torres Islands in the far north of Vanuatu, after their coastal homes were repeatedly exposed to storm surges, king tides, aggressive seas and flooding which has also provided standing water in which mosquitoes can breed (leading to more malaria, skin diseases, etc.). The relocation project, formally completed in August, 2005 by the United Nations Environment Program (UNEP) involved over 100 villagers living in the Lateu settlement: the villagers were moved 600 m inland where the ground is 15 m higher. The UNEP's "Capacity Building for the Development of Adaptation in Pacific Island Countries" relocation project underlines the increasingly drastic measures now underway to conserve low lying communities in the Pacific Island region and elsewhere, arguably linked to the release of greenhouse gas emissions to the atmosphere. However, it is challenging to keep resettling villages farther inland because the islands are narrow: the biggest island in the Torres is about 16 km (10 miles) long and 3 km (2 miles) wide.

Recognizing the importance of disaster risk reduction, the government of Vanuatu has taken steps to strengthen the nation's disaster management institutions in order to enhance disaster preparedness and increase the capacity of communities to cope with destructive calamities. For example, in 2013, a new disaster warning center (capable of monitoring volcanic, seismic, and tsunami activity) opened in Port Vila. As cyclone Pam drew near, this state of the art technology successfully issued disaster warnings to more than 80 % of the population. The relatively low death toll in Vanuatu is a result of improved communication technologies and the fact that Vanuatu was the first Pacific Island country to integrate disaster risk management into national planning in 2006. After cyclone Pam, the government of Australia began funding a long-term recovery package, developed in consultation with Vanuatu's government in order to restore health and education facilities, support livelihoods, promote economic recovery, support gender and disability inclusion, and repair and rebuild critical public infrastructure.

However, as one of the world's Least Developed Countries (LDCs), Vanuatu requires even more investment in education, safety and urban disaster risk reduction measures. It is proposed that the government improve the implementation of land use planning policies and that authorities apply and enforce the national building code in all structures, including private, commercial and public buildings.

8 Game Theory for Disaster Forensics and Complexity: The Climate Change Refugee Problem

Advances in game theory can help to address issues of complexity in disaster forensics problems. A game-theoretic tool known as conflict analysis [15, 17] is now applied. Game theory allows any number of DMs, each of whom may have a large number of options or alternatives to be taken. Here, a set of states is defined as possible combinations of all options of the DMs (where logically impossible states should be deleted). Once a set of states is determined, the preferences of each DM on every state must be determined. Reliance on ordinal preference ranking instead of cardinal evaluations has both advantages and disadvantages. Using ordinal preferences, a DM can express her preference order with less difficulty and more confidence than with cardinal preferences, although it is often difficult to express subtle differences in preference information. After the preference structure for each DM is determined, equilibrium solutions are obtained. Generally, it is possible to find one or more equilibrium solutions. The graph model for conflict resolution is now described. In a graph model of a conflict [15], the DMs and the possible states of the conflict are specified, along with the state transitions controlled by each DM. A graph model also includes each DM's ordinal ranking of all possible states where ties are allowed.

When a graph model is analyzed, each state is assessed for stability from the point of view of each DM. A state is stable for a DM if that DM would choose not to depart from it, should it arise. Solution concepts are models of the DM's thinking processes in deciding what would be the likely outcome of a move away from a given state. Note that a state may be stable under some solution concepts but not others. Of course, different DMs may have different solution concepts. A state that all DMs find stable is an equilibrium, and constitutes a possible resolution of the conflict model. Different solution concepts imply different levels of foresight, or measure a DM's ability to consider possible moves that could take place in the future. A DM with high foresight thinks further ahead. Nash stability (R) has low foresight, and the level of the foresight increases from low to high. Nonmyopic stability (NM) has the highest foresight and limited-move stability (L_h) has variable foresight level given by the parameter h. Some solution concepts, such as L_h and NM, allow strategic disimprovements, which occur when a DM (temporarily) moves to a worse state in order to reach a more preferred state eventually; other solution concepts, such as R and sequential stability (SEQ), never allow disimprovements; still others, general meta-rationality (GMR) and symmetric metarationality (SMR) permit strategic dis-improvements by opponents only. Different solution concepts also imply different levels of preference knowledge. Under R, GMR and SMR, a DM need only know its own preferences, while a DM must know the preference information for all DMs for the solution concepts SEQ, L_h and NM.

The decision support system GMCR II implements the Graph Model for Conflict Resolution within a Windows environment . The structure of GMCR II involves the following components:

- Modeling Subsystem (decision makers, options, feasible states, allowable state transitions, relative preferences)
- Analysis Engine (stability analysis, coalition analysis)
- Output Interpretation System (stability, equilibria, coalitional stability)

The modelling subsystem of GMCR II allows users to enter conflict models conveniently and expeditiously. Users input DMs and options, patterns of infeasible states, allowable transitions and preference information. Then, GMCR II will generate the required input for stability analysis, including feasible states, allowable state transitions, and ranking of states from most to least preferred, allowing ties, for each DM. Based on the information generated at modelling stage, the analysis engine performs a thorough stability analysis on the conflict model, including the stability results for every state, and for each DM, under the aforementioned solution concepts. The output interpretation subsystem presents the results from the analysis engine in a user-friendly manner. Information about individual stability, equilibria, and coalition stability is easily identified and compared. The Vanuatu Disaster Management issue is now modeled below for the point in time just after the March 2015 Cyclone Pam disaster. Here, a unique model of the disaster management situation is developed and detailed modelling and analytical results are presented and explained in the subsequent section.

The options for these three main DMs—Western countries, United Nations, and Vanuatu—are given below:

As shown in Table 1, Western countries (Western) have the option to increase funding for disaster forensics analysis, to implement policies that that allow for a drastic reduction in greenhouse gas emissions, and to take (or avoid) climate leadership globally. The United Nations (UN) can promote resilience within Vanuatu and increase disaster funding to the nation. Finally, the nation of Vanuatu can take measures to prevent its citizens from becoming climate refugees that must

Table 1 Decision makers and options in the Vanuatu disaster management model

1. Western nations
• Disaster forensics focus (forensics) • Radical reduction in greenhouse gas emissions (drastic) • Avoid global climate change leadership (avoid)
2. United nations
• Promote resilience within Vanuatu (promote) • Increase disaster aid to Vanuatu (aid)
3. Vanuatu
• Prevent climate refugees (prevent) • Mainstream climate adaptation into disaster mitigation (mainstream) • Enhance access to resources (enhance)

leave the country, to mainstream climate adaptation and enhance access of its citizens to health and educational resources. Other influential groups were not considered in this model because key strategic insights can be obtained from these players. The DMs and their options are displayed in Table 2, where a "Y" opposite to an option indicates "Yes" the option is selected by the DM controlling it, whereas an "N" corresponds to "No", the option is not taken. A *strategy* for a given DM is any feasible combination of its options. For example, in the status quo state shown in Table 2, the island nation of Vanuatu is selecting none of its three options. Similar explanations can be put forward for the other two DMs in Table 2. A *state* is formed when each DM selects a *strategy*. As an example, Table 2 shows the status quo state existing in 2015. Written horizontally in text, the status quo state (NNY NN NNN) is created by Western Nations, UN and Vanuatu following strategies (NNY), (NN), and (NNN), respectively. All DMs, options and feasible states are shown in Table 3.

Table 2 Status Quo in the Vanuatu disaster management model

DM and option	Status quo	
West		
1. Disaster forensics focus	N	Strategy for Western nations
2. Radical emissions reduction	N	
3. Avoid climate leadership	Y	
UN		
4. Promote disaster resilience	N	Strategy for Environmental groups
5. Increase disaster assistance	N	
Vanuatu		
6. Prevent climate refugees	N	Strategy for Vanuatu
7. Mainstream climate adaptation	N	
8. Improve access	N	

Table 3 Feasible states of the Vanuatu game

DM and option	1	2	3	4	5	6	7	8	9	10	11	12	13
Western countries													
1. Disaster forensics focus	N	N	N	N	N	N	Y	N	Y	N	Y	Y	Y
2. Radical reduction	N	N	N	Y	Y	N	N	Y	Y	N	N	Y	Y
3. Climate change leadership	Y	Y	Y	Y	Y	N	N	N	N	N	N	N	N
UN													
4. Disaster resilience	N	N	Y	Y	Y	N	N	N	N	Y	Y	Y	Y
5. Disaster assistance	N	N	N	N	Y	Y	Y	Y	Y	Y	Y	Y	Y
Vanuatu													
6. Prevent climate refugees	N	Y	N	N	N	Y	Y	Y	Y	Y	Y	N	Y
7. Mainstream adaptation	N	N	N	N	N	Y	Y	Y	Y	Y	Y	N	Y
8. Improve access to education	N	N	N	N	N	Y	Y	Y	Y	Y	Y	N	Y

In GMCR II, users go to the main menu "Conflict" to enter the description of the conflict model. Under this menu, users can input the title of the conflict, the date of analysis for the model, and a brief introduction of the conflict. Then they input information about DMs and their options by going to "Modelling → States → Generate Possible …". A pop-up window appears to guide users adding the DMs and their corresponding options in the form of "full title" and "short title".

8.1 *Feasible States and State Ranking for Each DM*

In the Vanuatu disaster management model, there are 8 options in total. Because each option can be either selected or rejected, $2^8 = 256$ states are mathematically possible. However, many of these states are infeasible in the real world, for a variety of reasons. In GMCR II, four types of infeasibilities are available to specify infeasible patterns: "Mutually Exclusive Options," "At Least One Option," "Option Dependence," and "Direct Specification." Users go to "Modeling → States → Remove Infeasible…" to specify four types of infeasibilities. In this conflict model, only the "Option Dependence" category of infeasible patterns is employed to remove infeasible states. After all these infeasible patterns are input in GMCR II, thirteen feasible states are left in the model as shown in Table 3. For identification purposes, they have been labeled as states 1–13.

Before carrying out a stability analysis, GMCR II requires that the feasible states be ranked from most to least preferred for each DM, where ties are allowed. GMCR II possesses two flexible approaches, called Option Weighting and Option Prioritization, for conveniently specifying preference information in terms of options for each DM. An internal algorithm then automatically orders the states for the DM based upon this preference information. Option Weighting allows users to assign a number or numerical weight to each of the options from the viewpoint of each DM, where a positive or negative number means the DM likes or does not like the option, and the magnitude of the number reflects the degree of preference. Option Prioritization provides an intuitive specification based on preference statements listed from most important at the top to least important at the bottom. In addition to these two means to specify the ranking of feasible states for each DM, GMCR II also allows users to fine-tune the preference ranking by directly re-ordering states, joining two or more states into an equally preferred group, and splitting an equally preferred group apart. Option Prioritization along with Direct Ranking is employed to come up with the preference ranking for the DMs in this conflict. Table 4 lists the preference statements using option numbers in order of priority for each DM.

As shown in Table 4, both conditional and unconditional preference statements are acceptable in GMCR II. In addition, two types of conditions, "IF" and "IFF (if and only if)", are permitted. Finally, the symbols "–", "&", and "|" represent "not," "and," and "or," respectively. From Table 4, one can easily interpret the preference statements given in order of priority from the top of the column to the bottom for

Table 4 Preference statements for the decision makers in the Waiahole conflict

Western nations	UN	Vanuatu
−1	−3	−1
3 IF (1&−2)	4&5	(6&7&8) IF (2)
2 IF 4|5	−1	4|5
−5 IF −4	6&7	−3

Table 5 Evolution of the disaster management problem from State 1 (Status Quo) to State 13 (Final State)

DM and option	1		3		4		5		12		13
Western nations											
1. Disaster forensics focus	N		N		N		N	→	Y		Y
2. Radical emissions reduction	N		N	→	Y		Y		Y		Y
3. Avoid climate leadership	Y		Y		Y		Y	→	N		N
UN											
4. Disaster resilience	N	→	Y		Y		Y		Y		Y
5. Increase disaster aid	N		N		N	→	Y		Y		Y
Vanuatu											
6. Prevent climate refugees	N		N		N		N		N	→	Y
7. Mainstream adaptation	N		N		N		N		N	→	Y
8. Improve access to education	N		N		N		N		N	→	Y

each DM. The option number −1 given at the top of the left column indicates that Western Nations most prefers that they do not enforce drastic reduction in greenhouse gases. Second most preferred for Western Nations is 3 IF (1&−2). The preference statements for UN and Vanuatu can also be easily interpreted from the prioritized list given in the second and third column in Table 4, respectively. After all preference statements are entered, GMCR II will generate the resulting preference ranking using an efficient algorithm and assuming transitivity.

In a stability analysis, GMCR II calculates the stability of every feasible state for each DM for all of the solution concepts discussed in Sect. 2. If a state is stable, according to a given solution concept, for all DMs, it constitutes an equilibrium under that solution concept. It is, therefore, a compromise resolution, since no DM has an incentive to unilaterally move away from it. By going to "Analysis → Run," one equilibria—state 13—is derived. Evolution of the conflict from the status quo state (state 1) to state 13 is shown in Table 5. Note that the dynamics of the disaster management problem show that over time Western Nations choose to invest more resources on disaster forensics analyses, begin to adopt policies to seriously reduce greenhouse game emissions and begin to accept global leadership for addressing climate change challenges. The United Nations in turn provides more resources to developing countries for disaster resilience and increase funding for disaster mitigation, preparedness, response and recovery activities. Finally, the nation of Vanuatu takes measures to mainstream disaster risk reduction into all sectors of the economy, improve access to education and healthcare to reach social development goals and reduce the risk of its citizens becoming internally displaced by climate change.

9 Multiple Participant Decision Making for Disaster Forensics: A Large-World Complex Systems Approach

The twentieth century has witnessed the development of powerful non-quantitative decision making approaches for the resolution of complex disaster forensics issues and it is valuable to harness these approaches to improve disaster decision making. Figure 4 illustrates that a disaster forensics analysis involves complexity, uncertainty and high decision stakes and hence constitutes a "Large World Decision Making" problem. This is compared to classic forensics analyses which fall within the realm of traditional "Small World Decision Making". The previous section considered traditional game theory which represents a small world approach. While this section takes an operations research approach to structuring, analyzing, and modeling the previously discussed Vanuatu disaster management challenge, the

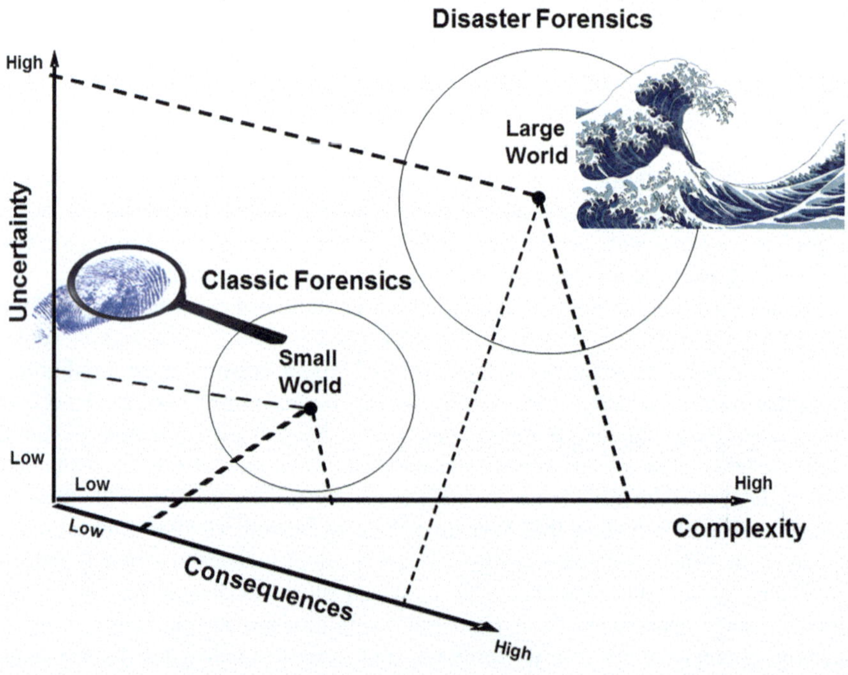

Fig. 4 Large world decision making (disaster forensics analysis) versus small world decision making (classical forensics analysis)

decision makers, options, and preferences are now not assumed to be fixed: players can interact to design the game they eventually play (by communicating with each other prior to playing a game) while simultaneously selecting a focal equilibrium set in that game. Accordingly, we capture the same climate change, disaster and forensics issues discussed in the previous section; however, we will now adopt a complex systems modeling approach.

Disaster management agencies, politicians, non-governmental organizations and policy makers involved in disaster management often find existing disaster forensics procedures and policies to be inadequate. Consequently, they may face severe challenges as they seek to reduce disaster risk and overcome previous guidelines: There are various tools available to operationalize the complex systems, large world approach. For example, Drama Theory II (DT II) as created by Levy et al. [34] allows for "large world" uncertainty in that it doesn't assume a Bayesian model in which all possible sequences of disaster forensics events (probabilistic or deterministic) and all preferences for them are known in advance (game theory can only account for "small world" uncertainty). Rather, it is assumed that emergency management personnel engage in a type of pre- play drama-theoretic "communication" (including words, explicit deeds, implicit acts etc.) in order to achieve the "common knowledge" necessary for responding to a disaster and defining the disaster forensics activities that are then carried out.

Large World decision making assumes that decision makers involved in a type of pre-play "communication" (including words, explicit deeds, implicit acts, etc.) in order to achieve the "common knowledge" necessary for analyzing the disaster. Hence, in Large World decision making a wider definition of rationality is employed than the deeply embedded "instrumental rationality" assumptions (i.e. static beliefs and preferences) of traditional game theory [34]. In contrast to Small World decision making (game theory), Large World Decision Making asserts that disaster forensics requires players to engage in a rational-emotional process of re-defining both the game and their "positions" in it until there emerges a satisfactory resolution on which they all agree. In this way, Large World decision making involves activities that are not included in the narrower (traditional) game theoretic interpretation, including emotional appeals, apparent irrationality, and argumentation between parties.

There are several key distinctions between Large World approaches (i.e. DT II) and Small World (i.e. game theory) with respect to disaster forensic analyses. First, DT II makes the analysis more relevant and applicable to real-world disasters by eliminating the need for the cardinal (von-Neumann Morgenstern) payoffs and mixed (i.e. probabilistic) strategies used in game theory (drama theory uses ordinal preferences and pure strategies). Second, DT II assumes any number of players, each of whom has a large number of options or alternatives to choose from. Third, game-theoretic analysis identifies equilibria within a fixed model which requires players to use artificially fixed preferences and options (and to make decisions based on moves in a model calibrated at the beginning of the analysis). This ignores the complex reality of disaster forensics analyses, where time is spent debating how to avoid cascading impacts by analyzing dynamic and time-contingent strategies, preferences, and opportunities. In contrast DT II constitutes a theory of mutual interactive game design in which players, by communicating with others prior to managing the disaster, interact to design a game that they will ultimately play (including players, options and beliefs).

Drama-theory constitutes a flexible and powerful decision support tool to help policy makers understand and model "wicked" problems at the complex level of decision making (often known as the "messes" or "swamp") such as managing the unknown disaster impacts of global environmental change. Borrowing from theatrical terminology, an episode in drama theory is defined as moving through phases of *scene-setting*, *build-up*, *climax* where players re-define a "moment of truth" by finding justifications for changing the parameters of a game, including the set of players involved as well as their positions, stated intentions and preferences The following disaster forensics case study is modeled using concepts from DT II with the assistance of the Decision Support System Confrontation Manager™ (which supports a previous version of drama theory). Option board diagrams were obtained using Windows Paint™ to adjust output from Confrontation Manager™.

In this case study we focus on improving a disaster forensics analysis in the Pacific islands region: particular attention is given to the island nation of Vanuatu, and its relationship to donors and the UN. Vanuatu seeks to reduce the risk of additional climate refugees and to meet social development goals and enhance disaster resilience. The initial disaster forensics model contains three decision makers, namely Australia (a major donor), the Western Countries (West) and Vanuatu. Note that the Vanuatu has three options it controls while all other parties have one (Australia can increase its aid expenditures to Vanuatu and the West can reduce its emissions of greenhouse gases (Fig. 5). The preferences, options, and intentions of these three key stakeholders are discussed together with the dilemmas facing all players and the short and long term implications of each dilemma. The options in Fig. 5 constitute the minimum needed to specify parties' positions and stated intentions. Acceptance or rejection of an option is always represented by a filled-in or empty shape. The three dilemmas of drama theory are defined in relation to each option: A party may have a trust, rejection or persuasion dilemma with any party in relation to any option owned by that party. The three dilemmas of DT II can be summarized as follows. Consider the situation in which actor A's intention contradicts actor B's position. If B has no doubts about A's intention, B has a persuasion dilemma with A over this option; however, if B does doubt it, then player A has a rejection/threat dilemma with B over this option. Hence the rejection/threat dilemma applies to options where player A intends to flout player B's position, and B doubts that A will. Finally, if A's intention is consistent with B's position for an option and B doubts it, then B has a trust dilemma with A over this option.

Let us now consider each dilemma in detail. The persuasion dilemma arises if either you won't say whether you will carry out my position, or you say you won't and I do not doubt it. More formally, player A has a persuasion dilemma with player B when A does not doubt B's intention to flout A's position (or threat). A implicitly "gives in" to this dilemma if it communicates no new reasons why B

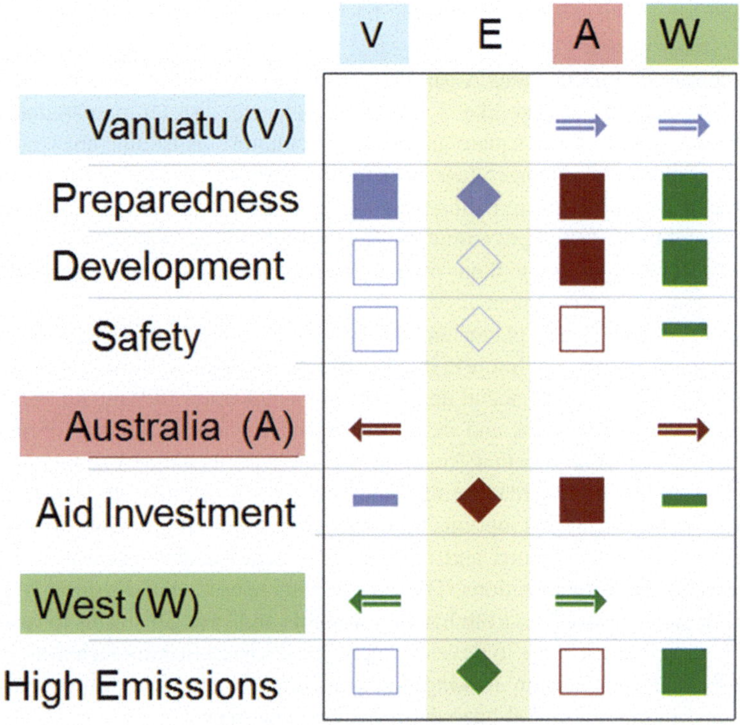

Fig. 5 Large World Disaster Forensics Options Board

should not flout its position (or threat); it "fights" the dilemma by trying to communicate new reasons. The trust dilemma can be simply defined as follows: I do not trust you to carry out your promise. Specifically, Player A has a trust dilemma with B if she doubts B's stated intention to meet A's position. A implicitly "gives in" to this dilemma if she communicates no new reasons for B to adhere to her intention; she "fights" by communicating new reasons. The final dilemma is the rejection dilemma: you do not believe that I will carry out a threat (or position) that conflicts with yours. Player A has a rejection dilemma to B when A's stated intention flouts B's position, but B doubts this intention. A implicitly "gives in" to this dilemma if it communicates now new reasons why B should believe its intention. On the other hand, A fights the rejection dilemma by communicating new reasons why B should believe its intention.

A dash is inserted if it is uncertain, conditional or left open whether an option should or will be adopted or rejected. In Fig. 5, the first column shows the position of Vanuatu (V). The country's position is to increase disaster preparedness and enhance disaster forensics analyses (solid square in row 1), and is not able to take extra steps to enhance urban development, improve the safety of women or meet other social development goals (solid square in row 2). Vanuatu's position also includes requesting Western nations to reduce emissions thereby reducing disaster risk (open square in row 1). The nation does not take a position on Australia's aid strategy (dash in row 1).

The third column illustrates Australia's (A) position. Australia wants to increase aid to Vanuatu and would like Vanuatu to enhance disaster preparedness and development goals (solid square in the third column). Australia does not want Western countries to change their policies on climate change or to reduce on emissions The fourth column illustrates the position of Western (i.e. developed) countries and their elected governments. They initially do not take a position on any option except for allowing their own high emissions and encouraging disaster preparedness and development in Vanuatu.

The intentions of each player are shown in the second column (entitled *E* for equilibrium or threatened future). If all positions remain as shown in Fig. 5 the result will be that adequate social development cannot occur in Vanuatu because of rising greenhouse emissions and increased natural hazards. In the equilibria situation (column 2) it is shown that Australia will increase its aid to Vanuatu however Western countries will increase emissions. One way to improve the situation is to include a new player and options into the analysis. It is valuable to consider involving the United Nations and other Pacific regional development and emergency management organizations. The current climate crisis and the threat of global environmental catastrophes, from harsher droughts and stronger storms to rising sea levels, forces policy makers to develop improved decision making techniques at the "wicked" level of decision making for modeling the implications of complex, urgent, and important global hazards, risks, and disasters.

Reducing disaster risk in Vanuatu and the rest of the Pacific Island region will require transformative actions include promoting renewable energy sources on an unprecedented scale, generating a deep public and private sector awareness of disaster forensics tools and initiatives, and advancing green economic development in the local context. Among these elements, education plays a significant role because it serves as the foundation for the success of the other elements and the subsequent implementation of comprehensive sustainability plans. The future of sustainability in Vanuatu rests in hands of business leaders, government officials and local communities that are willing to engage in the planning process, demand regulatory action, and embrace stewardship.

The state Hawaii offers some valuable lessons for other Pacific island nations. In the 1970s, the State of Hawaii was a leader in long-range strategic planning. Conceived in the mid-1970s, the visionary "Hawaii State Plan" was among the first planning documents in the US that provided visionary and holistic policies for the economic, social and environmental future of an entire state. Building upon the Hawaii State Plan the "2050 Hawaii Sustainability Plan" involves respecting the state's island communities; striking a balance between economic, social, and environmental goals; meeting the needs of the current generation without compromising the ability of future generations to meet their own; and respecting key Hawaiian values including Auamo Kuleana (collective transformation through individual excellence), 'Ike 'Āina (knowledge learned from connection to land) and Aloha 'Āina (love and commitment to land). Currently, Hawaii produces about

21 % of its electricity from renewable energy and depends upon imported fossil fuel. However, in 2015, Hawaii became the first state in the nation to adopt a 100 % renewable energy portfolio standard: the clean energy initiative directs the state's utilities to generate 100 % of their electricity from renewable sources by 2045. Achieving the 100 % renewable energy portfolio standard involves a two pronged strategy. The first involves conservation by establishing energy efficient building codes and promoting energy efficient lifestyles. The second emphasizes modernizing the power grid, establishing alternative fuels, wisely harnessing renewable energy (from solar, wind, ocean, geothermal, hydroelectric, and biomass resources) and reducing the use of fossil fuels. For example, a significant step towards energy independence was taken when the Hawaii Public Utilities Commission approved the development of a $41 million solar farm on Kaua'i, which will provide 6 % of the island's energy needs each day.

Vanuatu also may wish to embrace the risk management advances found in the ISO 31000 international risk standard for developing their own risk-related standards and policies, especially in the areas of disaster forensics, disaster risk reduction and the management of hazards. The first editions of ISO 31000 and ISO Guide 73 were published in 2009. Reducing, anticipating and managing risk are essential tasks for the government of Vanuatu. ISO 31000 has been adopted by national governments, UN agencies, as well as by multinational corporations and

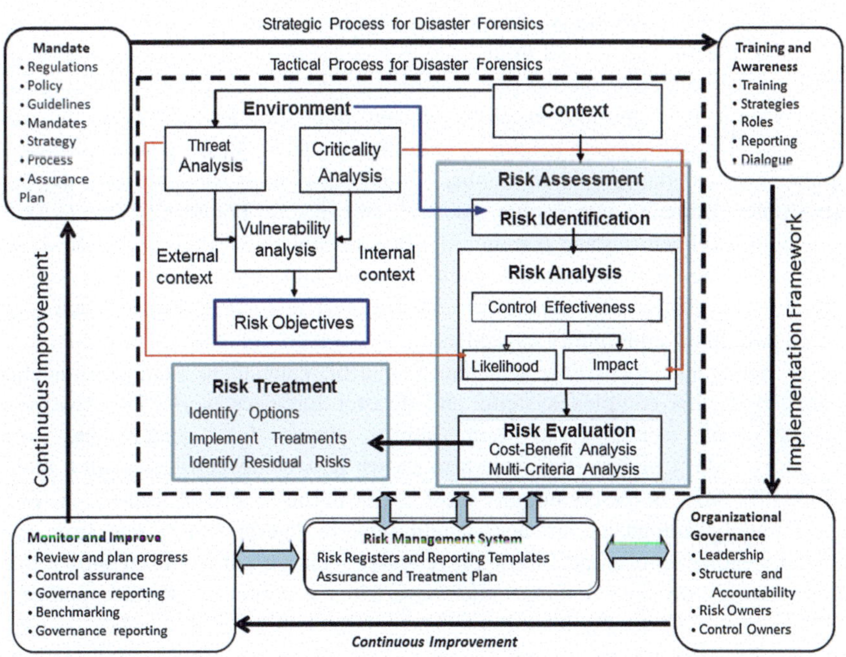

Fig. 6 Strategic and tactical processes for disaster forensics and ISO 31000 (modified from ISO 31000, 2009)

more than 50 national standards bodies (for their own national standards) covering over 70 % of the global population. Specifically, Fig. 6 highlights tactical and strategic disaster forensics processes that Vanuatu can use in the context of ISO 31000. The overarching ISO 31000 principle is that risk management should have net value to stakeholders decision by helping them to understand the effects of uncertainty on renewable energy objectives and to devise cost-effective risk treatments, reduce disaster risk, enhance public safety, improve sustainability, promote a low carbon energy future, enhance public benefits and reduce harm.

10 Conclusions

Complexity theory for disaster forensics, with roots in the fields of statistical physics, information theory, non-linear dynamics and complex systems literature, seeks to reduce disaster risk for individuals, communities and nations by promoting positive transformations in the co-evolving and mutually dependent human-environmental condition, by capitalizing on opportunities provided by human creativity, diplomatic openings, technologic capacities and environmental change. It is shown that complexity theory and disaster forensics are effective tools for designing and implementing effective disaster risk reduction initiatives that equitably and responsibly meet the biophysical needs of human communities, while reducing disaster vulnerability, maintaining long-term security, respecting financial constraints, meeting ecological limits, and improving institutional arrangements for transparent and effective disaster risk governance. Large world decision making has been shown to help promote sustainable hazard mitigation by identifying the preconditions of instability and helping to proactively address them in an increasingly complex and uncertain world. A number of complexity tools were put forth that cut across policy domains, geographic, political, and sectoral boundaries. These tools were applied to Pacific Island communities that are among the most at risk on earth. Collectively, they demonstrate the quality, breadth and depth of complex systems methodologies that can be used to promote disaster resilience, ensure sustainable security and build community capacities.

The author also spent time with the people of Vanuatu to share the insights gleaned from this complex systems and disaster forensics work. This involved teaching people techniques to secure houses, prepare food gardens and store emergency food. Communities were also shown how to develop evacuation plans and identify those in the community most at risk during a time of disaster. A team of community volunteers are also taught how to interpret warnings from the national weather bureau and follow preventative measures. It has been shown that "Large World" decision making under uncertainty constitutes a flexible and efficient approach for reducing disaster risk and coping with the climate-related natural hazards. The disaster management lessons learned in Vanuatu are relevant for nations confronting the mass migration of its citizens from climate change. For example, low-lying Kiribati has purchased land in Fiji, which Kiribati President

Anote Tong says is an investment and also a guarantee in case the entire nation needs to move and becomes climate-refugees.

Disaster preparedness investments have saved countless lives in the Pacific Island region. For example, before Tropical Cyclone Pam the government of Vanuatu used an SMS warning system to alert people of the approaching cyclone. In addition, text messages, containing condensed versions of warnings from the national meteorology service, were sent every 3 h as the cyclone intensified. The messages were then sent hourly as the cyclone came closer to making landfall. This was particularly critical for many in the outer islands—and many Vanuatu chiefs used smartphones for the first time. While technical problems at the national broadcaster meant emergency radio bulletins were only reaching some of Vanuatu's 65 inhabited islands many citizens (especially in Shefa and the outer islands) obtained critical information from the SMS alerts. Penetration of mobile phones in Vanuatu was shown to be quite high (estimated to be 85 % or more of the population) so this is a vital means of communicating with the public.

References

1. Abel N (1999) Resilient rangeland regions. In: Eldridge D, Freudenberger D (eds) People and rangelands: building the future, vol 1. VI International Rangeland Conference Inc., Townsville, Australia, pp 21–30
2. Abel N, Ross H, Walker P (1998) Mental models in rangeland research, communication and management. Rangeland J 20(1):77–91
3. American Rivers (2014) Map of US dam removals 1936–2014. Retrieved from http://www.americanrivers.org/initiatives/dams/dam-removals-map/
4. Arrow KJ (1971) Essays in theory of risk bearing. Markham, Chicago
5. Beckerman W (1992) Economic growth and natural capital. World Dev 481:175 183
6. Boulding KJ (1966) The economics of the coming spaceship earth. In: Jarrett HE (ed) Environmental quality in a growing economy. John Hopkins Press, Baltimore, pp 3–14
7. Christensen NL (1997) Managing for heterogeneity and complexity on dynamic landscapes. In: Shachak M, Ostfeld RS, Likens GE (eds) The ecological basis of conservation: heterogeneity, ecosystems, and biodiversity. Chapman and Hall, New York, pp 167–186
8. Cohen WM, Levinthal DA (1990) Absorptive capacity: a new perspective on learning and motivation. Adm Sci Q 25:125–152
9. Coppock DL (1993) Vegetation and pastoral dynamics in the southern Ethiopian rangelands: implications for theory and management. In: Benke RH, Scoones I, Kerven C (eds) Range ecology at disequilibrium. Overseas Development Institute, the International Institute for Environment and Development, and the Commonwealth Secretariat, London, pp 31–41
10. Craik KJW (1952) The nature of explanation. Cambridge University Press, Cambridge
11. Department of Foreign Affairs and Trade (DFAT) (2016) Government of Australia. http://dfat.gov.au/about-us/publications/Pages/aid-investment-plan-aip-vanuatu-2015-16-to-2018-19.aspx
12. Dregne H, Kasas M, Rozanov B (1991) A new assessment of the world status of desertification. Desert Control Bulletin 20(5):6–18
13. East A, Pess G, Bountry J, Magirl C, Ritchie A, Logan J, Shafroth P (2015) Large-scale dam removal on the Elwha River, Washington, USA: River channel and floodplain geomorphic change. Geomorphology 228(1):765–786. doi:10.1016/j.geomorph.2014.08.028
14. Ellis JE, Coughenour MB, Sn DM (1993) Climate change, variability, ecosystem stability, and the implications for range and livestock development. In: Benke RH, Scoones I, Kerven C

(eds) Range ecology at disequilibrium. Overseas Development Institute, the International Institute for Environment and Development, and the Commonwealth Secretariat, London, pp 31–41
15. Fang L, Hipel KW, Kilgour D (1993) Interactive decision making: the graph model for conflict resolution. Wiley, New York
16. Forensic Investigations of Disasters (FORIN) project (2015) Integrated research on disaster risk. http://www.irdrinternational.org
17. Fraser NM, Hipel KW (1984) Conflict analysis: models and resolutions. North-Holland, New York
18. Gard TC (1988) Introduction to stochastic differential equations. Marcel Dekker, New York
19. Gomez-Pompa A, Vazquez-Yanes C, Guevara S (1972) The tropical rain forest: a non-renewable resource. Science 177:762–765
20. Grant G, Lewis S (2015) The remains of the dam: what have we learned from 15 years of US dam removals. In: Lollino G, Arattano M, Rinaldi M, Giustolisi O, Marechal J, Grant G (eds) River basins, reservoir sedimentation and water resources, vol 3. Springer, Basel, pp 31–35
21. Gunderson LH, Holling CS, Light SS (1995) Barriers and bridges to the renewal of ecosystems and institutions. Columbia University Press, New York
22. Hart P, Rosenthal U, Kouzmin A (1993) Crisis decision making: the centralization thesis revisited. Admin Soc 25(1):12–45
23. Harvey OJ (1966) Experience, structure, and adaptability. Springer, New York
24. Havas M, Hutchison T, Likens G (1984) Scientific consensus for resource management. Environ Technol 18:176–190
25. Hilgard EW (1899) Irrigation risks in arid lands. Research Bulletin RB-04, University of California, Irvine, California
26. Holling CS (1973) Resilience and stability of ecological systems. Annu Rev Ecol Syst 4:1–23
27. Holling CS (1995) What barriers? What bridges? In: Gunderson LH, Holling CS, Light SS (eds) Barriers and bridges to the renewal of ecosystems and institutions. Columbia University Press, New York, pp 3–36
28. Johnson-Laird PN (1983) Mental models. Cambridge University Press, Cambridge
29. Karl TR, Trenberth KE (1999) The human impact on climate. Sci Am 281:100–105
30. Kay JJ (1991) The concept of ecological integrity, alternative theories of ecology, and implications for decision-support indicators. In: Victor PA, Kay JJ, Ruitenbeek HJ (eds) Economic, ecological, and decision theories: indicators of ecologically sustainable development. Ministry of Supply and Services Canada, Ottawa, pp 23–58
31. Kelly GA (1955) The psychology of personal constructs, vol II. W.W. Norton, New York
32. Kuhn SS (1962) The structure of scientific revolutions. University of Chicago Press, Chicago
33. Landau M (1991) Multi organizational systems in public administration. J Public Admin Theory 1(1):5–18
34. Levy JK, Hipel KW, Howard N (2009) Advances in drama theory for managing global hazards and disasters. Part I: theoretical foundation, Part II: Coping with Global Climate Change and Environmental Catastrophe Group Decision and Negotiation, vol 18, No (4), pp 303–334
35. Lister NM (1997) PhD doctoral comprehensive examination. University of Waterloo, Waterloo, Canada
36. Ludwig D (1996) The distribution of population survival times. Am Naturalist 147:506–526
37. Mackay H (1994) Why people don't listen? Pan, Sydney
38. Major J, O'Connor J, Podolak C, Keith M, Grant G, Spicer K, et al (2012) Geomorphic response of the Sandy River, Oregon, to Removal of Marmot Dam. U.S. Geological Survey Professional Paper 1792. Retrieved from http://pubs.usgs.gov/pp/1792/
39. Maruyama M (1980) Mindscapes and science theories. Curr Anthropol 21:589–599. Reprinted in excepted form as: Most frequently found mindscape types. In: Caley MT, Sawada D (eds) Mindscapes: the epistemology of Magoroh Maruyama. Gordon and Breach Science Publishers, Langhorne, Pennsylvania, pp 1–30

40. May RM (1977) Thresholds and breakpoints in ecosystems with a multiplicity of stable states. Nature 269:471–477
41. Nabokov V (1974) Look at the Harlequins. McGraw-Hill, New York
42. Natural Hazards Risk Atlas (2015) Verisk Maplecroft, identifies the threat posed to more than 1,300 cities. https://maplecroft.com/themes/nh/
43. Needham J (1969) Science and Society in East and West: the grand titration. Allen and View, London
44. Patten BC (1975) Ecosystem linearization: an evolutionary design problem. Am Naturalist 109 (969):529–539
45. Pearson AJ, Snyder NP, Collins MJ (2011) Rates and processes of channel response to dam removal with a sand-filled impoundment. Water Resour Res 47:W08504. doi:10.1029/2010WR009733
46. Perrow C (1984) Normal accidents: living with high risk technologies. Basic Books, New York
47. Prato T (1999) Multiattribute decision analysis for ecosystem management. Ecol Econ 30:207–222
48. Regier HA (1993) Indicators of ecosystem integrity. In: McKenzie DH, Hyatt DE, McDonald VJ (eds) Ecological indicators, vol 1. Elsevie Press, New York, pp 183–200
49. ReliefWeb (2016) http://reliefweb.int/report/vanuatu/one-year-tropical-cyclone-pam-red-cross-reaches-44000-people-humanitarian-assistance
50. Sagan S (1993) The limits of safety: organizations, accidents, and nuclear weapons. Princeton University Press, Princeton
51. Scheffer M, Hosper SH, Meijer ML, Moss B, Jeppesen E (1993) Alternative equilibria in shallow lakes. Trends Ecol Evol 8(8):275–279
52. Scholes RJ, Walker BH (1993) An African savanna: synthesis of the Nylsvley study. Cambridge University Press, Cambridge
53. Shafik N, Bandyopadhay S (1992) Economic growth and environmental quality: time series and cross-country evidence. World development background report, Washington, DC
54. Simon KA (1969) Sciences of the artificial. MIT Press, Cambridge
55. Simon JL, Kahn H (1984) The resourceful earth: a response to Global 2000. Basil Blackwell, Oxford
56. Stoddart LA, Smith AD, Box TW (1975) Range management, 3rd edn. McGraw-Hill, New York
57. Sutphen S, Bott V (1990) Issue salience and preparedness as perceived by city managers. In: Sylves RT, Waugh Jr, WL (eds) Cities and disaster: North American studies in emergency management. Charles C. Thomas, Springfield
58. Thom R (1969) Topological models in biology. Topology 8:10–24
59. Timmerman P (1986) Myths and paradigms of interactions between development and environment. In: Clark WC, Munn RE (eds) Sustainable development of the biosphere. Cambridge University Press, Cambridge, pp 292–320
60. UNESCAP (2011) United Nations Economic and Social Commission for Asia and the Pacific. http://www.unescap.org/resources/review-developments-transport-asia-and-pacific-2011World
61. United Nations University (2014) Institute for Environment and Human Security. Risk Report http://ehs.unu.edu/news/news/world-risk-report-2014.html#info
62. Vreede M (1977) Deserts and men. Government Printing Office, The Hague
63. Walker KJ (1994) The political economy of environmental policy: an Australian introduction. University of New South Wales Press, Kensington
64. Young SR (1991) A metaphysics for the post-modern. Soc Dynamicist 22:5–7
65. Zarfl C, Lumsdon AE, Berlekamp J, Tydecks L, Tockner K (2015) A global boom in hydropower dam construction. Aquat Sci 77(1):161–170

The manufacturer's authorised representative in the EU is Springer Nature Customer Service Centre GmbH, Europaplatz 3, 69115 Heidelberg, Germany. If you have any concerns regarding our products, please contact ProductSafety@springernature.com

Printed and bound by CPI Group (UK) Ltd, Croydon, CR0 4YY
05/10/2026
02228127-0012